Fluid Mechanics

FLUID MECHANICS

Hilary D. Brewster

Oxford Book Company

Jaipur ▪ New Delhi

Published By

Oxford Book Company

Regd. Off. "Bony Residency", Gate No. 2
Opp. Tilak Public School, Vishweswariya Nagar,
Gopalpura Road, Jaipur - 302018, India
Ph: 0141 - 2761280, Telefax: 2761381
email - oxfordbookjpr@gmail.com
website - oxfordbookcompany.com

Delhi Office: 102, 1st Floor, Satyam House,
4327/3, Ansari Road, Darya Ganj,
New Delhi - 110002, India
Ph: 011-45652440

Published in India, 2025

ISBN : 9788189473983

Price : ₹ 2295/-

Typeset by Shivangi Computers, Jaipur

Printed and bound in India by Thomson Press Ltd.,
New Delhi

Preface

Fluid Mechanics, understanding and applying the principles of how motions and forces act upon fluids such as gases and liquids, is introduced and comprehensively covered in this widely adopted text. This book *'Fluid Mechanics'* continues the tradition of precision, accuracy, accessibility and strong conceptual presentation. The author balances three separate approaches integral, differential and experimental to provide a foundation for fluid mechanics concepts and applications. The application of theory in fluid mechanics and enables students new to the science to grasp fundamental concepts in the subject. Despite dramatic advances in numerical and experimental methods of fluid mechanics, the fundamentals are still the starting point for solving flow problems. This textbook introduces the major branches of fluid mechanics of incompressible and compressible media, the basic laws governing their flow, and gas dynamics. Fluid Mechanics demonstrates how flows can be classified and how specific engineering problems can be identified, formulated and solved, using the methods of applied mathematics.

The concepts of fluid mechanics, covering both the physical and mathematical aspects of the subject. The text aims to bridge the gap between civil and mechanical engineering courses, and hence covers a wide variety of topics. This book remains one of the most comprehensive and useful texts on fluid mechanics available today, with applications going from engineering to geophysics, and beyond to biology and general science. This book features the applications of essential concepts as well as the coverage of topics in the this field.

Hilary D. Brewster

Contents

Chapter 1

Fluid Mechanics

What is fluid mechanics? As its name suggests it is the branch of applied mechanics concerned with the statics and dynamics of fluids (both liquids and gases). The analysis of the behaviour of fluids is based on the fundamental laws of mechanics which relate continuity of mass and energy with force and momentum together with the familiar solid mechanics properties. Even among fluids which are accepted as fluids there can be wide differences in behaviour under stress.

Fluids obeying Newton's law where the value of μ is constant are known as Newtonian fluids. If μ is constant the shear stress is linearly dependent on velocity gradient. This is true for most common fluids. Fluids in which the value of μ is not constant are known as non-Newtonian fluids

FLUID MECHANICS IN CHEMICAL ENGINEERING

A knowledge of fluid mechanics is essential for the chemical engineer because them ajority of chemical-processing operation sarecon ducted either partly or totally in the fluid phase. Examples of such operations abound in the biochemical, chemical, energy, fermentation, materials, mining, petroleum, pharmaceuticals, polymer, and waste-processing industries.

There are two principal reasons for placing such an emphasis on fluids. First, at typical operating conditions, an enormous number of materials normally exist as gases or liquids, or can be transformed into such phases. Second, it is usually more efficient and cost-effective to work with fluids in contrast to solids.

Even some operations with solids can be conducted in a quasi-fluidlike manner; examples are the fluidized-bed catalytic refining of hydrocarbons, and the long-distance pipelining of coal particles using water as the agitating and transporting medium.

Although there is inevitably a significant amount of theoretical development, almost all the material in this book has some application to chemical processing and other important practical situations. Throughout, we shall endeavor to present an understanding of the physical behaviour involved; only then is it really possible to comprehend the accompanying theory and equations.

GENERAL CONCEPTS OF A FLUID

We must begin by responding to the question, "What is a fluid?" Broadly speaking, a fluid is a substance that will deform continuously when it is subjected to a tangential or shear force, much as a similar type of force is exerted when a water-skier skims over the surface of a lake or butter is spread on a slice of bread.

The rate at which the fluid deforms continuously depends not only on the magnitude of the applied force but also on a property of the fluid called its viscosity or resistance to deformation and flow. Solids will also deform when sheared, but a position of equilibrium is soon reached in which elastic forces induced by the deformation of the solid exactly counterbalance the applied shear force, and further deformation ceases.

A simple apparatus for shearing a fluid is shown in figure. The fluid is contained between two concentric cylinders; the outer cylinder is stationary, and the inner one (of radius R) is rotated steadily with an angular velocity I. This shearing motion of a fluid can continue indefinitely, provided that a source of energy—supplied by means of a torque here—is available for rotating the inner cylinder. The diagram also shows the resulting velocity profile; note that the velocity in the direction of rotation varies from the peripheral velocity RI of the inner cylinder down to zero at the outer stationary cylinder, these representing typical no-slip conditions at both locations. However, if the intervening space is filled with a solid—even one with obvious elasticity, such as rubber—only a limited rotation will be possible before a position of equilibrium is reached, unless, of course, the torque is so high that slip occurs between the rubber and the cylinder.

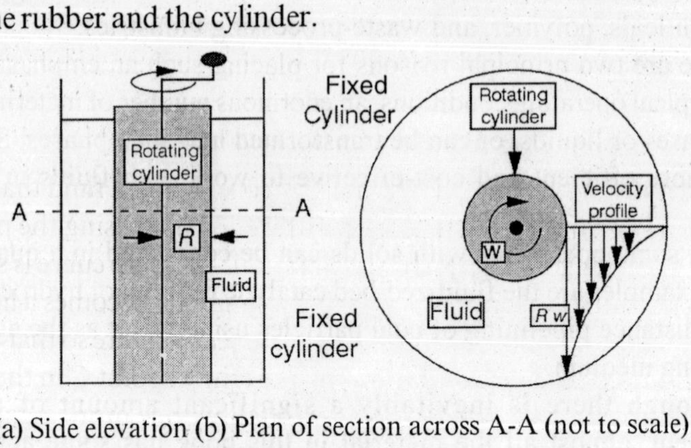

(a) Side elevation (b) Plan of section across A-A (not to scale)

Fig. Shearing of a fluid

There are various classes of fluids. Those that behave according to nice and obvious simple laws, such as water, oil, and air, are generally called Newtonian fluids.

These fluids exhibit constant viscosity but, under typical processing conditions, virtually no elasticity. Fortunately, a very large number of fluids of interest to the chemical engineer exhibit Newtonian behaviour, which is devoted to the study of non-Newtonian fluids. A fluid whose viscosity is not constant (but depends, for example, on the intensity to which it is being sheared), or which exhibits significant elasticity, is termed non-Newtonian.

For example, several polymeric materials subject to defor-mation can "remember" their recent molecular configurations, and in attempting to recover their recent states, they will exhibit elasticity in addition to viscosity. Other fluids, such as drilling mud and toothpaste, behave essentially as solids and will not flow when subject to small shear forces, but will flow readily under the influence of high shear forces.

Fluids can also be broadly classified into two main categories—liquids and gases. Liquids are characterized by relatively high densities and viscosities, with molecules close together; their volumes tend to remain constant, roughly independent of pressure, temperature, or the size of the vessels containing them. Gases, on the other hand, have relatively low densities and viscosities, with molecules far apart; generally, they will rapidly tend to fill the container in which they are placed. However, these two states—liquid and gaseous—represent but the two extreme ends of a continuous spectrum of possibilities.

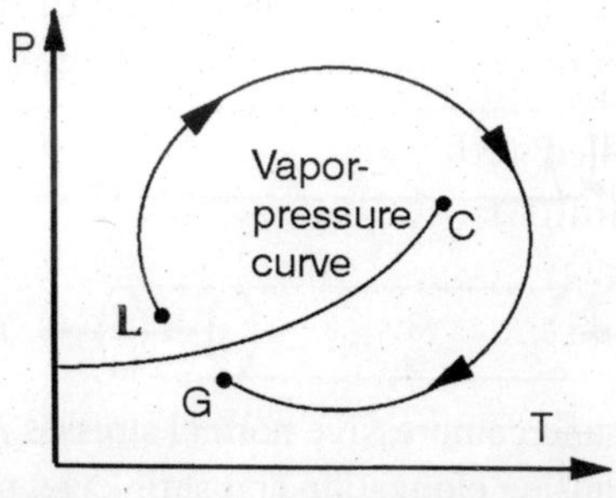

Fig. When does a liquid become a gas?

The situation is readily illustrated by considering a fluid that is initially a gas at point G on the pressure/temperature. By increasing the pressure, and perhaps lowering the temperature, the vapour-pressure curve is soon reached and crossed, and the fluid condenses and apparently becomes a liquid at point L. By continuously adjusting the pressure and temperature so that the clockwise path is followed, and circumnavigating the critical point C in the process, the fluid is returned to G, where it is presumably once more a gas. But where does the transition from liquid at L to gas at G occur? The answer is at no single point, but rather that the change is a continuous and gradual one, through a whole spectrum of intermediate states.

STRESSES, PRESSURE, VELOCITY, AND THE BASIC LAWS

Stresses. The concept of a force should be readily apparent. In fluid mechanics, a force per unit area, called a stress, is usually found to be a more convenient and versatile quantity than the force itself. Further, when considering a specific surface, there are two types of stresses that are particularly important.

- The first type of stress, acts perpendicularly to the surface and is therefore called a normal stress; it will be tensile or compressive, depending on whether it tends to stretch or to compress the fluid on which it acts. The normal stress equals F/A, where F is the normal force and A is the area of the surface on which it acts. The dotted outlines show the volume changes caused by deformation. In fluid mechanics, pressure is usually the most important type of compressive stress.
- The second type of stress, acts tangentially to the surface; it is called a shear stress τ, and equals F/A, where F is the tangential force and A is the area on which it acts. Shear stress is transmitted through a fluid by interaction of the molecules with one another. A knowledge of the shear stress is very important when studying the flow of viscous Newtonian fluids. For a given rate of deformation, measured by the time derivative $d\gamma/dt$ of the small angle of deformation γ, the shear stress τ is directly proportional to the viscosity of the fluid.

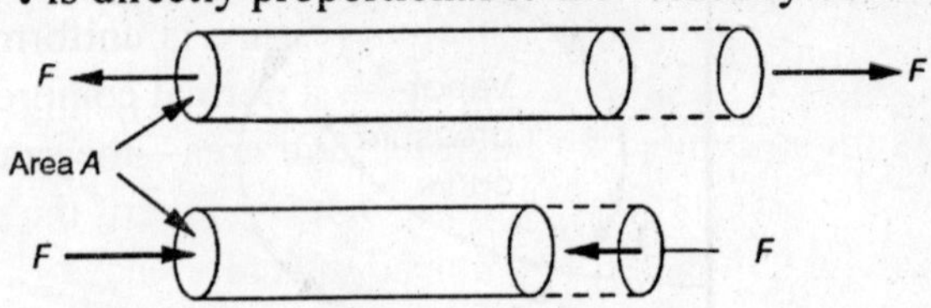

Fig. (a) Tensile and compressive normal stresses F/A, act-ing on a cylinder, causing elongation and shrinkage, respectively

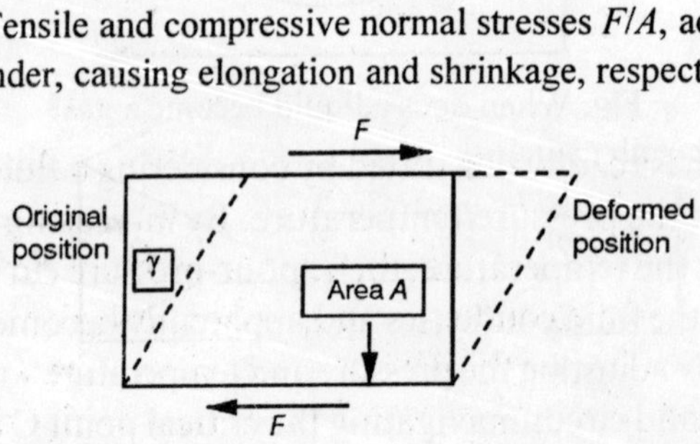

Fig. (b) Shear stress $\tau = F/A$, acting on a rectangular parallelepiped, shown in cross section, causing a deformation measured by the angle γ

Pressure: In virtually all hydrostatic situations—those involving fluids at rest—the fluid molecules are in a state of compression. For example, for the swimming pool whose cross section, this compression at a typical point P is caused by the downwards gravitational weight of the water above point P. The degree of compression is measured by a scalar, p—the pressure.

A small inflated spherical balloon pulled down from the surface and tethered at the bottom by a weight will still retain its spherical shape (apart from a small distortion at the point of the tether), but will be diminished in size. It is apparent that there must be forces acting normally inward on the surface of the balloon, and that these must essentially be uniform for the shape to remain spherical.

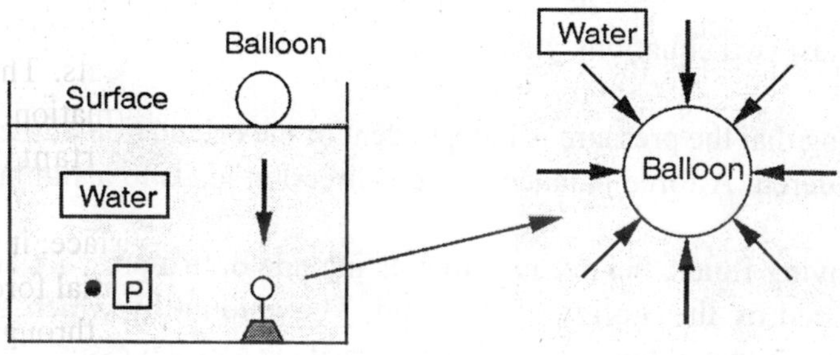

(a) (b)

Fig. (a) Balloon submerged in a swimming pool; (b) enlarged view of the compressed balloon, with pressure forces acting on it

Although the pressure p is a scalar, it typically appears in tandem with an area A (assumed small enough so that the pressure is uniform over it). By definition of pressure, the surface experiences a normal compressive force $F = pA$. Thus, pressure has units of a force per unit area—the same as a stress.

The value of the pressure at a point is independent of the orientation of any area associated with it, as can be deduced with reference to a differentially small wedge-shaped element of the fluid.

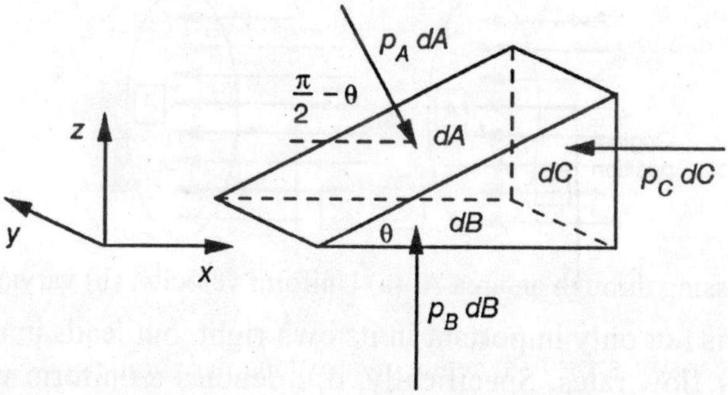

Fig. Equilibrium of a Wedge of Fluid

Due to the pressure there are three forces, $p_A dA, p_B dB$, and $p_C dC$, that act on the three rectangular faces of areas dA, dB, and dC. Since the wedge is not moving, equate the two forces acting on it in the horizontal or x direction, noting that $p_A dA$ must be resolved through an angle $(\pi/2 - \theta)$ by multiplying it by $\cos(\pi/2 - \theta) = \sin\theta$:

$$p_A dA \sin\theta = p_C dC.$$

The vertical force $p_B dB$ acting on the bottom surface is omitted from Eqn. because it has no component in the x direction. The horizontal pressure forces acting in the y direction on the two triangular faces of the wedge are also omitted, since again these forces have no effect in the x direction. From geometrical considerations, areas dA and dC are related by:

$$dC = dA \sin\theta.$$

These last two equations yield:

$$p_A = p_C,$$

verifying that the pressure is independent of the orientation of the surface being considered. A force balance in the z direction leads to a similar result, $p_A = p_B$.

For moving fluids, the normal stresses include both a pressure and extra stresses caused by the motion of the fluid. The amount by which a certain pressure exceeds that of the atmosphere is termed the gauge pressure, the reason being that many common pressure gauges are really differential instruments, reading the difference between a required pressure and that of the surrounding atmosphere. Absolute pressure equals the gauge pressure plus the atmospheric pressure. Velocity. Many problems in fluid mechanics deal with the velocity of the fluid at a point, equal to the rate of change of the position of a fluid particle with time, thus having both a magnitude and a direction. In some situations, particularly those treated from the macroscopic viewpoint, it sometimes suffices to ignore variations of the velocity with position. In other cases—particularly those treated from the microscopic viewpoint, it is invariably essential to consider variations of velocity with position.

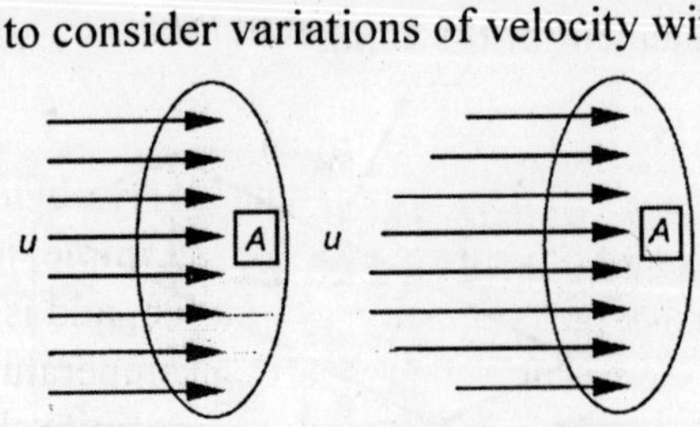

Fig. Fluid passing through an area A: (a) Uniform velocity, (b) varying velocity.

Velocity is not only important in its own right, but leads immediately to three fluxes or flow rates. Specifically, if u denotes a uniform velocity (not varying with position):

- If the fluid passes through a plane of area A normal to the direction of the velocity, the correspond-ing volumetric flow rate of fluid through the plane is $Q = uA$.
- The corresponding mass flow rate is $m = \rho Q = \rho uA$, where ρ is the (constant) fluid density. The alternative notation with an overdot, $\dot{m}$, is also used.
- When velocity is multiplied by mass it gives momentum, a quantity of prime importance in fluid mechanics. The corresponding momentum flow rate pass-ing through the area

A is $\dot{M} = mu = \rho u^2 A$.

pressions will be seen later to involve integrals over the area A: $Q = \int_A udA, m \int_A \rho u^2 dA.$

Basic laws. In principle, the laws of fluid mechanics can be stated simply, and—in the absence of relativistic effects—amount to conservation of mass, energy, and momentum. When applying these laws, the procedure is first to identify a system, its boundary, and its surroundings; and second, to identify how the system interacts with its surroundings. Let the quantity X represent either mass, energy, or momentum. Also recognize that X may be added from the surroundings and transported into the system by an amount X_{in} across the boundary, and may likewise be removed or transported out of the system to the surroundings by an amount X_{out}.

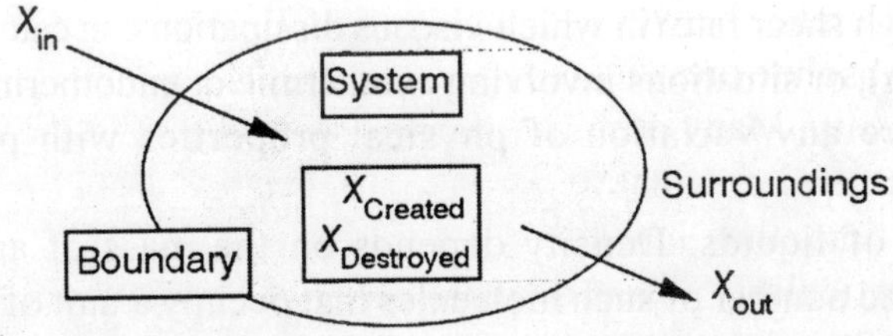

Fig. A system and transports to and from it.

The general conservation law gives the increase "X_{system} in the X-content of the system as:

$$X_{in} - X_{out} = \Delta X_{system}.$$

Although this basic law may appear intuitively obvious, it applies only to a very restricted selection of properties X. For example, it is not generally true if X is another extensive property such as volume, and is quite meaningless if X is an intensive property such as pressure or temperature.

In certain cases, where X^i is the mass of a definite chemical species i, we may also have an amount of creation $X^i_{created}$ or destruction $X^i_{destroyed}$ due to chemical reaction, in which case the general law becomes:

$$X^i_{in} - X^i_{out} + X^i_{created} - X^i_{destroyed} = \Delta X^i_{system}.$$

The conservation law, and such fundamental importance that in various guises it will find numerous applications throughout all of this text.

To solve a physical problem, the following information concerning the fluid is also usually needed:

- The physical properties of the fluid involved.
- For situations involving fluid flow, a constitutive equation for the fluid, which relates the various stresses to the flow pattern.

PHYSICAL PROPERTIES—DENSITY, VISCOSITY, AND SURFACE TENSION

There are three physical properties of fluids that are particularly important: density, viscosity, and surface tension. Each of these will be defined and viewed briefly in terms of molecular concepts, and their dimensions will be examined in terms of mass, length, and time (M, L, and T). The physical properties depend primarily on the particular fluid. For liquids, viscosity also depends strongly on the temperature; for gases, viscosity is approximately proportional to the square root of the absolute temperature. The density of gases depends almost directly on the absolute pressure; for most other cases, the effect of pressure on physical properties can be disregarded.

Typical processes often run almost isothermally, and in these cases the effect of temperature can be ignored. Except in certain special cases, such as the flow of a compressible gas (in which the density is not constant) or a liquid under a very high shear rate (in which viscous dissipation can cause significant internal heating), or situations involving exothermic or endothermic reactions, we shall ignore any variation of physical properties with pressure and temperature.

Densities of liquids. Density depends on the mass of an individual molecule and the number of such molecules that occupy a unit of volume. For liquids, density depends primarily on the particular liquid and, to a much smaller extent, on its temperature. Representative densities of liquids are given in table. The accuracy of the values given in tables is adequate for the calculations needed in this text. However, if highly accurate values are needed, particularly at extreme conditions, then specialized information should be sought elsewhere.

Density: The density ρ of a fluid is defined as its mass per unit volume, and indicates its inertia or resistance to an accelerating force. Thus:

$$\rho = \frac{\text{mass}}{\text{volume}} [=] \frac{M}{L^3},$$

in which the notation "[=]" is consistently used to indicate the dimensions of a quantity. It is usually understood in Equation. that the volume is chosen so that it is neither so small that it has no chance of containing a representative selection of molecules nor so large that (in the case of gases) changes of pressure cause significant changes of density throughout the volume. A medium characterized by a density is called a continuum, and follows the classical laws of mechanics— including Newton's law of motion.

Table: Specific Gravities, Densities, and Thermal Expansion Coefficients of Liquids at 20 °C

Liquid	*Sp. Gr.* s	*Density,* kg/m^3	ρ lb_m/ft^3	α $°C^{-1}$
Acetone	0.792	792	49.4	0.00149
Benzene	0.879	879	54.9	0.00124
Crude oil, 35°API	0.851	851	53.1	0.00074
Ethanol	0.789	789	49.3	0.00112
Glycerol	1.26 (50 °C)	1,260	78.7	—
Kerosene	0.819	819	51.1	0.00093
Mercury	13.55	13,550	845.9	0.000182
Methanol	0.792	792	49.4	0.00120
n-Octane	0.703	703	43.9	—
n-Pentane	0.630	630	39.3	0.00161
Water	0.998	998	62.3	0.000207

Degrees A.P.I. (American Petroleum Institute) are related to specific gravity s by the formula:

$$°\text{A.P.I.} = \frac{141.5}{s} - 131.5$$

Note that for water, °A.P.I. = 10, with correspondingly higher values for liquids that are less dense. Thus, for the crude oil listed in Table, equation. indeed gives 141.5/0.851 – 131.5 = 35°A.P.I.

Densities of gases. For ideal gases, $pV = nRT$, where p is the absolute pressure, V is the volume of the gas, n is the number of moles (abbreviated as "mol" when used as a unit), R is the gas constant, and T is the absolute temperature. If M_w is the molecular weight of the gas, it follows that:

$$\rho = \frac{nM_w}{V} = \frac{M_w p}{RT}.$$

Thus, the density of an ideal gas depends on the molecular weight, absolute pres-sure, and absolute temperature. Values of the gas constant R are given in Table for various systems of units. Note that degrees Kelvin, formerly represented by " °K," is now more simply denoted as "K."

Table: Values of the Gas Constant, R

Value	*Units*
8.314	J/g-mol K
0.08314	liter bar/g-mol K
0.08206	liter atm/g-mol K
1.987	cal/g-mol K
10.73	psia ft^3 /lb-mol °R
0.7302	ft^3 atm/lb-mol °R
1,545	ft lb_f /lb-mol °R

For a nonideal gas, the compressibility factor Z (a function of p and T) is introduced into the denominator of equation, giving:

$$\rho = \frac{nM_w}{V} = \frac{M_w p}{ZRT}.$$

Thus, the extent to which Z deviates from unity gives a measure of the nonideality of the gas.

The isothermal compressibility of a gas is defined as:

$$\beta = \frac{1}{V}\left(\frac{\partial V}{\partial p}\right)_T,$$

and equals—at constant temperature—the fractional decrease in volume caused by a unit increase in the pressure. For an ideal gas, $\beta = 1/p$, the reciprocal of the absolute pressure.

The coefficient of thermal expansion α of a material is its isobaric (constant pressure) fractional increase in volume per unit rise in temperature:

$$\alpha = \frac{1}{V}\left(\frac{\partial V}{\partial T}\right)_p,$$

Since, for a given mass, density is inversely proportional to volume, it follows that for moderate temperature ranges (over which α is essentially constant) the density of most liquids is approximately a linear function of temperature:

$$\rho = \rho_0[1 - \alpha(T - T_0)],$$

where ρ_0 is the density at a reference temperature T_0. For an ideal gas, $\langle = 1/T$, the reciprocal of the absolute temperature.

The specific gravity s of a fluid is the ratio of the density ρ to the density ρ_{SC} of a reference fluid at some standard condition:

$$s = \frac{\rho}{\rho_{SC}}.$$

For liquids, ρ_{SC} is usually the density of water at 4 °C, which equals 1.000 g/ml or 1,000 kg/m^3. For gases, ρ_{SC} is sometimes taken as the density of air at 60 °F and 14.7 psia, which is approximately 0.0759 lb_m/ft^3, and sometimes at

0 °C and one atmosphere absolute; since there is no single standard for gases, care must obviously be taken when interpreting published values. For natural gas, consisting primarily of methane and other hydrocarbons, the gas gravity is defined as the ratio of the molecular weight of the gas to that of air (28.8 lb_m /lb-mol).

Values of the molecular weight M_w are listed in Table for several commonly occurring gases, together with their densities at standard conditions of atmospheric pressure and 0 °C.

Table: Gas Molecular Weights and Densities (the Latter at Atmospheric Pressure and 0 °C)

Gas	*Mw*	*Standard* *kg/m³*	*Density* *lb_m/ft^3*
Air	28.8	1.29	0.0802
Carbon dioxide	44.0	1.96	0.1225
Ethylene	28.0	1.25	0.0780
Hydrogen	2.0	0.089	0.0056
Methane	16.0	0.714	0.0446
Nitrogen	28.0	1.25	0.0780
Oxygen	32.0	1.43	0.0891

Viscosity: The viscosity of a fluid measures its resistance to flow under an applied shear stress. There, the fluid is ideally supposed to be confined in a relatively small gap of thickness *h* between one plate that is stationary and another plate that is moving steadily at a velocity *V* relative to the first plate.

In practice, the situation would essentially be realized by a fluid occupying the space between two concentric cylinders of large radii rotating relative to eatch other. A steady force *F* to the right is applied to the upper plate (and, to preserve equilibrium, to the left on the lower plate) in order to maintain a constant motion and to overcome the viscous friction caused by layers of molecules sliding over one another.

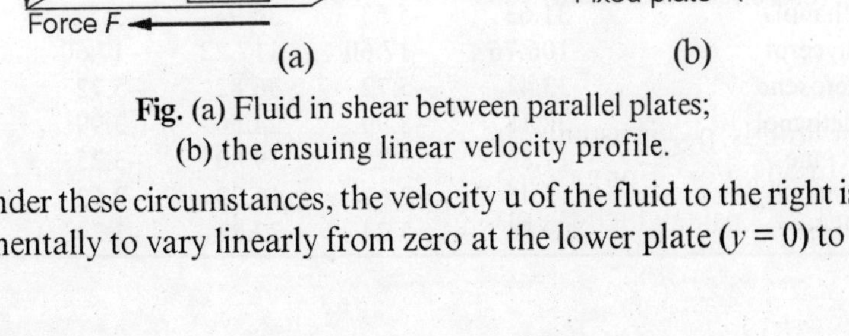

Fig. (a) Fluid in shear between parallel plates; (b) the ensuing linear velocity profile.

Under these circumstances, the velocity u of the fluid to the right is found experimentally to vary linearly from zero at the lower plate ($y = 0$) to *V* itself at the upper plate, corresponding to no-slip conditions at each plate. At any intermediate distance *y* from the lower plate, the velocity is simply:

$$u = \frac{y}{h}V.$$

Recall that the shear stress Δ is the tangential applied force F per unit area:

$$\tau = \frac{F}{A},$$

in which A is the area of each plate. Experimentally, for a large class of materials, called Newtonian fluids, the shear stress is directly proportional to the velocity gradient:

$$\tau = \mu\frac{du}{dy} = \mu\frac{V}{h}.$$

The proportionality constant μ is called the viscosity of the fluid; its dimensions can be found by substituting those for F (ML/T^2), A (L^2), and du/dy (T^{-1}), giving:

$$\mu \;[=]\; \frac{M}{LT}.$$

Representative units for viscosity are g/cm s (also known as poise, designated by P), kg/m s, and lb_m/ft hr. The centipoise (cP), one hundredth of a poise, is also a convenient unit, since the viscosity of water at room temperature is approximately 0.01 P or 1.0 cP. Table gives viscosity conversion factors. The viscosity of a fluid may be determined by observing the pressure drop when it flows at a known rate in a tube. More sophisticated methods for determining the rheological or flow properties of fluids—including viscosity; such methods often involve containing the fluid in a small gap between two surfaces, moving one of the surfaces, and measuring the force needed to maintain the other surface stationary.

Table: Viscosity Parameters for Liquids

Liquid	*a* (*T* in *K*)	*b* (*T* in *K*)	*a* (*T* in °*R*)	*b* (*T* in °*R*)
Acetone	14.64	–2.77	16.29	–2.77
Benzene	21.99	–3.95	24.34	–3.95
Crude oil, 35° API	53.73	–9.01	59.09	–9.01
Ethanol	31.63	–5.53	34.93	–5.53
Glycerol	106.76	–17.60	117.22	–17.60
Kerosene	33.41	–5.72	36.82	–5.72
Methanol	22.18	–3.99	24.56	–3.99
Octane	17.86	–3.25	19.80	–3.25
Pentane	13.46	–2.62	15.02	–2.62
Water	29.76	–5.24	32.88	–5.24

The kinematic viscosity ν is the ratio of the viscosity to the density:

$$\nu = \frac{\mu}{\rho},$$

and is important in cases in which significant viscous and gravitational forces coexist. The reader can check that the dimensions of ν are L^2/T, which are identical to those for the diffusion coefficient $\mathcal{D}$ in mass transfer and for the thermal diffusivity $\alpha = k/\rho c_p$ in heat transfer. There is a definite analogy among the three quantities—indeed, as seen later, the value of the kinematic viscosity governs the rate of "diffusion" of momentum in the laminar and turbulent flow of fluids.

Viscosities of liquids. The viscosities μ of liquids generally vary approximately with absolute temperature T according to:

$$\ln \mu = a + b \ \ln T \text{ or } \mu = e^{a+b\ln T},$$

and—to a good approximation—are independent of pressure. Assuming that μ is measured in centipoise and that T is either in degrees Kelvin or Rankine, appro-priate parameters a and b are given in Table for several representative liquids. The resulting values for viscosity are approximate, suitable for a first design only.

Viscosities of gases. The viscosity μ of many gases is approximated by the formula:

$$\mu = \mu_0 \left(\frac{T}{T_0} \right)^n,$$

in which T is the absolute temperature (Kelvin or Rankine), μ_0 is the viscosity at an absolute reference temperature T_0, and n is an empirical exponent that best fits the experimental data. The values of the parameters μ_0 and n for atmospheric pressure are given in Table ; recall that to a first approximation, the viscosity of a gas is independent of pressure. The values μ_0 are given in centipoise and correspond to a reference temperature of T_0 = 273 K.= 492 °R.

Table: Viscosity Parameters for Gases

Gas	μ_0, cP	n
Air	0.0171	0.768
Carbon dioxide	0.0137	0.935
Ethylene	0.0096	0.812
Hydrogen	0.0084	0.695
Methane	0.0120	0.873
Nitrogen	0.0166	0.756
Oxygen	0.0187	0.814

Surface tension: Surface tension is the tendency of the surface of a liquid to behave like a stretched elastic membrane. There is a natural tendency for liquids to minimize their surface area. The obvious case is that of a liquid droplet on a horizontal surface that is not wetted by the liquid—mercury on glass, or water on a surface that also has a thin oil film on it. For small droplets, the droplet adopts a shape that is almost perfectly spherical, because in this configuration there is the least surface area for a given volume.

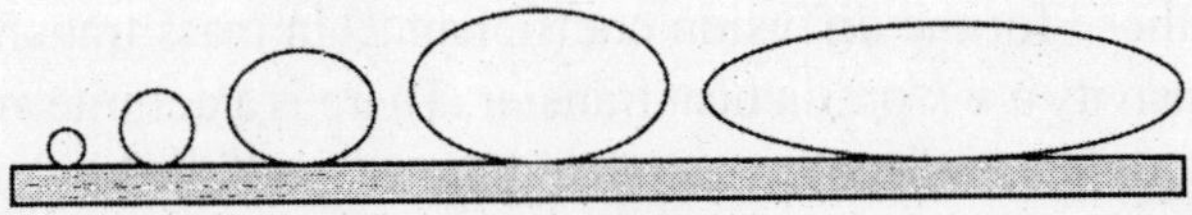

Fig. The larger droplets are flatter because gravity is becoming more important than surface tension

For larger droplets, the shape becomes somewhat flatter because of the increasingly important gravitational effect, which is roughly proportional to a^3, where a is the approximate droplet radius, whereas the surface area is proportional only to a^2. Thus, the ratio of gravitational to surface tension effects depends roughly on the value of $a^3/a^2 = a$, and is therefore increasingly important for the larger droplets. Overall, the situation is very similar to that of a water-filled balloon, in which the water accounts for the gravitational effect and the balloon acts like the surface tension.

A fundamental property is the surface energy. A molecule I, situated in the interior of the liquid, is attracted equally in all directions by its neighbors. However, a molecule S, situated in the surface, experiences a net attractive force into the bulk of the liquid. (The vapour above the surface, being comparatively rarefied, exerts a negligible force on molecule S.) Therefore, work has to be done against such a force in bringing an interior molecule to the surface. Hence, an energy σ, called the surface energy, can be attributed to a unit area of the surface.

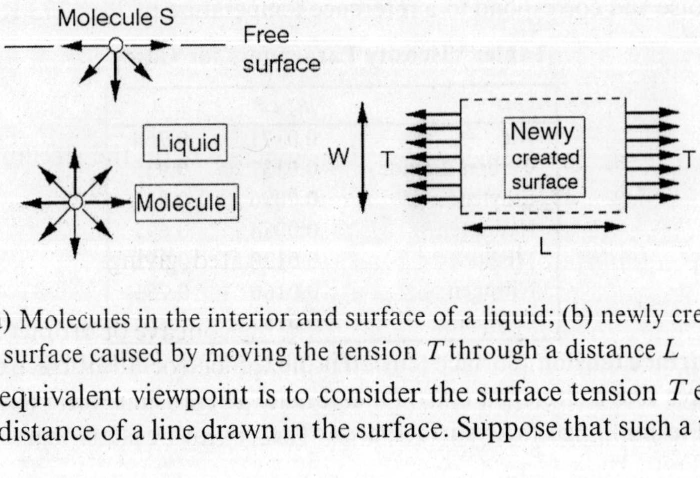

Fig. (a) Molecules in the interior and surface of a liquid; (b) newly created surface caused by moving the tension *T* through a distance *L*

An equivalent viewpoint is to consider the surface tension *T* existing per unit distance of a line drawn in the surface. Suppose that such a tension has moved a distance L, thereby creating an area WL of fresh surface. The work done is the product of the force, TW, and the distance L through which it moves, namely TWL, and this must equal the newly acquired surface energy σWL. Therefore, $T = \sigma$; both quantities have units of force per unit distance, such as N/m, which is equivalent to energy per unit area, such as J/m^2. We next find the amount $p_1 - p_2$ by which the pressure p_1 inside a liquid droplet of radius r, exceeds the pressure p_2 of the surrounding vapour. The equilibrium of the upper hemisphere of the droplet, which is also surrounded by an imaginary cylindrical "control surface" ABCD, on which forces in the vertical direction will soon be equated. Observe that the internal pressure p_1 is trying to blow apart the two hemispheres (the lower one is not shown), whereas the surface tension σ is trying to pull them together.

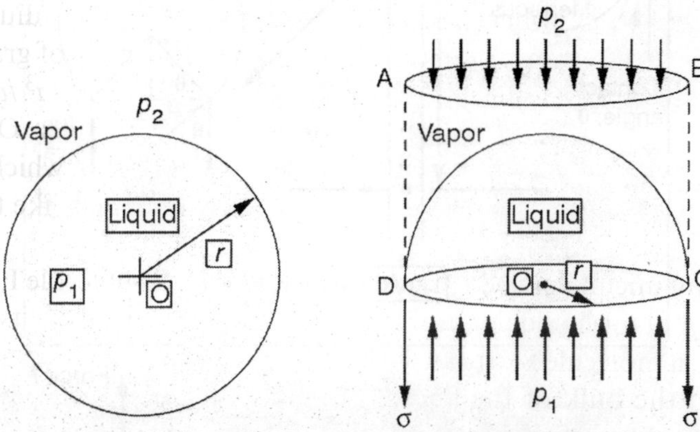

(a) Liquid droplet (b) Forces in equilibrium

Fig. Pressure change across a curved surface

In more detail, there are two different types of forces to be considered:

- That due to the pressure diverence between the pressure inside the droplet and the vapour outside, each acting on an area πr^2 (that of the circles CD and AB):

$$(p_1 \ - p_2)\pi r^2.$$

- That due to surface tension, which acts on the circumference of length $2\pi r$:

$$2\pi r\sigma.$$

At equilibrium, these two forces are equated, giving:

$$\Delta p \ = p_1 - p_2$$

That is, there is a higher pressure on the concave or droplet side of the interface. What would the pressure change be for a bubble instead of a droplet? Why? More generally, if an interface has principal radii of curvature r_1 and r_2, the increase in pressure can be shown to be:

$$p_1 - p_2 = \left(\frac{1}{r_1} + \frac{1}{r_2}\right)$$

For a sphere of radius r, both radii are equal, so that $r_1 = r_2 = r$, and $p_1 - p_2 = 2\sigma/r$.

The radii r_1 and r_2 will have the same sign if the corresponding centers of curvature are on the same side of the interface; if not, they will be of opposite sign.

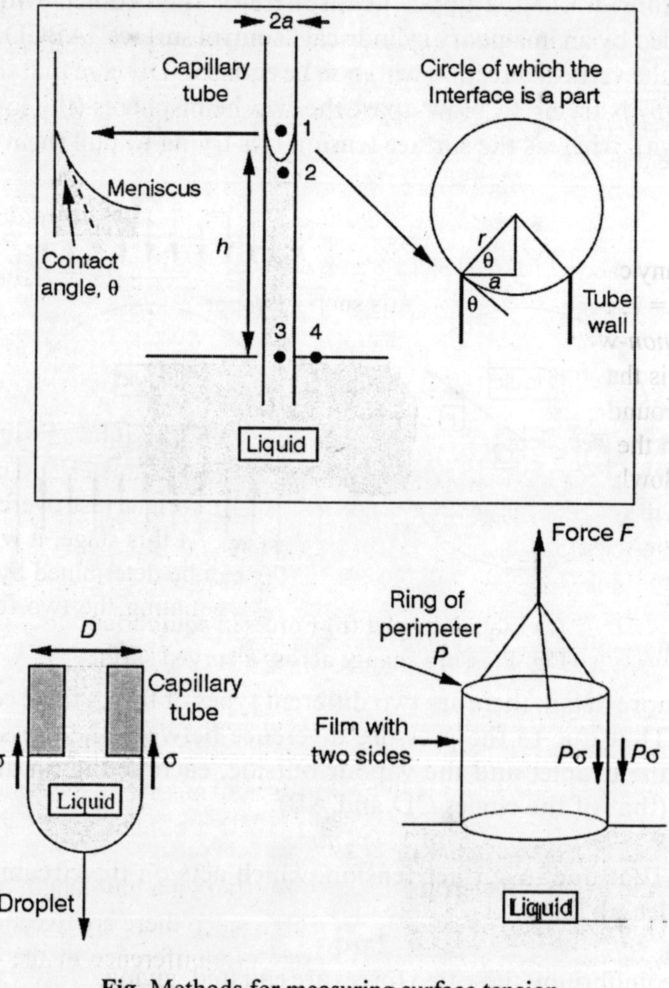

Fig. Methods for measuring surface tension

A brief description of simple experiments for measuring the surface tension σ of a liquid:

- In the *capillary-rise* method, a narrow tube of internal radius a is dipped vertically into a pool of liquid, which then rises to a height h inside the tube; if the *contact angle* (the angle between the free surface and the wall) is θ, the meniscus will be approximated by part of the surface of a sphere; from the geometry shown in the enlargement on the right-hand side of Fig the radius of the sphere is seen to be $r = a/\cos\theta$. Since the surface is now concave on the air side, the reverse of equation occurs, and $p_2 = p_1 - 2\sigma/r$, so that p_2 is *below* atmospheric pressure p_1. Now follow the path 1–2–3–4, and observe that $p_4 = p_3$ because points 3 and 4 are at the same elevation in the same liquid. Thus, the pressure at point 4 is:

$$p_4 = p_1 - \frac{2\sigma}{r} + \rho gh.$$

However, $p_4 = p_1$ since both of these are at atmospheric pressure. Hence, the surface tension is given by the relation:

$$\sigma = \frac{1}{2}\rho ghr = \frac{\rho gha}{2\cos\theta}.$$

In many cases—for complete wetting of the surface—θ is essentially zero and $\cos\theta = 1$. However, for liquids such as mercury in glass, there may be a complete *non*-wetting of the surface, in which case $\theta = \pi$, so that $\cos\theta = -1$; the result is that the liquid level in the capillary is then *depressed* below that in the surrounding pool.

- In the *drop-weight* method, a liquid droplet is allowed to form very slowly at the tip of a capillary tube of outer diameter D. The droplet will eventually grow to a size where its weight just overcomes the surface-tension force $\pi D\sigma$ holding it up. At this stage, it will detach from the tube, and its weight $w = Mg$ can be determined by catching it in a small pan and weighing it. By equating the two forces, the surface tension is then calculated from:

$$\sigma = \frac{w}{\pi D}$$

- In the *ring tensiometer,* a thin wire ring, suspended from the arm of a sensitive balance, is dipped into the liquid and gently raised, so that it brings a thin liquid film up with it. The force F needed to support the film is measured

by the balance. The downward force exerted on a unit length of the ring by one side of the film is the surface tension; since there are two sides to the film, the total force is $2P\sigma$, where P is the circumference of the ring. The surface tension is therefore determined as:

$$\sigma = \frac{F}{2P}.$$

In common with most experimental techniques, all three methods described above require slight modifications to the results expressed in equations because of imperfections in the simple theories.

Surface tension generally appears only in situations involving either free surfaces (liquid/gas or liquid/solid boundaries) or interfaces (liquid/liquid boundaries); in the latter case, it is usually called the *interfacial* tension. Representative values for the surface tensions of liquids at 20°C, in contact either with air or their vapour (there is usually little difference between the two),are given in Table.

Table: Surface Tensions

Liquid	σ *dynes/cm*
Acetone	23.70
Benzene	28.85
Ethanol	22.75
Glycerol	63.40
Mercury	435.5
Methanol	22.61
n-Octane	21.80
Water	72.75

UNITS AND SYSTEMS OF UNITS

Mass, Weight, and Force.

The *mass* M of an object is a measure of the amount of matter it contains and will be constant, since it depends on the number of constituent molecules and their masses. On the other hand, the *weight* w of the object is the gravitational force on it, and is equal to Mg, where g is the local gravitational acceleration. Mostly, we shall be discussing phenomena occurring at the surface of the earth, where g is approximately 32.174 ft/s^2 = 9.807 m/s^2 = 980.7 cm/s^2. These values are simply taken as 32.2, 9.81, and 981, respectively.

Table: Representative Units of Force

System	*Units of Force*	*Customary Name*
SI	kg m/s^2	Newton
CGS	g cm/s^2	Dyne
FPS	lbm ft/s^2	Poundal

Newton's second law of motion states that a force F applied to a mass M will give it an acceleration a:

$$F = ma,$$

From which is apparent that force has *dimensions* ML/T^2. Table gives the corresponding *units* of force in the SI (meter/kilogram/second), CGS (centimeter/ gram/second), and FPS (foot/pound/second) systems.

The poundal is now an archaic unit, hardly ever used. Instead, the *pound force*, lb_f , is much more common in the English system; it is defined as the gravitational force on 1 lb_m, which, if left to fall freely, will do so with an acceleration of 32.2 ft/s^2. Hence:

$$1\ lb_f = 32.2 lb_m \frac{ft}{s^2} = 32.2 \text{ poundals.}$$

When using lb_f in the ft, lb_m, s (FPS) system, the following conversion factor, commonly called "g_c," will almost invariably be needed:

$$g_c = 32.2 \frac{lb_m ft/s^2}{lb_f} = 32.2 \frac{lb_m ft}{lb_f s^2}.$$

Some writers incorporate g_c into their equations, but this approach may be confusing since it virtually implies that one particular set of units is being used, and hence tends to rob the equations of their generality.

Why not, for example, also incorporate the conversion factor of 144 in^2/ft^2 into equations where pressure is expressed in lbf/in_2? We prefer to omit all conversion factors in equations, and introduce them only as needed in evaluating expressions numerically. If the reader is in any doubt, units should *always* be checked when performing calculations.

SI Units.

Table: SI Units

Physical Quantity	*Name of Unit*	*Symbol for Unit*	*Definition of Unit*
Basic Units			
Length	meter	m	–
Mass	kilogram	kg	–
Time	second	s	–
Temperature	degree Kelvin	K	–
Supplementary Unit			
Plane angle	radian	rad	—
Derived Units			
Acceleration			m/s^2
Angular velocity			rad/s
Density			kg/m^3
Energy	joule	J	kg m^2/s^2
Force	newton	N	kg m/s^2
Kinematic viscosity			m^2/s
Power	watt	W	kg m^2/s^3 (J/s)
Pressure	pascal	Pa	kg/m s^2 (N/m^2)
Velocity			m/s
Viscosity			kg/m s

The most systematically developed and universally accepted set of units occurs in the *SI* units or *Systeme International d'Unit'es*; the subset we mainly need is shown in Table.

The basic units are again the meter, kilogram, and second (m, kg, and s); from these, certain *derived* units can also be obtained. Force (kg m/s^2) has already been discussed; energy is the product of force and length; power amounts to energy per unit time; surface tension is energy per unit area or force per unit length, and so on. Some of the units have names, and these, together with their abbreviations. Tradition dies hard, and certain other "metric" units are so well established that they may be used as *auxiliary* units; these are shown in Table. The *gram* is the classic example. Note that the basic SI unit of mass (kg) is even represented in terms of the gram, and has not yet been given a name of its own.

Table: Auxiliary Units Allowed in Conjunction with SI Units

Physical Quantity	*Name of Unit for*	*Symbol Unit*	*Definition of Unit*
Area	hectare	ha	10^4 m^2
Kinematic viscosity	stokes	St	10^{-4} m^2/s
Length	micron	μm	10^{-6} m
Mass	tonne	t	10^3 kg = Mg
	gram	g	10^{-3} kg = g
Pressure	bar	bar	10^5 N/m^2
Viscosity	poise	P	10^{-1} kg/m s
Volume	liter	l	10^{-3} m^3

Table shows some of the acceptable prefixes that can be used for accommodating both small and large quantities. For example, to avoid an excessive number of decimal places, 0.000001 s is normally better expressed as 1 μs (one microsecond). Note also, for example, that 1 μkg should be written as 1 mg—one prefix being better than two.

Table: Prefixes for Fractions and Multiples

Factor	*Name*	*Symbol*	*Factor*	*Name*	*Symbol*
10^{-12}	pico	p	10^3	kilo	k
10^{-9}	nano	n	10^6	mega	M
10^{-6}	micro	μ	10^9	giga	G
10^{-3}	milli	m	10^{12}	tera	T

Some of the more frequently used conversion factors are given in Table.

HYDROSTATICS

Variation of pressure with elevation.

Here, we investigate how the pressure in a stationary fluid varies with elevation *z*. The result is useful because it can answer questions such as "What forces are exerted on the walls of an oil storage tank?" Consider a hypothetical differential cylindrical element of fluid of cross-sectional area *A*, height *dz*, and volume *Adz*, which is also surrounded by the same fluid. Its weight, being the downwards gravitational force on its mass, is $dW = \rho A\, dz\, g$. Two completely equivalent approaches will be presented:

Table: Commonly Used Conversion Factors

Area	1 mile^2	=	640 acres
	1 acre	=	0.4047 ha
Energy	1 BTU	=	1,055 J
	1 cal	=	4.184 J
	1 J	=	0.7376 ft lbf
	1 erg	=	1 dyne cm
Force	1 lb_f	=	4.448 N
	1 N	=	0.2248 lbf
Length	1 ft	=	0.3048 m
	1 m	=	3.281 ft
	1 mile	=	5,280 ft
Mass	1 lbm	=	0.4536 kg
	1 kg	=	2.205 lb_m
Power	1 HP	=	550 ft lb_f/s
	1 kW	=	737.6 ft lb_f/s
Pressure	1 atm	=	14.696 lb_f/in^2
	1 atm	=	1.0133 bar
	1 atm	=	1.0133 × 105 Pa
Time	1 day	=	24 hr
	1 hr	=	60 min
	1 min	=	60 s
Viscosity	1 cP	=	2.419 lb_m/ft hr
	1 cP	=	0.001 kg/ms
	1 cP	=	0.000672 lb_m/ft s
	1 lb_f s/ft^2	=	4.788 × 10^4 cP
Volume	1 ft^3	=	7.481 U.S. gal
	1 U.S. gal	=	3.785 l
	1 m^3	=	264.2 U.S. gal

Method 1.

Let *p* denote the pressure at the base of the cylinder; since *p* changes at a rate *dp/dz* with elevation, the pressure is found either from Taylor's expansion or the definition of a derivative to be $p + (dp/dz)dz$ at the top of the cylinder. Hence, the fluid exerts an *upward* force of *pA* on the base of the cylinder, and a *downward* force of $[p+(dp/dz)dz]A$ on the top of the cylinder. Next, apply

Newton's second law of motion by equating the net upward force to the mass times the acceleration—which is zero, since the cylinder is stationary:

$$\underbrace{pA - \left(p + \frac{dp}{dz}dz\right)A}_{\text{Net pressure force}} - \underbrace{\rho A\,dz\,g}_{\text{Weight}} = \underbrace{(\rho A\,dz)}_{\text{mass}} \times 0 = 0.$$

Cancellation of pA and division by Adz leads to the following *differential equation,* which governs the rate of change of pressure with elevation:

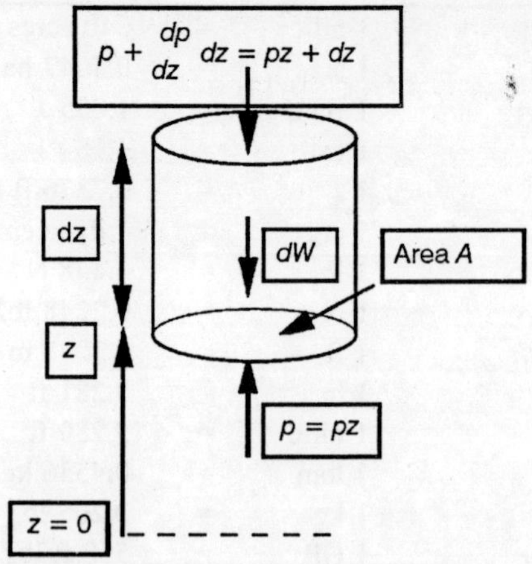

Fig. Forces acting on a cylinder of fluid

Method 2.

Let pz and $pz+dz$ denote the pressures at the base and top of the cylinder, where the elevations are z and $z+dz$, respectively. Hence, the fluid exerts an *upward* force of pzA on the base of the cylinder, and a *downward* force of $pz+dzA$ on the top of the cylinder. Application of Newton's second law of motion gives:

Isolation of the two pressure terms on the left-hand side and division by Adz gives:

$$\underbrace{p_z A - p_{z+dz}}_{\text{Net pressure force}}\ \underbrace{A - \rho A\,dz\,g}_{\text{Weight}} = \underbrace{(\rho A\,dz)}_{\text{mass}} \times 0 = 0.$$

Isolation of the two pressure terms on the left-hand side and division by Adz gives:

$$\frac{p_{z+dz} - p_z}{dz} = -\rho g.$$

As dz tends to zero, the left-hand side of equation becomes the derivative dp/dz, leading to the same result as previously

$$\frac{dp}{dz} = -\rho g.$$

The same conclusion can also be obtained by considering a cylinder of *finite* height Δz and then letting Δz approach zero.

Note that equation predicts a pressure *decrease* in the vertically upward direction at a rate that is proportional to the local density. Such pressure variations can readily be detected by the ear when traveling quickly in an elevator in a tall building, or when taking off in an airplane. The reader must thoroughly understand *both* the above approaches. For most of this book, we shall use Method 1, because it eliminates the steps of taking the limit of $dz \to 0$ and invoking the definition of the derivative.

Pressure in a liquid with a free surface.

In figure the pressure is ps at the free surface, and we wish to find the pressure p at a depth H below the free surface—of water in a swimming pool, for example.

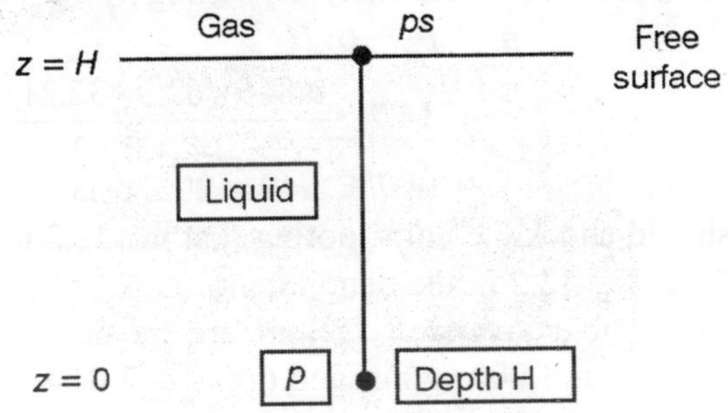

Fig. Pressure at a depth H

Separation of variables in equation and integration between the free surface ($z = H$) and a depth H ($z = 0$) gives:

$$\int_{ps}^{p} dp = -\int_{H}^{0} \rho g\, dz.$$

Assuming– quite reasonably—that ρ and g are constants in the liquid, these quantities may be taken outside the integral, yielding:

$$p = p_s + \rho g H,$$

which predicts a *linear* increase of pressure with distance downward from the free surface. For large depths, such as those encountered by deep-sea divers, very substantial pressures will result.

EXAMPLE PRESSURE IN AN OIL STORAGE TANK

What is the absolute pressure at the bottom of the cylindrical tank of figure, filled to a depth of H with crude oil, with its free surface exposed to the atmosphere? The specific gravity of the crude oil is 0.846. Give the answers for (a) H= 15.0 ft (pressure in lbf/in2), and (b) H= 5.0 m (pressure in Pa and bar). What is the purpose of the surrounding dike?

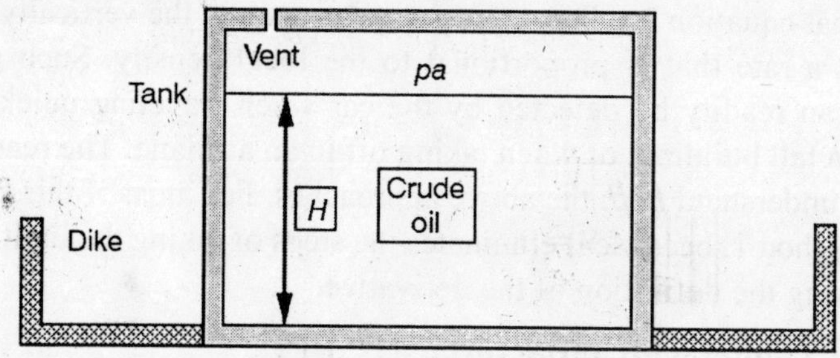

Fig. Crude oil storage tank.

Solution

- The pressure is that of the atmosphere, *pa*, plus the increase due to a column of depth H = 15.0 ft. Thus, setting $p_s = p_a$, equation gives:

$$p = p_a + \rho g H$$

$$= 14.7 + \frac{0.846 \times 62.3 \times 32.215.0}{144 \times 32.2}$$

$$= 14.7 + 5.49 = 20.2 \text{ psia.}$$

The reader should check the units, noting that the 32.2 in the numerator is g [=] ft/s^2, and that the 32.2 in the denominator is gc [=] lbm ft/lbf s^2.

- For SI units, no conversion factors are needed. Noting that the density of water is 1,000 kg/m^3, and that p_a = 1.01 × 10_5 Pa absolute:

$$p = 1.01 \times 10^5 + 0.846 \times 1{,}000 \times 9.81 \times 5.0 = 1.42 \times 10^5 \text{ Pa} = 1.42 \text{ bar.}$$

In the event of a tank rupture, the dike contains the leaking oil and facilitates prevention of spreading fire and contamination of the environment.

Epilogue: When he arrived at work in an oil refinery one morning, the author saw firsthand the consequences of an inadequately vented oil-storage tank.

Rain during the night had caused partial condensation of vapour inside the tank, whose pressure had become sufficiently lowered so that the external atmospheric pressure had crumpled the steel tank just as if it were a flimsy tin can. The refinery manager was not pleased.

EXAMPLE MULTIPLE FLUID HYDROSTATICS

The U-tube shown in figure contains oil and water columns, between which there is a long trapped air bubble. For the indicated heights of the columns, find the specific gravity of the oil.

Solution: The pressure p^2 at point 2 may be deduced by starting with the pressure p^1 at point 1 and adding or subtracting, as appropriate, the hydrostatic pressure changes due to the various columns of fluid. Note that the width of the U-tube (2.0 ft) is irrelevant, since there is no change in pressure in the horizontal leg. We obtain:

$$p_2 = p_1 + \rho_o g h_1 + \rho_a g h_2 + \rho_w g h_3 - \rho_w g h_4,$$

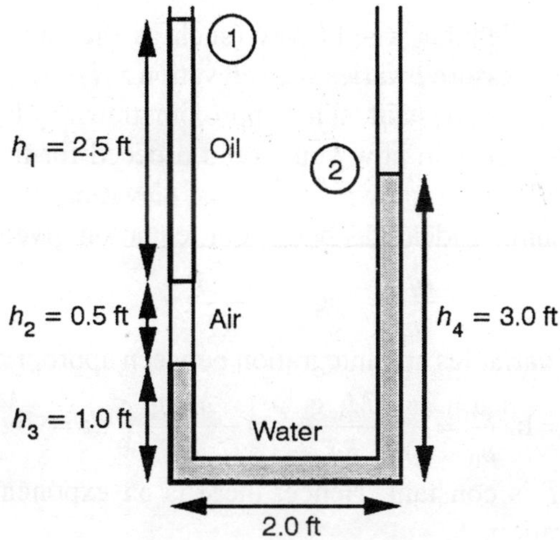

Fig. Oil/air/water system.

in which ρ_o, ρ_a, and ρ_w denote the densities of oil, air, and water, respectively. Since the density of the air is very small compared to that of oil or water, the term containing ρ_a can be neglected. Also, $p_1 = p_2$, because both are equal to atmospheric pressure. Equation can then be solved for the specific gravity *so* of the oil:

$$s_o = \frac{\rho_o}{\rho_w} = \frac{h_4 - h_3}{h_1} = \frac{3.0 - 1.0}{2.5} = 0.80.$$

Pressure Variations in a Gas

For a gas, the density is no longer constant, but is a function of pressure (and of temperature—although temperature variations are usually less significant than those of pressure), and there are two approaches:

- For *small* changes in elevation, the assumption of constant density can still be made, and equations similar to equation are still approximately valid.
- For moderate or large changes in elevation, the density in equation is given by, $\rho = M_w p / RT$ or| $\rho = M_w p / ZRT$, depend-ing on whether the gas is ideal or nonideal. It is understood that *absolute* pressure and temperature must *always* be used whenever the gas law is involved. A separation

of variables can still be made, followed by integration, but the result will now be more complicated because the term *dp/p* occurs, leading—at the simplest (for an isothermal situation)—to a decreasing *exponential* variation of pressure with elevation.

EXAMPLE PRESSURE VARIATIONS IN A GAS

For a gas of molecular weight *M*w (such as the earth's atmosphere), investigate how the pressure p varies with elevation z if $p = p_0$ at $z = 0$. Assume that the temperature T is constant. What approximation may be made for *small* elevation increases? Explain how you would proceed for the nonisothermal case, in which $T = T(z)$ is a known function of elevation.

Solution: Assuming ideal gas behaviour, equation give:

$$\frac{dp}{dz} = -\rho g = -\frac{M_w p}{RT} g.$$

Separation of variables and integration between appropriate limits yields:

$$\int_{p_0}^{p} \frac{dp}{p} = \text{In}\frac{p}{p_0} = -\int_0^z \frac{M_w g}{RT} dz = -\frac{M_w g}{RT}\int_0^z dz = -\frac{M_w gz}{RT}$$

Since M_{wg}/RT is constant. Hence, there is an exponential decrease of pressure with elevation.

$$p = p_0 \exp\left(-\frac{M_w g}{RT} z\right).$$

Since a Taylor's expansion gives $e^{-x} = 1 - x + x_2/2 - \ldots$, the pressure is *approximated* by $e^{-x} = 1 - x + x^2/2 - \ldots$ the pressure is *approximated* by:

$$p = p_0\left[1 - \frac{M_w g}{RT} z + \left(\frac{M_w g}{RT}\right)^2 \frac{z^2}{2}\right]$$

For *small* values of *Mwgz/RT*, the last term is an insignificant second-order effect (compressibility effects are unimportant), and we obtain:

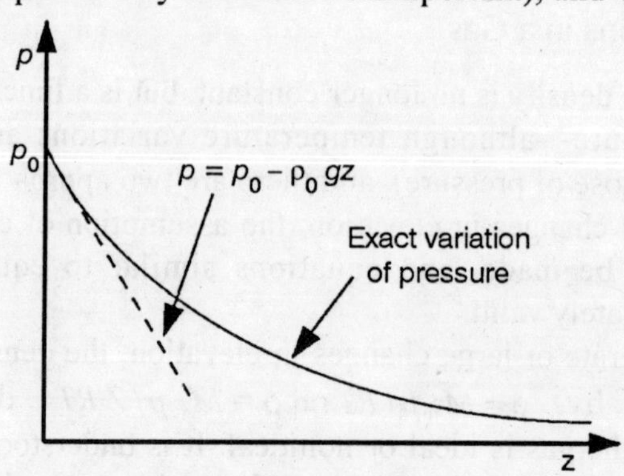

Fig. Variation of gas pressure with elevation

$p = p_0 1 - \dfrac{M_w p_0}{RT} gz = p_0 - \rho_0 gz$, in which ρ_0 is the density at elevati-on

$z = 0$; this approximation—essentially one of constant density and is clearly applicable only for a small change of elevation changes the upper limit on z for which this linear approximation is realistic.

If there are significant elevation changes the approximation of equation *cannot* be used with any accuracy. Observe with caution that the Taylor's expansion is only a vehicle for demonstrating what happens for small values of $M_w gz/RT$. Actual calculations for larger values of $M_w gz/RT$ should be made using equation.

For the case in which the temperature is not constant, but is a known function $T(z)$ of elevation (as might be deduced from observations made by a meteorological balloon), it must be included *inside* the integral:

$$\int_{p_1}^{p_2} \frac{dp}{p} = -\frac{M_w g}{R} \int_0^z \frac{dz}{T(z)}.$$

Since $T(z)$ is unlikely to be a simple function of z, a numerical method will probably have to be used to approximate the second integral of equation.

Total force on a dam or lock gate.

Figure shows the side and end elevations of a dam or lock gate of depth D and width W. An expression is needed for the total horizontal force F exerted by the liquid on the dam, so that the latter can be made of appropriate strength. Similar results would apply for liquids in storage tanks. Gauge pressures are used for simplicity, with $p = 0$ at the free surface and in the air outside the dam. Absolute pressures could also be employed, but would merely add a constant atmospheric pressure everywhere, and would eventually be canceled out. If the coordinate z is measured from the bottom of the liquid upward, the corresponding *depth* of a point below the free surface is $D - z$. Hence, from equation, the differential horizontal force dF on an infinitesimally small rectangular strip of area $dA = Wdz$ is:

$$dF = pWdz = \rho g(D - z)Wdz.$$

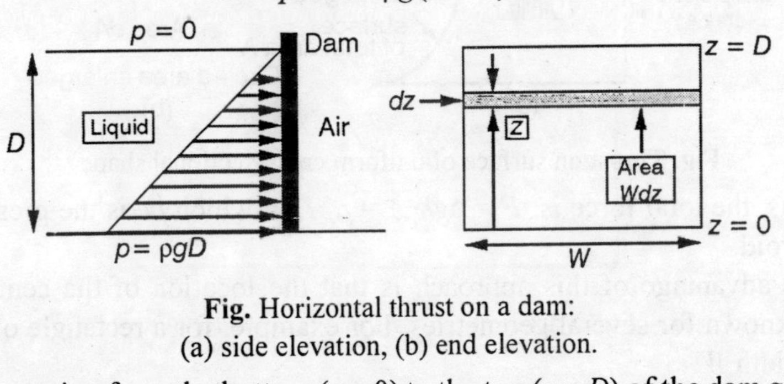

Fig. Horizontal thrust on a dam:
(a) side elevation, (b) end elevation.

Integration from the bottom ($z = 0$) to the top ($z = D$) of the dam gives the total horizontal force:

$$F = \int_0^F dF = \int_0^D \rho g W(D - z)dz = \frac{1}{2}\rho g W D^2.$$

Horizontal Pressure Force on an Arbitrary Plane Vertical Surface.

The preceding analysis was for a regular shape. A more general case, figure shows a *plane* vertical surface of *arbitrary* shape. Note that it is now slightly easier to work in terms of a *downward* coordinate *h*.

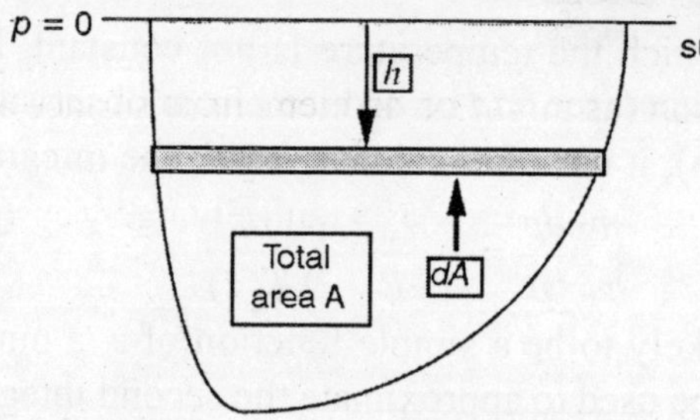

Fig. Side view of a pool of liquid with a submerged vertical surface

Again taking gauge pressures for simplicity (the gas law is not involved), with $p = 0$ at the free surface, the total horizontal force is:

$$\int_A p dA = \int_A \rho g h dA = \rho g A \frac{\int_A h dA}{A}.$$

But the depth *hc* of the centroid of the surface is defined as:

$$h_c \equiv \frac{\int_A h dA}{A}.$$

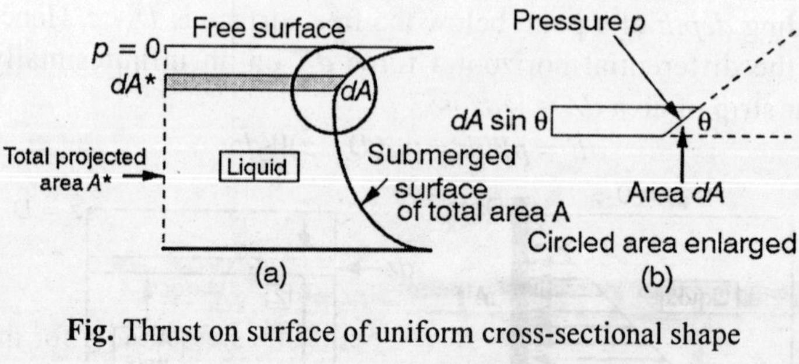

Fig. Thrust on surface of uniform cross-sectional shape

Thus, the total force is: $F = \rho g h_c A = p_c A$, in which *pc* is the pressure at the centroid.

The advantage of this approach is that the location of the centroid is already known for several geometries. For example, for a rectangle of depth *D* and width *W*:

$h_c = \frac{1}{2}D$ and $F = \frac{1}{2}\rho g W D^2$, in agreement with the earlier result of equation. Similarly, for a vertical circle that is just submerged, the depth of the centroid equals its radius. And, for a vertical triangle with one edge coincident with the surface of the liquid, the depth of the centroid equals one-third of its altitude.

Horizontal Pressure Force on a Curved Surface

Figure shows the cross section of a submerged surface that is no longer plane. However, the shape *is* uniform normal to the plane of the diagram. In general, the local pressure force pdA on an element of surface area dA does not act horizontally; therefore, its horizontal component must be obtained by projection through an angle of $(\pi/2 - \theta)$, by multiplying by $\cos(\pi/2 - \theta) = \sin\theta$. The total horizontal force F is then:

$$F = \int_A p \sin\theta \, dA = \int_{A^*} p \, dA^*,$$

in which $dA_* = dA \sin\theta$ is an element of the projection of A onto the hypothetical vertical plane A*.

HYDROSTATIC FORCE ON A CURVED SURFACE

A submarine, whose hull has a circular cross section of diameter D, is just submerged in water of density ρ. Derive an equation that gives the total horizontal force Fx on the left half of the hull, for a distance W normal to the plane of the diagram.

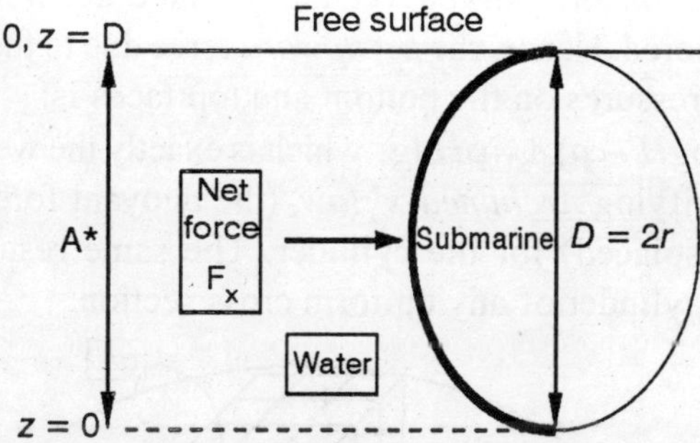

Fig. Submarine just submerged in seawater

If $D = 8$ m, the circular cross section continues essentially for the total length $W = 50$ m of the submarine, and the density of sea water is $\rho = 1{,}026$ kg/m$_3$, determine the total horizontal force on the left-hand half of the hull.

Solution: The force is obtained by evaluating the integral of equation, which is identical to that for the rectangle:

$$F_x = \int_{A^*} p \, dA = \int_{z=0}^{z=D} \rho g W (D - z) dz = \frac{1}{2} \rho g W D^2.$$

Insertion of the numerical values gives:

$$F_x = \frac{1}{2} \times 1{,}026 \times 9.81 \times 50 \times 8.0^2 = 1.61 \times 10^7\, N.$$

Thus, the total force is considerable—about 3.62×10^6 lb_f.

Buoyancy Forces.

If an object is submerged in a fluid, it will experience a net upward or *buoyant* force exerted by the fluid. To find this force, first examine the buoyant force on a submerged circular cylinder of height *H* and cross-sectional area *A*.

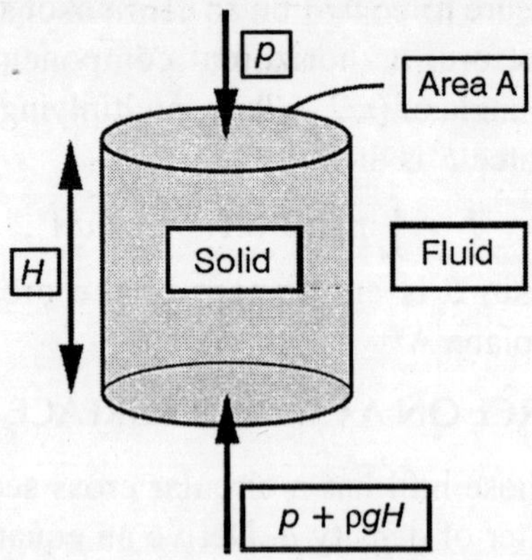

Fig. Pressure forces on a submerged cylinder

The forces on the curved vertical surface act horizontally and may therefore be ignored. Hence, the net *upward* force due to the difference between the opposing pressures on the bottom and top faces is:

$F = (p + \rho gH - p)A = \rho HAg$, which is exactly the weight of the displaced liquid, thus verifying *Archimedes' law*, (the buoyant force equals the weight of the fluid displaced) for the cylinder. The same result would clearly be obtained for a cylinder of any uniform cross section.

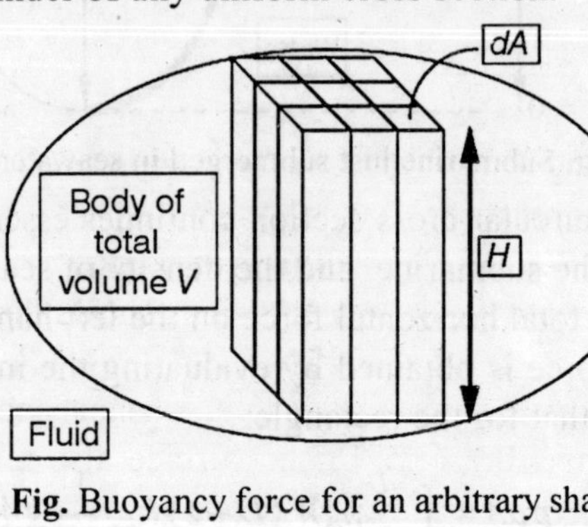

Fig. Buoyancy force for an arbitrary shape.

Figure shows a more general situation, with a body of arbitrary shape. However, Archimedes' law still holds since the body can be decomposed into an infinitely large number of vertical rectangular parallelepipeds or "boxes" of infinitesimally small cross-sectional area *dA*. The effect for one box is then summed or "integrated" over all the boxes, and again gives the net upward buoyant force as the weight of the liquid displaced.

APPLICATION OF ARCHIMEDES' LAW

Consider the situation in figure, in which a barrel rests on a raft that floats in a swimming pool. The barrel is then pushed off the raft, and may either float or sink, depending on its contents and hence its mass. The cross-hatching shows the volumes of water that are displaced. For each of the cases, determine whether the water level in the pool will rise, fall, or remain constant, relative to the initial level in (a).

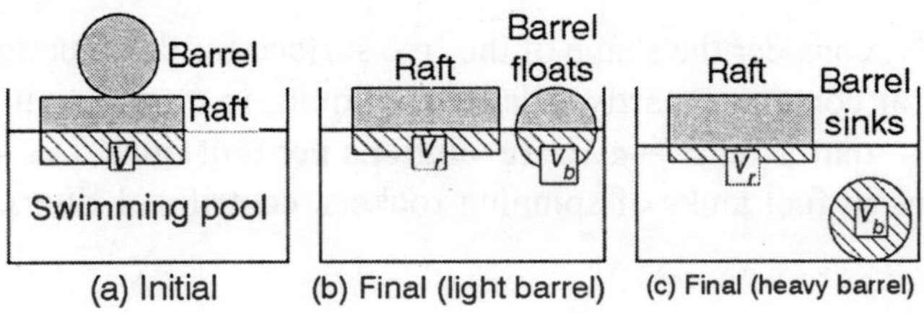

(a) Initial (b) Final (light barrel) (c) Final (heavy barrel)

Fig. Raft and barrel in swimming pool: (a) initial positions, (b) light barrel rolls off and floats, (c) heavy barrel rolls off and sinks. The cross-hatching shows volumes below the surface of the water

Solution:

Initial state. Let the masses of the raft and barrel be *Mr* and *Mb*, respectively. If the volume of displaced water is initially *V* in (a), Archimedes' law requires that the total weight of the raft and barrel equals the weight of the displaced water, whose density is

$$\rho : (M_r + M_b)g = V\rho g.$$

Barrel floats. If the barrel floats, as in (b), with submerged volumes of *Vr* and *Vb* for the raft and barrel, respectively, Archimedes' law may be applied to the raft and barrel separately:

Raft: $m_r g = V_r \rho g,$ Barrel: $m_b g = V_b \rho g.$

Addition and compare of the two equations shows that:

$$V_r + V_b = V.$$

Therefore, since the volume of the water is constant, and the total displaced volume does not change, the level of the surface also remains *unchanged.* *Barrel sinks.* Archimedes' law may still be applied to the raft, but the weight of the water displaced by the barrel no longer suffices to support the weight of the barrel, so that:

Raft: $m_r g = V_r \rho g$, Barrel: $m_b g > V_b \rho g$.

Addition of the two relations and comparison with equation shows that:

$$V_r + V_b < V.$$

Therefore, since the volume of the water in the pool is constant, and the total displaced volume is *reduced*, the level of the surface *falls*. This result is perhaps contrary to intuition: since the *whole* volume of the barrel is submerged in (c), it might be thought that the water level will rise above that in (b). However, because the barrel must be heavy in order to sink, the load on the raft and hence V_r are substantially reduced, so that the total displaced volume is also reduced. This problem illustrates the need for a complete analysis rather than jumping to a possibly erroneous conclusion.

PRESSURE CHANGE CAUSED BY ROTATION

Finally, consider the shape of the free surface for the situation, in which a cylindrical container, partly filled with liquid, is rotated with an angular velocity ω—that is, at $N = \omega/2\pi$ revolutions per unit time. The analysis has applications in fuel tanks of spinning rockets, centrifugal filters, and liquid mirrors.

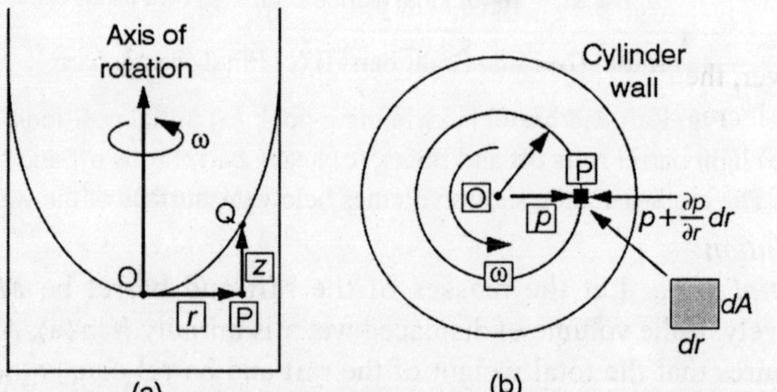

Fig. Pressure changes for rotating cylinder: (a) elevation, (b) plan

Point O denotes the origin, where $r = 0$ and $z = 0$. After a sufficiently long time, the rotation of the container will be transmitted by viscous action to the liquid, whose rotation is called a *forced vortex.* In fact, the liquid spins as if it were a *solid body*, rotating with a uniform angular velocity ω, so that the velocity in the direction of rotation at a radial location r is given by $v\theta = r\omega$. It is therefore appropriate to treat the situation similar to the hydrostatic investigations already made.

Suppose that the liquid element P is essentially a rectangular box with crosssectional area dA and radial extent dr. (In reality, the element has slightly tapering sides, but a more elaborate treatment taking this into account will yield identical results to those derived here.) The pressure on the inner face is

p, whereas that on the outer face is $p + (\partial p/\partial r)dr$. Also, for uniform rotation in a circular path of radius r, the acceleration toward the centre O of the circle is $r\omega 2$. Newton's second law of motion is then used for equating the net pressure force toward O to the mass of the element times its acceleration:

$$\underbrace{\left(p+\frac{\partial p}{\partial r}dr-p\right)dA}_{\text{Net pressure force}}=\underbrace{\rho(dA\,dr)}_{\text{Mass}}r\omega^2.$$

Note that the use of a *partial* derivative is essential, since the pressure now varies in both the horizontal (radial) *and* vertical directions. Simplification yields the variation of pressure in the radial direction:

$$\frac{\partial p}{\partial r}=\rho r\omega^2,$$

so that pressure *increases* in the radially outward direction. Observe that the gauge pressure at all points on the interface is zero; in particular, $p_O = p_Q = 0$. Integrating from points O to P (at constant z):

$$\int_{p=0}^{p_P} dp=\rho\omega^2\int_0^r r\,dr,$$

$$p_P=\frac{1}{2}\rho\omega^2r^2$$

However, the pressure at P can also be obtained by considering the usual hydrostatic increase in traversing the path QP:

$$p_P = \rho g z.$$

Elimination of the intermediate pressure p_P between equations relates the elevation of the free surface to the radial location:

$$z = \frac{\omega^2 r^2}{2g}.$$

Thus, the free surface is *parabolic* in shape; observe also that the density is not a factor, having been canceled from the equations.

There is another type of vortex—the *free* vortex—that is also important, in cyclone dust collectors and tornadoes, for example. There, the velocity in the angular direction is given by $vE = c/r$, where c is a constant, so that vE is inversely proportional to the radial position.

OVERFLOW FROM A SPINNING CONTAINER

A cylindrical container of height H and radius a is initially half-filled with a liquid. The cylinder is then spun steadily around its vertical axis Z-Z. At what value of the angular velocity ω will the liquid just start to spill over the top of the container? If $H = 1$ ft and $a = 0.25$ ft, how many rpm (revolutions per minute) would be needed?

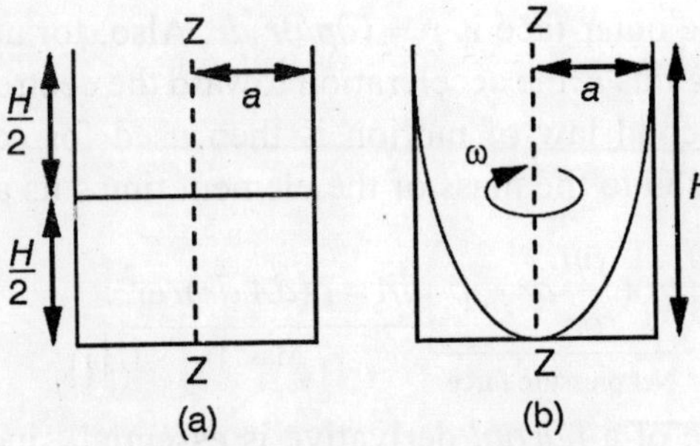

Fig. Geometry of a spinning container: (a) at rest, (b) on the point of overflowing

Solution: From equation. the shape of the free surface is a parabola. Therefore, the air inside the rotating cylinder forms a paraboloid of revolution, whose volume is known from calculus to be exactly one-half of the volume of the "circumscribing cylinder," namely, the container.8 Hence, the liquid at the centre reaches the bottom of the cylinder *just* as the liquid at the curved wall reaches the top of the cylinder. In equation, therefore, set $z = H$ and $r = a$, giving the required angular velocity:

$$w = \sqrt{\frac{2gH}{a^2}}.$$

For the stated values:

$$\omega = \sqrt{\frac{2\times 32.2\times 1}{0.25^2}} = 32.1\frac{\text{rad}}{\text{s}}, \qquad N = \frac{\omega}{2\pi} = \frac{32.1\times 60}{2\pi} = 306.5 \text{ rpm}.$$

Chapter 2

Physical Basics of Fluid

SOLIDS AND FLUIDS

All substances of our natural and technical environment can be subdivided into solid, liquid and gaseous media, on the basis of aggregation. This subdivision is considered in many fields of engineering in order to point out important differences concerning the properties of the substances. This could also be applied to fluid mechanics, however, this would not be particularly advantageous. It is rather recommended to employ fluid mechanics aspects to achieve a subdivision of media appropriate for the treatment of fluid flow processes. To this end, the term fluid is introduced for designating all those substances that cannot be classified clearly as solids.

From the point of view of fluid mechanics, all media can be subdivided into solids and fluids, the difference between both groups being that solids possess elasticity as an important property, while fluids have viscosity as a characteristic property. Shear stresses imposed on to a solid from outside lead to inner elastic shear forces which prevent irreversible changes of position of molecules of the solid. When, on the contrary, external shear stresses are imposed on to fluids, they react with the build-up of velocity gradients, the build-up of the gradient occurring through the molecule-dependent momentum transport, i.e. through the fluid viscosity. Thus elasticity (solids) and viscosity (liquids) are the properties of matters that are employed in fluid mechanics for subdividing media. However there are few exceptions to this subdivision: such as in the case of some of the matters in rheology exhibit mixed properties to such an extent that for small deformations they behave like solids and behave like liquids in the case of large deformations.

At this point, attention is drawn to another important fact regarding the characterisation of fluid properties. A fluid tries to evade smallest external shear stresses by starting to flow. Hence it can be inferred that a fluid at rest is characterized by a state which is free of external shear stresses. Each area in a fluid at rest can therefore experience normal forces only. When shear stresses occur in a medium at rest, this medium is assigned to solids. The viscous forces accompanied by external motion observed in a fluid should not be mistaken with the elastic forces in solids. The viscous force cannot be analogously addressed as the elastic force. This is the case for all liquids and gases which take part in fluid motion. The present book is dedicated to such a treatment of fluid flows. On the basis of the above mentioned treatments of fluid flows, the fluids in motion can simply be classified as media free from stresses and distinguished from solids. The "shear stresses" that are often introduced when treating fluid flows of common liquids and gases represent molecule-dependent momentum-transport terms in reality.

Neighboring layers of a flowing fluid, having a velocity gradient, do not interact with each another through "shear stresses" but through an exchange of momentum due to the molecular motion. This can be explained by simplified derivations aiming at the physical understanding of the molecular processes. The derivations are carried out for an ideal gas, since they can be understood particularly well in this case. The results from these derivations can therefore not be transferred in all aspects to fluids with more complex properties.

For further subdivision of the fluids, it is recommended to make use of their response to normal stresses or pressure on fluid elements. When a fluid reacts to pressure changes by changing its volume and consequently density, the fluid is called compressible. When no volume or density changes occur with pressure differences, the fluid is regarded as incompressible. Although strictly speaki ng, incompressible fluids do not exist. However, such a subdivision is reasonable and moreover useful and this will also be shown in following derivations of basic fluid mechanics equations. Indeed, this subdivision -distinguishes liquids from gases.

In general, fluids can be further classified into liquid and gases. Liquids and some plastic materials show very small expansion coeffcients (typical values for isobaric expansion are $\beta_P = 10 \cdot 10^{-6}\ /K$, while gases have much larger expansion coeffcients (typical values are $\beta_P = 1000 \cdot 10^{-6}/K$). A comparison of both subgroups of fluids shows that liquids fulfill the condition of incompressibility with a precision that is adequate for the most of flow problems.

On this assumption, the basic equations of fluid mechanics can be simplified, as the following derivations show; in particular the number of equations needed for the general description of fluid flow processes being reduced from 6 to 4. The simplifications of the basic equations for incompressible media allow a considerable reduction in the complexity of the flow solutions in simple and complex geometries, as in the case of problems without heat transfer the energy equation does not have to be solved.

The simplified basic equations of fluid mechanics derived for incompressible media can occasionally also be applied to the flows of compressible fluids, such as cases where the density variations occurring in the entire flow field are small as compared to the fluid density.

For further characterization of a fluid, it is referred to the well-known fact that solids conserve their form, while a fluid volume has no form of its own, but assumes the form of the container in which it is kept. Liquids differ from gases in terms of the area taken by the fluids constituting only part of the container, while the remaining part is either not filled or contains a gas, there exists a free surface between them. Such a surface does not exist when the container is filled only with a gas. The gas takes up the entire container volume.

Finally, it can be concluded that there is a number of media those can only be categorized in a limited way according to the above classification. They are e.g. media that consist of two phase mixtures. These have properties that cannot be classified so easily. This holds also for a number of other media that can, as per the above classification, be assigned neither to the solids nor to the fluids and they start to flow only above a certain value of the "shear stress". Media of this kind would be excluded in this book, so that the above indicated classifications of media into solids and fluids remain valid. Further restrictions to the fluid properties that are applied in dealing with flow problems in this book are clearly indicated in the respective sections.

In this way it should be possible to avoid mistakes that often arise from the derivations of fluid mechanics equations for simplified fluid properties and/or simplified flow cases.

MOLECULAR PROPERTIES AND QUANTITIES OF CONTINUUM MECHANICS

As all matter consists of molecules or aggregations of molecules, all macroscopic properties of matter can be described by molecular properties. Thus it is possible to evaluate all properties of fluids that are of importance for considerations in fluid mechanics linked to properties of molecules, i.e. to describe the macroscopic properties of fluids by molecular properties. However, such a description of the state of matter requires much efforts due to necessary formalism and moreover would be unclear. A molecular-theoretical presentation of fluid properties would hardly be appropriate to supply practice-oriented fluid mechanics information useful for an engineer in easily comprehensible (and also applicable) form. For this reason, it is more advantageous to introduce quantities of continuum mechanics for describing fluid properties.

The connection between continuum mechanics quantities, introduced in fluid mechanics and the molecular properties should be considered as the most important links between the two different ways of description and presentation of fluid properties.

Some state parameters such as density r, pressure P, temperature T are essential for the description of fluid mechanics processes and can be expressed in terms of molecular quantities for ideal gases. From the following derivations one can infer that the "effects" of molecules or molecular properties on fluid elements or control volumes are taken into consideration by introducing the properties, density r, pressure P, temperature T, viscosity μ etc. in an "integral form" and it is sufficient for fluid mechanics considerations. Therefore, continuum mechanics considerations do not neglect the molecular structure of the fluids, but take them into account in integral form, i.e. averaged over several molecules.

The mass per unit volume is called specific density ρ of a matter. For a fluid element this quantity depends on its position in space, i.e. $x_i = (x_1, x_2, x_3)$, and also on time t, so that generally

$$\rho(x_i, t) = \lim_{\Delta V \to \delta V_{\Re}} \frac{\Delta M}{\Delta V} = \frac{\delta m \Re_1}{\delta V_{\Re}}$$

holds. If n is considered to be the mean number of the molecules existing per unit volume and with m the mass available per molecule, the following connection holds:

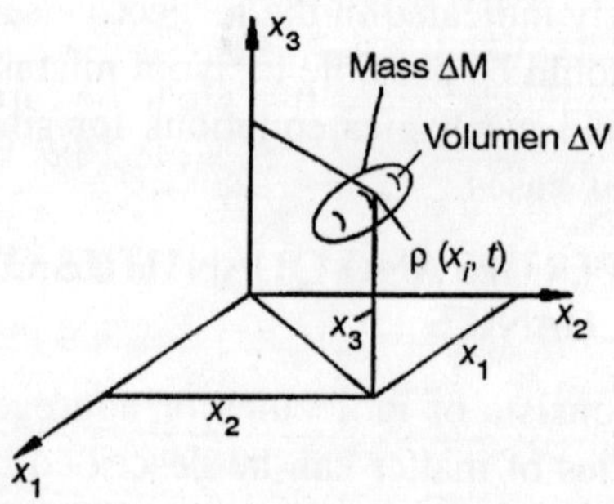

Fig. Definition of the Fluid Density $\rho(x_i, t)$.

$$\rho(x_i, t) = mn(x_i, t)$$

The density of the matter is thus identical with the number of molecules avail-able per unit volume, multiplied by the mass of a single molecule. Therefore, changes in density in space and in time correspond to spatial and temporal changes of the mean number of the molecules available per unit volume.

By stochastic consideration of thermal molecular motions in a fluid volume having a large number of molecules under normal conditions, a mean number of molecules can be specified at time t with sufficient clarity. Volumes in the order of magnitude of 10^{-18} - 10^{-20}m^3 are considered as sufficiently large for arriving at clear definitions of density. The treatments of flow processes in fluid mechanics are usually carried out in a much larger volume therefore, the specification of "mean" number of molecules in order to designate the available mass in the considered central volume or density is appropriate.

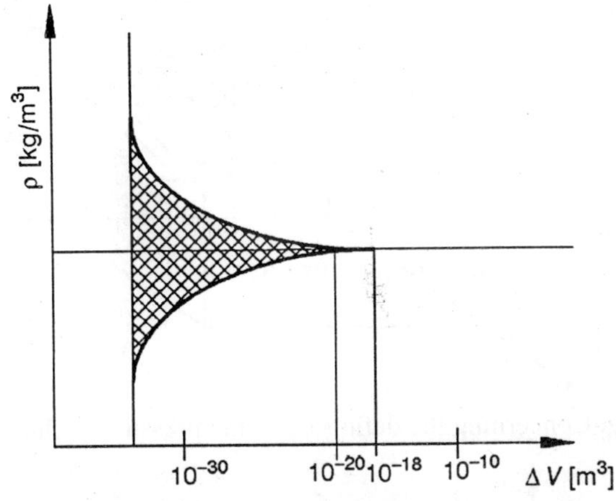

Fig. Fluctuations while determining the density of fluids

The local density $\rho(x_i, t)$ therefore describes a property of matter that is essential for fluid mechanics with a precision that is almost always sufficient. The control volume in the fluid mechanics considerations is selected to such a large extent that the determination of a local density value completely fulfills the requirements of the considerations that are to be carried out from the flow mechanics point of view, in spite of the molecular basic structure of the considered fluids.

Similar considerations can also be made for the pressure that occurs in a fluid at rest and which is defined as the force acting per area unit, i.e.

$$P\,(x_i, t) = \lim_{\Delta F_j \to \delta F_j} \frac{\Delta K_j}{\Delta F_j}$$

From the molecular-theoretical point of view, the pressure effect is defined as the temporal momentum change occurring per unit area, i.e. the force which the molecules experience and exert when colliding in an elastic way with the considered area. The following relation holds :

$$P = \frac{1}{3} mn\overline{u^2} = \frac{1}{3}\rho\overline{u^2}$$

In equation m is the molecular mass, n the number of the molecules per unit volume and u the thermal speed of the molecules.

Analogous to the above volume dimensions, it can be stated that most of the fluid mechanics considerations do not require area resolutions that fall below 10^{-12} bis $10^{-13} m^2$ and therefore the mean numbers of molecules are suffcient to have the force effect of the molecules on the areas. This, however, corresponds to the indication of a pressure, $P\,(x_i, t)$ for the fluid.

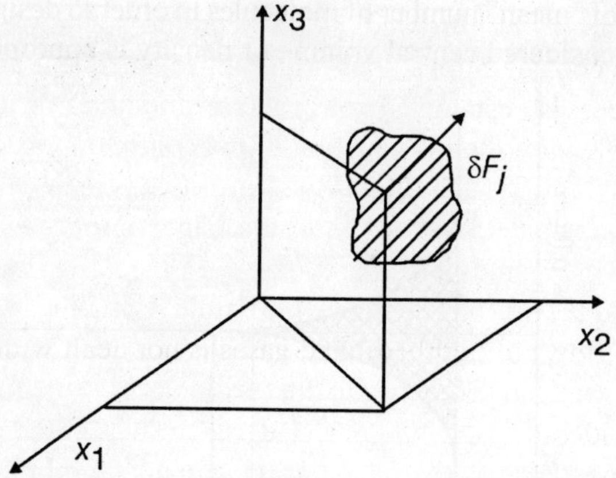

Fig. Concerning the definition of the pressure in fluid $P(x_i, t)$

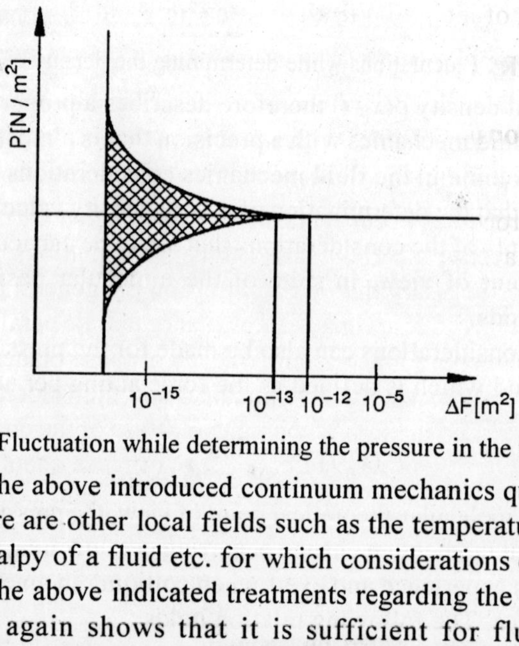

Fig. Fluctuation while determining the pressure in the fluid

Similar to the above introduced continuum mechanics quantities $\rho(x_i, t)$ and $P(x_i, t)$ there are other local fields such as the temperature, the internal energy and enthalpy of a fluid etc. for which considerations can be repeated analogously to the above indicated treatments regarding the density and the pressure. This again shows that it is sufficient for fluid mechanics considerations to neglect the complex molecular properties and to introduce continuum mechanics quantities into the fluid mechanics considerations that correspond to mean values of corresponding molecular parameters. fluid mechanics considerations can therefore be carried out on the basis of continuum mechanics quantities.

However, there are some important domains in fluid mechanics where continuum considerations are not appropriate, e.g. the investigation of flows in highly diluted gas systems. No clear continuum mechanics quantities can be defined there for the volume and the areas with which fluid mechanics processes are to be resolved, as the required spatial resolution of the flow mechanics considerations does not promise sufficient numbers of molecules for the necessary establishment of the mean values of the parameters which are available with the introduction of the continuum mechanics quantities.

When treating such fluid flows, priority has to be given to the molecular-theoretical considerations of fluid mechanics processes as compared to the continuum mechanics considerations.

In the present introduction of the fluid mechanics of viscous media, the domain of flows of highly diluted gases is not dealt with, so that all required considerations can take place in the terminology of continuums mechanics. In these considerations, molecular effects, e.g. within the conservation law for mass, momentum and energy are presented in integral form, i.e. the molecular structure of the considered fluids is not neglected but taken into consideration in the form of integral quantities.

TRANSPORT PROCESSES IN NEWTONIAN FLUIDS

General Considerations

When treating fluids with the transport of heat and molecular mass transport processes occur that cannot be neglected and that hence have to be taken into account in the general transport equations. A physically correct treatment is necessary that orients itself on the general representations and is indicated below. These figures show planes that lie parallel to the $x_1 - x_3$ plane of a cartesian coordinate system. In each of these planes the temperature $T =$ const (a) the concentration $c =$ const (b) and the velocity $(U_j) =$ const (*c*) are such that when taking into account the increase of the quantities in x_2-direction $= x_i$-direction, a positive gradient in each of these quantities exists. It is these gradients that result in molecular transports of heat, chemical species and momentum.

The heat transport occurring as a consequence of the molecular motion is given by the Fourier law of heat conduction and the mass transport occurring analogously given by the Fick's law of diffusion. Fourier law of heat conduction:

$$\dot{q}_i = -\lambda \frac{\partial T}{\partial x_i}$$

λ = coefficient of heat conduction Fick law of diffusion

$$\dot{m}_i = -D \frac{\partial c}{\partial x_i}$$

D = diffusion coefficient

In an analogous way, the molecule-dependent momentum transport also has to be described by the Newtonian law which in the presence of only one velocity component U_j can be stated as follows here.

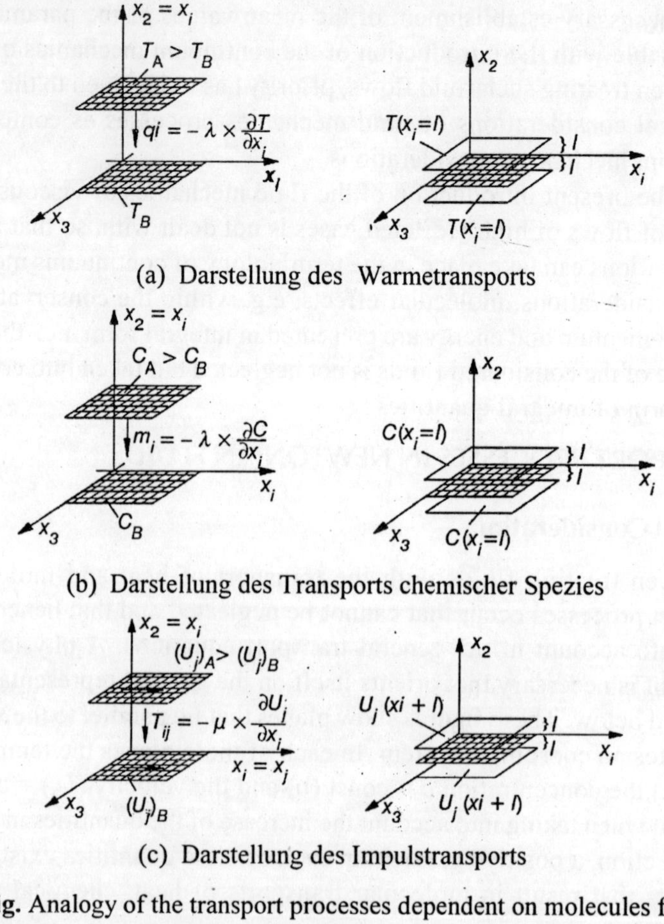

Fig. Analogy of the transport processes dependent on molecules for (a) heat transport, (b) matter transport and (c) momentum transport.

$$\tau_{ij} = -\mu \frac{\partial U_j}{\partial x_i}$$

Newtonian law of momentum transport:

μ = dynamic viscosity.

In τ_{ij}, $\dot{q}_i$ and m_i the direction i indicates the "molecular transport direction", and j indicates the components of the velocity vector for which momentum transport considerations are carried out. The complete equation for τ_{ij}, in the presence of a Newtonian medium can be indicated as follows:

$$\tau_{ij} = \left(\frac{\partial U_j}{\partial x_i} + \frac{\partial U_i}{\partial x_j}\right) + \frac{2}{3}\mu\delta_{ij}\left(\frac{\partial U_k}{\partial x_k}\right)^2$$

τ_{ij} represents momentum transport per unit area for unit time or "stress" i.e. force per unit area. It is therefore often designated as "shear stress" and the sign before μ is chosen positive. This has to be taken into account when comparing representations in this book with corresponding statements in other books. The existing differences in the viewpoints are considered in following two annotations.

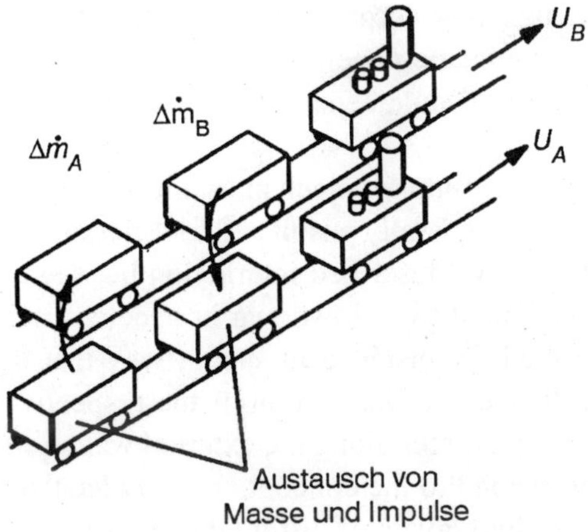

Fig. Exchange of mass and momentum Illustrative explanation of tij as momentum transport

Annotation: The following illustrative example shall show how the viscosity-dependent momentum transport introduced in the continuum mechanics is reflected through the motion of molecules.

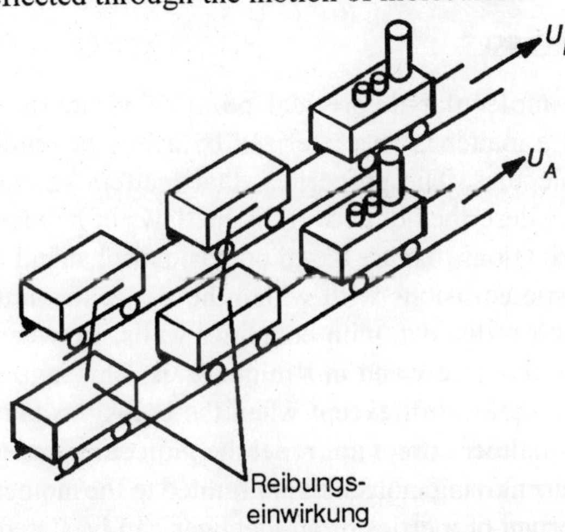

Fig. Influence of friction Illustrative representation of tij as friction terms

Two passenger trains may run next to one another with different speeds. In each of the trains, persons carry sacks along with them. These sacks are being thrown by the passengers of the one train to the passengers in the other train, so that a momentum transfer takes place; it should be noted that the masses m_A and *mB* of the trains do not change. Due to the fact that the persons in the quicker train catch the sacks that are being thrown to them from the slower train, the quicker train is slowed down. In an analogous way the slower train is accelerated. Momentum transfer in the direction of travel takes place by an momentum transport perpendicular to the direction of travel. This idea, transferred to the molecule-dependent momentum transport in fluids, is in accordance with the molecule dependent transport processes that were stated above.

Annotation: In continuums mechanics, the viscous-dependent interaction between fluid layers is generally postulated as "friction forces" between layers. This would, in the above described interaction between trains running along each another, correspond to a slowdown or acceleration by frictional forces that could be applied for instance in such a way that the passengers in the trains exert an influence on the motion of the respective other train by bars with which the friction forces along the external wagon wall are induced. This idea does not correspond to the conception of molecular dependent transport processes between fluid layers of different speeds.

If one carries out physically correct considerations regarding the molecular dependent momentum transport t_{ij}. In addition, considerations are presented on the following pages concerning pressure, heat exchange and diffusion in gases in order to show the connection between molecular and continuum-mechanics quantities.

PRESSURE IN GASES

From the molecular-theoretical point of view, the gaseous state of aggregation of a matter is characterized by a free or random motion of the atoms and molecules. The properties, that matters assume in this state of aggregation, are described quite well by the laws of an ideal gas. These laws result from derivations that are based on basic mechanical laws and that start from ideal elastic collisions with which the molecules interact among each another and with walls, e.g. with container walls. Between these collisions, the molecules move freely and in straight lines. This is to say that no forces act between the molecules, except when the collisions take place. Likewise container walls neither attract nor repel the molecules and the interactions of the walls with the moving molecules are limited to the moment of the collision. The most important properties of an ideal gas can be stated as follows:

- The volume of the molecules and the atoms is extremely small as com- pared to their distance from one another so that the molecules can be regarded as material points;

- The molecules exert, except for the moment of the collision, neither at-tractive nor repulsive forces on each another;
- For the collisions between two molecules or a molecule and a wall, the laws of the perfect elastic impact hold. (Collisions of two molecules take place exclusively.)

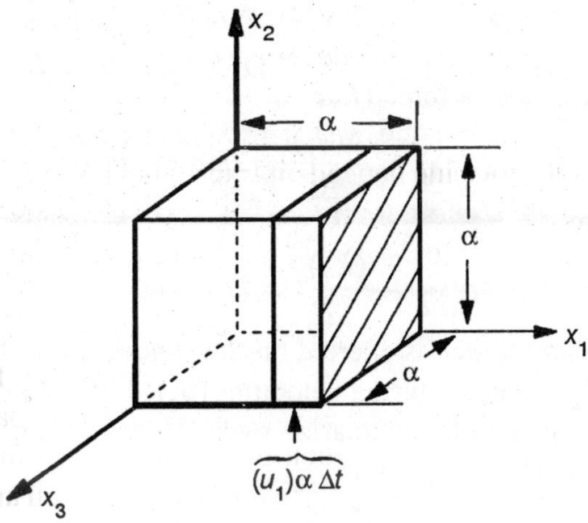

Fig. Control volume for derivations concerning the pressure effect of molecules

When one takes into account the characteristic properties of an ideal gas listed in points, the following derivations indicated can be formulated to derive the pressure that represents a characteristic continuum mechanics quantity of the gas, by taking molecular-theoretical considerations into account. These derivations only consider the known basic laws of mechanics and the properties indicated above in points.

In order to derive the "pressure effect" of the molecules on an area as a consequence of the molecular motion, the derivations carried out by considering a control volume with edge length. Regarding this control volume, the area standing perpendicular to the axis x_1 is shown shaded. All considerations are made for this area. For other areas of the control volume derivations have to be carried out in an analogous way, so that the considerations for the shaded area in figure can be considered as generally valid.

In the control volume, N molecules are present altogether. By the introduction of the number of molecules per m^3 (molecular density) n, this number N is given by:

$$N = na^3.$$

From n_α molecules per m^3, n_i molecules with a velocity component $(u_1)_i$ may move in the direction of the axis x_1 and interact with the shaded area. In time Δt all molecules will hit the wall area, which have a distance of $(u_1)_\alpha \Delta t$ from it. These are:

$$z_\alpha = n_\alpha a^2 (u_1)_\alpha \Delta t.$$

Each of z_α molecules exerts a momentum on the wall that is formulated by the law of the ideal elastic impact:

$$\Delta(i_1)_\alpha = -m\Delta(u_1)_\alpha = 2m(u_1)_\alpha.$$

For the total momentum transferred by z_a molecules to the wall, it holds:

$$\Delta(J_1)_\alpha = z_\alpha \Delta(i_1)_\alpha = n_\alpha a^2 (u_1)_i \Delta t[2m(u_1)_\alpha],$$

$$\Delta(J_1)_\alpha = 2ma^2 \Delta t n_\alpha (u^2{}_1)_\alpha.$$

The wall experiences a force $(K_1)_\alpha$

$$(\mathrm{K}_1)_a = \frac{\Delta(J_1)i}{\Delta t} = 2ma^2 n_\alpha (u_1^2)_\alpha$$

or the pressure

$$(P_1)_\alpha = \frac{(K_1)_\alpha}{a^2} = 2mn_\alpha (u_1^2)_\alpha$$

The total pressure which is exerted on the shaded area, summarizes the pressure shares $(P_1)_\alpha$ of all differing velocities $(u_1)_i$. If one wants to calculate the total pressure, one has to summarize over all these contributions. Then one obtains from the above equation:

$$P_1 = \sum_{a=1}^{N_x} (P_1)_\alpha = 2m \sum_{\alpha=1}^{N_x} n_\alpha (u_1^2)_\alpha$$

The summation occurring in the above relation can be substituted by the following definition of the mean value of the velocity square, $u^2{}_1$:

$$\sum_{\alpha=1}^{N_x} n_\alpha (u_1^2)_\alpha = n_x \overline{(u_1^2)}$$

Here n_x is the total number of an average molecules per m^3 moving in the positive direction x_1, i.e.

$$n_x = \frac{1}{6} n$$

where n represents the mean number of molecules present per m^3. $\overline{u_1^2}$ represents the square of the "effective value" of the molecular velocity, which according to the above derivations can be defined as follows:

$$\overline{(u_1^2)} = \frac{1}{n_x} \sum_{i=1}^{N_x} n_\alpha (u_1^2)_\alpha = \frac{6}{n} \sum_{i=1}^{N_x} n_\alpha (u_1^2)_\alpha$$

The thermodynamic pressure P in a free fluid flow is defined generally as the mean value of the sum of the pressures in all three direction:

$$P = \frac{1}{3}(P_1 + P_2 + P_3) = \frac{1}{3} mn \left[\sum_{\alpha=1}^{N_x} n_\alpha (u_1^2)_\alpha + \sum_{\beta=1}^{N_y} n_\beta (u_2^2)_\beta \right.$$

$$+\sum_{\beta=1}^{N_z} n_\gamma (u_3^2)_\gamma \Big]$$

$$=$$

$$\frac{1}{3}mn\Big[\sum_{\alpha=1}^{N_x}\sum_{\beta=1}^{N_y}\sum_{\beta=1}^{N_z} n_\alpha \beta_\gamma \left(u_1^2 + u_2^2 + u_3^2\right)_{\alpha\beta\gamma}$$

$$= \frac{1}{3}mn\overline{(u^2)}$$

The pressure with which the shaded area is embossed is thus

$$P = \frac{1}{3}mn\overline{(u^2)}$$

As m is the mass of a single molecule and n the mean number of the molecules per m^3, the expression (mn) corresponds to the density ρ in the terminology of continuum mechanics:

$$P = \frac{1}{3}mn\overline{(u^2)} = \frac{1}{3}\rho\overline{(u^2)}$$

$$\mathrm{P} = = \frac{1}{3}\rho\overline{(u^2)}$$

This relation contains a molecular quantity and also the mean molecular velocity square. The mean velocity square can be eliminated by another quantity of continuum mechanics, namely the temperature T of an ideal gas.

The mean kinetic energy of a molecule can be written according to the equipartition law of statistical physics where u represents the gas volume per kmol.

$$e_k = \frac{1}{2}m\overline{(u^2)} = \frac{3}{2}kT$$

where $k = 1,380658 \cdot 10^{-23} \frac{J}{K}$ represents the Boltzmann constant. From and follows:

$$P = \frac{1}{3}\rho\left(3\frac{ki}{m}T\right) = \rho\frac{k}{m}T$$

Further it holds:

$$k = \frac{\Re}{L} = \frac{\text{universal Gas Constant}}{\text{Loschmidts's number}}$$

then:

$$P = \frac{\Re T}{Lm}\rho$$

$M = Lm$ is the mass per kmol of an ideal gas, so that $u = M/\rho$ can be written.

$$Pu = \Re T$$

Strictly speaking, the above derivations can only be stated for a monatomic gas, with the assumption of ideal gas properties. However, the above law can be transferred to polyatomic gases with "ideal gas properties" if the additional degrees of freedom present in polyatomic gases and the corresponding constitutes of the internal energy of a gas are taken into account. Generally the energy content of a gas can be stated as follows:

$$e_{ges} = \frac{a}{2}kT$$

a indicates the degrees of freedom of the molecular motion:

$a = 3$ with monatomic gases,

$a = 5$ with biatomic gases,

$a = 6$ with triatomic- and polyatomic gases.

The above derivations have shown, that the laws that are known from continuum mechanics can be derived from molecular-theoretical considerations.

This means that the laws of continuum mechanics, at least for the pressure derived here, with the introduction of density and temperature, are consistent with the corresponding considerations of mechanical theory of molecular motion.

MOLECULAR-DEPENDENT MOMENTUM TRANSPORT

Transport processes that are caused by the thermal motion of the molecules were referred in an introductory way. Attention was drawn to the analogy between momentum heat and mass transport and it was pointed out that the τ_{ij} terms used in fluid-mechanics are not considered to be as caused by friction, i.e. physically they represent no "shear stress" but molecular-dependent momentum transports occurring per unit area and time, the index *i* representing the considered transport direction and *j* is the direction of the considered momentum. In order to give an introduction into a physically correct consideration of the molecular-dependent momentum transport, the below indicated considerations for an ideal gas are made, where only a x_1-momentum transport in the direction x_2 is considered, i.e. the term τ_{21}.

For the following derivations a velocity distribution has to be used that does not correspond to the equilibrium distribution (Maxwell distribution), as with this distribution $\tau_{ij} = 0$ and $q_j = 0$. Therefore the following simple model of a non-equilibrium distribution is used. 1/6 of all molecules at a time moves with a velocity of $(-\overline{u}, 0, 0)$, $(\overline{u}, 0, 0)$, $(0, -\overline{u}, 0)$, $(0, \overline{u}, 0)$, $(0, 0, -\overline{u})$, $(0, 0, \overline{u})$ with the amount $\overline{u}$ in parallel to the axis of the coordinates.

When one assumes a molecular concentration per unit volume, i.e. *n* molecules per m^3, one third of them on an average move with a velocity of $\overline{u}$ in direction x_2 and of these again one half, i.e. each *n*/6 molecules per unit volume, move in negative and positive direction x_2. On an average $\frac{1}{3}n\bar{u}$ molecules per time and unit area unit traverse through the area of plane x_2 = constant, which is indicated in figure.

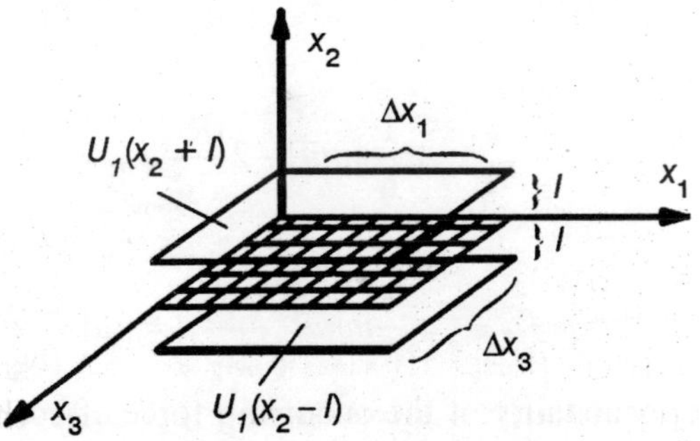

Fig. Molecular motion and sear stress

The molecules which traverse the plane x_2 =const in positive x_2 direction have on an average collided the last time with a distance of l with molecules below the plane, where l represents the mean free path of the molecular motion. The molecules coming from below thus possess on an average the mean velocity, which the flowing medium has in the plane $(x_2 - l)$. Consequently, the molecular transport in the positive direction x_2 an x_1 momentum, which on averaging can be stated as follows:

$$\Delta J_1 = \frac{1}{6}n\bar{u}[mU_1(x_2 - l)]\Delta t \Delta x_1 \Delta x_3$$

This is connected to an effect of forces τ_{21} quantifiable per time and unit area, i.e. with an effect of forces arising as a consequence of a x_1 momentum that comes about owing to a molecular stream in the positive direction x_2

$$\tau_{21}^{+} = \frac{\Delta J_1}{\Delta t}\frac{1}{\Delta x_1 \Delta x_3} = \frac{1}{6}mn\bar{u}U_1(x_2 - l)$$

In an analogous way, for the molecular stream which traverses the plane x_2 = const in the negative direction x_2 can be stated as x_1 momentum transport. For latter an effect per unit area can be calculated, which is indicated below:

$$\tau_{21}^{-} = -\frac{1}{6}mn\bar{u}U_1(x_2 + l)$$

Thus the entire momentum exchange per area unit, which the plane x_2 = const experiences, is:

$$\tau_{21} = \tau_{21}^{+} + \tau_{21}^{-} = \frac{1}{6}um\bar{u}[U_1(x_2 - l) - U_1(x_2 + l)]$$

For the velocities U_1 $(x_2 - l)$ and U_1 $(x_2 + l)$ can be stated by means of a Taylor series expansion:

$$U_1(x_2 - l) = U_1(x_2) - \left(\frac{\partial U_1}{\partial x_2}\right) l + \cdots$$

$$U_1(x_2 + l) = U_1(x_2) + \left(\frac{\partial U_1}{\partial x_2}\right) l + \cdots$$

Thus one obtains

$$\tau_{21} = \frac{1}{6} mn\bar{u} \left(-2 \frac{\partial U_1}{\partial x_2} l\right)$$

i.e.

$$\tau_{21} = \frac{1}{3} mn\bar{u}l \left(\frac{\partial U_1}{\partial x_2}\right) = -\mu \left(\frac{\partial U_1}{\partial x_2}\right)$$

Thus the proportionality of the occurring force effect as a consequence of the molecular motion with the present velocity gradient of the flow field was derived via approximation considerations by means of molecular-theoretical statements. The derivations indicate that one can understand the molecular motion as the cause of momentum transport and thus the force effect. If one attributes the viscosity μ as a material property. For an ideal gas holds:

$$\mu = \frac{1}{3} mn\bar{u}l .$$

If one takes into account the following relationship:

$$\bar{u} = \sqrt{\frac{8kT}{\pi m}}; \; l = \frac{1}{\sqrt{2} d^2 \pi n}.$$

where d is a dimension for the molecular diameter, one obtains:

$$\mu = \frac{2}{3\pi^{3/2}} \frac{\sqrt{mkT}}{d^2}.$$

This relationship tells us that for an ideal gas $\mu \sim \sqrt{T}$, i.e. the viscosity gas increases with the molecular mass $\mu \sim \sqrt{m}$, while the viscosity decreases with increasing molecular size, $\mu \sim (1/d^2)$.

The above indicated considerations were carried out to serve as an introduction into the derivations of continuum-mechanical properties of fluids, using average molecular sizes.

Only one transport direction was taken into account and only the x_1 -momentum was included in the considerations. The complete term τ_{ij} is derived previously for Newtonian media with complete considerations on the momentum transport in ideal gases.

It is shown that it comprises essentially three terms which can all be described physically with the considerations on the momentum transport in ideal gases. A generalization of the considerations for momentum-transport processes in fluids is possible.

MOLECULAR TRANSPORT OF HEAT AND MASS IN GASES

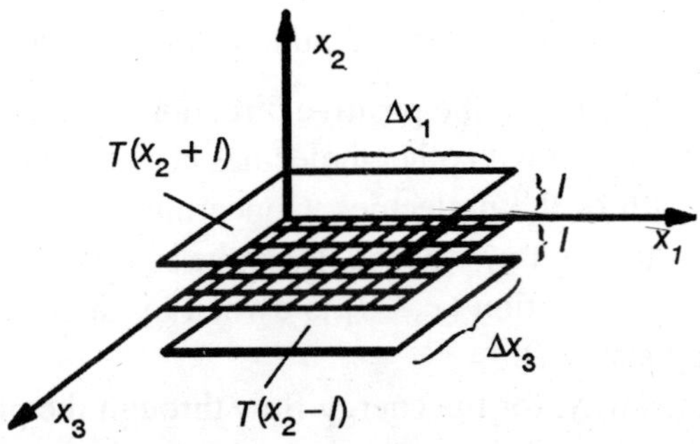

Fig. Heat transport through a plane, caused by molecules (principle sketch for derivation)

When the temperature in a system is not constant spatially, this system is thermally not homogeneous and energy (heat) will be transferred from areas of higher temperature to areas of lower temperature. For the one-dimensional problem, a heat flux of $\dot{q}_{x_2} = \dot{Q}/(\Delta t \Delta x_1 \Delta x_3)$ per unit area and unit time will take place, which is proportional to the temperature gradient existing on position $x_2 = 0$; this is known from experiments.

$$\dot{q}_{x_2 = \dot{q}_2} \quad = -\lambda \frac{\partial T}{\partial x_2}$$

The proportionality constant λ is designated as the thermal conductivity of the fluid (of the substance) or as the thermal conductivity coeffcient of the fluid. In this section considerations shall be made that are suitable for understanding the physical causes of heat conductivity from the point of view of the molecular theory of matter. The derivations are again given for a model of an ideal gas that is best suitable because its molecular motion for the intended derivations can be presented in a simple way.

If one considers a plane x_2 = const. in an ideal gas in which a temperature gradient exists that can be stated as derivation of $T(x_2)$., the heat conduction through the plane x_2 = const; can be explained such that the molecules in both directions traverse the plane and thus carry the "thermal energy" with them. When $(\partial T/\partial x_2) > 0$, the molecules which traverse the plane from top to bottom, have a higher mean energy than the molecules which traverse the plane in the opposite direction. The heat flow through the plane x_2 = const. can now be explained as the difference between the energy transports, which stem from the opposite-sense molecular flow. The following equations can be stated/ derived: Energy flow in the positive x_2 -direction:

$$\dot{q}_2^+ \quad = \frac{1}{6} n\overline{u}e(x_2 - l)$$

Here n is the number of the molecules per unit volume, $\overline{u}$ is the molecular velocity and $\left(\frac{1}{6}n\overline{u}\right)$ is the molecular number that traverses the considered area per unit area and time in the positive direction x_2. These molecules had, on average, the last contact with other molecules in a plane that has the distance of the mean free path of the molecules. Concerning the "energy content" of the molecular flow through x_2 = const., it can be said that the molecules hold energy there which at position $(x_2 - l)$ is owned by the elements of an ideal gas, i.e. the energy $e(x_2 - l)$.

In an analogous way, for the energy flow through the plane x_2 = const in the negative x_2 -direction, holds:

$$\dot{q}_2^- = -\frac{1}{6}n\overline{u}e(x_2 + l)$$

The heat flow results from the difference of the molecular-dependent energy streams, i.e. it holds:

$$\dot{q}_2 = \dot{q}_2^+ - \dot{q}_2^- = \frac{1}{6}n\overline{u}[e(x_2 + l) - e(x_2 + l)]$$

By means of the Taylor series expansion one obtains:

$$e(x_2 - l) = e(x_2) - \left(\frac{\partial e}{\partial x_2}\right)l + \left(\frac{1}{2}\frac{\partial^2 e}{\partial x_2^2}\right)l^2 - \cdots$$

$$e(x_2 + l) = e(x_2) + \left(\frac{\partial e}{\partial x_2}\right)l + \frac{1}{2}\left(\frac{\partial^2 e}{\partial x_2^2}\right)l^2 + \cdots$$

For the difference $e(x_2 - l) - e(x_2 + l)$, one obtains in a first approximation:

$$e(x_2 - l) - e(x_2 + l) = -2l\left(\frac{\partial e}{\partial x_2}\right) + \cdots$$

and thus for the heat flow:

$$\dot{q}_2 = -\frac{1}{3}n\overline{u}le\left(\frac{\partial e}{\partial x_2}\right) = -\frac{1}{3}n\overline{u}l\left(\frac{\partial e}{\partial T}\right)\left(\frac{\partial T}{\partial x_2}\right)$$

From the derivations for the "heat energy" of a molecule holds:

$$e = e_k = \frac{3}{2}kT \qquad \frac{\partial e}{\partial T} = \frac{3}{2}k = c_u$$

The Boltzmann constant k is understood to be a measure of the heat capacity of a molecule. When one considers again:

$$\overline{u} = \sqrt{\frac{8kT}{\pi m}}; \qquad l = \frac{1}{\sqrt{2}d^2\pi n}$$

one obtains

$$\lambda = \frac{1}{d^2}\sqrt{\left(\frac{k}{\pi}\right)^3 \frac{T}{m}}$$

However, it also increases with the molecular mass $\lambda \sim (1/\sqrt{m})$.

Similar to the considerations of the heat conductivity, where spatially different temperatures lead to temperature-smoothing processes, spatially different concentrations of a certain particle type cause concentration-smoothing processes that are to be understood analogously and to which the term "diffusion" is assigned. In order to follow up such processes, some gaseous radioactive particles could be used as trasers.

In equilibrium these marked particles are distributed evenly over the available volume. However, when the concentration of the marked parts is position dependent where the entire number of the particles per unit volume is constant, this state represents a non-equilibrium which will try to smooth concentration in the course of time by diffusion. This smoothing is possible by the temperature-dependent motion of the molecules. For the mathematical description of diffusion processes the equation can be employed.

$$\dot{m}_2 = -D\frac{\partial c}{\partial x_2}$$

Here $\dot{m}_2$ is the mass flow per unit time and area that runs parallel to the direction x_2 through a plane x_2 = const., D is the diffusion constant and c is

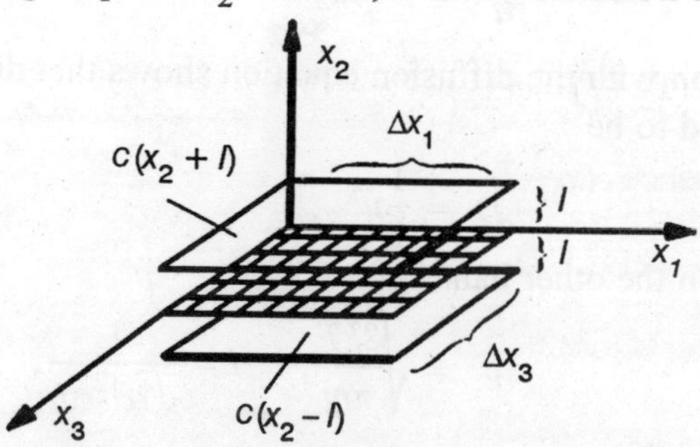

Fig. Transport of marked Molecules through a plane

The space-dependent concentration of the marked substance. The minus sign expresses that the particles move from the position of higher concentration to the position of lower concentration.

Analog to the molecular-theoretical considerations of the heat conduction, the considerations indicated below lead to a derivation of the diffusion equation and to a relationship which states the diffusion constants in molecular sizes. It holds again that the flow of the particles through a plane x_2 = const can be expressed as difference of the particle flows in positive or negative x_2 direction. In the positive x_2 direction, the area $\Delta x_1\ \Delta x_3$ of the considered plane is traversed by the particles, whose distance from the plane is not larger than $\bar{u}\Delta t$, i.e. $\frac{1}{6}$ $\Delta x_1\ \Delta x_3\ \bar{u}\Delta t c(x_2)$. by the particles.

If one considers the particle flow per unit time and area, it holds,

$$\dot{m}_2^+ = \frac{1}{6}\bar{u}c(x_2 - l)$$

The particle concentration that exists in the distance l from the considered plane is of relevance.

Accordingly, for the flow in negative x_2 direction holds:

$$\dot{m}_2^- = \frac{1}{6}\bar{u}c(x_2 + l).$$

With

$$c(x_2 - l) = c(x_2) - \left(\frac{\partial c}{\partial x_2}\right)l + \cdots$$

$$c(x_2 + l) = c(x_2) + \left(\frac{\partial c}{\partial x_2}\right)l + \cdots$$

yields for the desired quantity:

$$\dot{m}_2 = \dot{m}_2^+ + \dot{m}_2^-$$

$$\dot{m} = -\frac{1}{3}\bar{u}l\frac{\partial c}{\partial x_2}$$

A comparison with the diffusion equation shows that diffusion constant D was determined to be

$$D = \frac{1}{3}\bar{u}l$$

If one sets on the other hand

$$\bar{u} = \sqrt{\frac{8kT}{\pi m}}; \quad l = \frac{1}{\sqrt{2}d^2\pi n},$$

then one obtains:

$$D = \frac{2}{3}\frac{1}{nd^2}\sqrt{\frac{kT}{\pi^3 m}}$$

$1/\sqrt{m}$. On the other hand, there is a decrease with the molecular size, $D \sim (1/d^2)$, and also with the density of the gas ρ – nm, $D \sim (1/\rho)$.

VISCOSITY OF FLUIDS

The moleculare momentum transport in Newtonian fluid flow is given by:

$$\tau_{ij} = -\mu\left[\frac{\partial U_j}{\partial x_i}+\frac{\partial U_i}{\partial x_j}\right]-\mu\delta_{ij}\frac{\partial U_k}{\partial x_k}$$

The material property μ in the above presentation of τ_{ij} is defined as dynamic shear viscosity of a Newtonian medium, or as the shear coeffcient of viscosity of the fluid. The second coeffcient μ′ is defined as dynamic expansion viscosity coeffcient and can be formulated for a Newtonian medium as follows:

$$\mu' = -\frac{2}{3}\mu$$

with same physical units for μ and μ′, i.e. [μ] = [μ′]. The dynamic shear viscosity is a thermodynamic property of a fluid and is thus dependent on temperature and pressure. For a Newtonian medium, μ is independent of $\in_{ij} = \frac{1}{2}\left(\frac{\partial U_j}{\partial x_i}+\frac{\partial U_i}{\partial x_j}\right)$., i.e. t_{ij} is linear with the velocity gradients occurring in a flow, or is connected to local fluid-element deformations.

When the connection t_{ij} ($\in_{ij}$) is nonlinear, one speaks of non-Newtonian fluid viscosities. Some of these possible non-Newtonian fluid properties with pseudo-plastic behaviour, i.e. with increasing shear rate the fluid tends to have lower viscosities. Dilatant fluids on the other hand show an increase of viscosity with the increase of the rate of deformation and one defines them therefore as "shear thickening fluids". A Bingham fluid is shown that is characterized by a basic value t_{ij}. The treatment of Newtonian media rather than the fluids with properties of non-Newtonian media. The non-Newtonian fluid are only presented to point out that fluids with more complex fluid properties exist in nature.

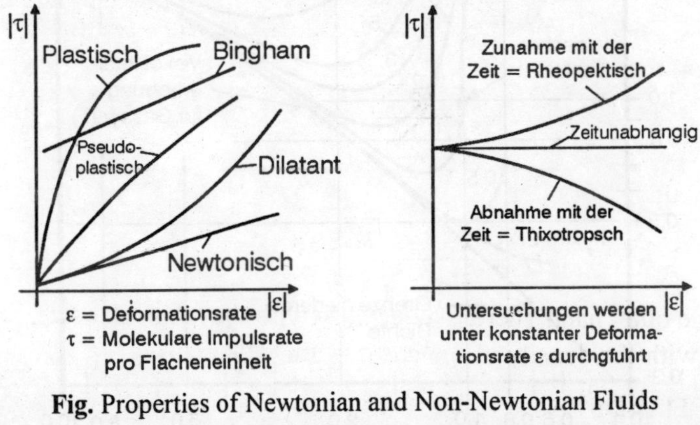

Fig. Properties of Newtonian and Non-Newtonian Fluids

The dynamic viscosity of a Newtonian fluid depends indirectly on the molecular interactions and can therefore be regarded as a thermodynamic property that varies with temperature and pressure. A complete theory of this viscosity as a transport property in gases and liquids is still under development and it can be looked up in the book of Hirschfelder. For an entire class of fluids, the function μ[*T*, *P*] can be presented in a description that was presented by Keenan, and which makes use of a normalised expression such that all values are normalised with the corresponding quantities at their critical state and following expression is obtained:

$$\frac{\mu}{\mu_c} = f\left[\left(\frac{T}{T_c}\right), \left(\frac{P}{P_c}\right)\right]$$

It shows, that the viscosity increases with pressure. The viscosity of liquids decreases with temperature. For gases, there is a very weak dependence of viscosity on pressure and it is generally neglected in gas dynamic consideration.

- The viscosity of liquids decreases rapidly with the temperature.
- The viscosity of gases increases with temperature under moderate pressure values.
- The viscosity of all fluids increases with pressure, independent of the aggregate state.
- The pressure dependency of gases is negligible.

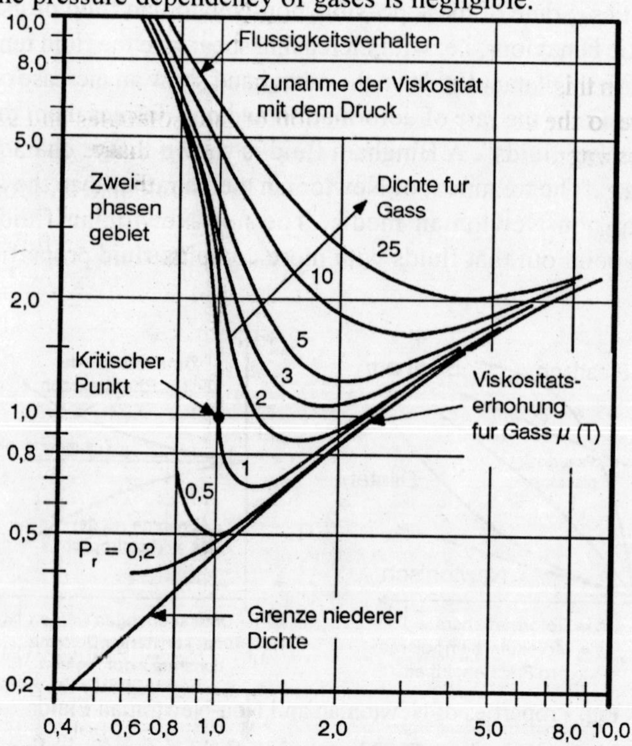

Fig. Standardized viscosity as a function of the pressure and temperature values standardized with the critical values

The above mentioned property is based on the fact that for the most fluids the critical pressure is higher than 10atm and hence conditions for small density are fulfilled very well under atmospheric pressures.

The theory of physical properties of gases under pressure conditions $P < P_c$ is very well developing and has been developed further until today on the basis of theories by Maxwell (1831 – 1879). All of these theories are based on the considerations. In accordance, the measured dynamic viscosity of a fluid results from the statistical average of molecule-dependent momentum transport of the motion of fluids. In case of gases, the dynamic viscosity reads:

$$\mu = \frac{2}{3}\rho l c$$

where ρ is the density of the fluid, l the mean free path length of the molecular motion and c = the sound velocity of the gas. For gases under normal pressure conditions $\rho l \approx$ const. However, more precise considerations show that ρl increases slightly with temperature and that this happens because of the so-called collision integral Ω_s. According to Chapman & Cowling / follows:

$$\mu = \frac{210^{-3}\sqrt{MT}}{\sigma^2 \Omega_s}$$

In this formula M is the molecular weight of the gas, T the absolute temperature, σ the collision cross section of the molecules and $\Omega_s = 1$ for the molecules interacting only in connection with the collision. When more complex molecular interactions exist, Ω_s has to be calculated according to following formula:

$$\Omega_s \approx 1.147\left(\frac{T}{T_c}\right)^{-0.145} + \left(\frac{T}{T_c} + 0.5\right)^{-2,0}$$

$T^* = T/T_c$ Ω_s	Ω_s Gl.
0.32.840	2.928
1.01.593	1.591
3.01.039	1.060
10.00.82440.8305	
30.00.70100.7015	
100.00.58870.5884	
400.00.48110.4811	

Table: Data for Stockmayer collision integral values for determining the viscosity of gases

The values determined from the Stockmayer potential, were compared in Table with the values of the above approximation relationship. For routine calculations, the following equations can be used

$$\frac{\mu}{\mu_0} \approx \left(\frac{T}{T_0}\right)^n$$

where μ_0 and T_0 are corresponding reference values that were obtained from measurements or calculations.

In general the value of ***n*** is around 0.7 More precise values of ***n*** are contained in Table for different gases.

T0 *Gas*	μ_0 *[K] [m Pa * s]*	*n*	*error [%]*	*temperature range [K]*	*S [K]*	*temperature range for ±2% error [K]*
Air	887.650.01716	0.666	± 4	745.65-4548.15	521.90	648.15-4548.15
Ar	887.65 0.02125	0.72	± 3	723.15-3648.15	598.15	548.15-3648.15
CO_2	887.65 0.01370	0.79	±5	744.4 -4098.15	773.15	700.65-4098.15
CO	887.65 0.01657	0.71	± 2	790.65-3648.15	579.40	565.65-3648.15
N_2	887.65 0.01663	0.67	± 3	773.15-3648.15	513.15	498.15-3648.15
O_2	887.65 0.01919	0.69	± 2	790.65-4773.15	585.65	691.90-4773.15
H_2	887.65 0.008411	0.68	± 2	453.15-2748.15	490.65	778.15-2748.15
Water vapour	210.65 0.01703	1.04	± 3	903.15-3648.15	2210.65	1085.65-3648.15

Table: Values for the Calculation of the Dynamic Viscosity of Gases

More extensive considerations were conducted by Sutherland, which were based on an intermolecular potential of forces with an attractive part. The resulting Sutherland formula is

$$\frac{\mu}{\mu_0} \approx \left(\frac{T}{T_0}\right)^{\frac{3}{2}} \frac{T_0 + S}{T + S}$$

In this relation S is an effective temperature, the so-called "Sutherland constant".

BALANCE CONSIDERATIONS AND CONSERVATION LAWS

Before we conduct detailed considerations on fluid mechanics processes, some remarks have to be made on the acquisition of information in fluid mechanics, especially on the knowledge in analytical fluid mechanics which is treated in this manuscript. Starting out from conservation laws, analytical fluid me-chanics employs deductive methods to solve various unsolved problems, i.e. to make statements on existing flow problems.

Here one makes use of derived relations that are based on balance considerations, as the reader knows them from other fields of natural and engineering science or also from everyday observations. In many domains of daily life one acquires, starting from in-tuitive knowledge on the existence of conservation laws, useful information from balance considerations which one conducts on defined fields, domains, periods, etc. The way by which the changes in quantities of our interest take place in detail is often not of interest rather only the "initial and end states" of the considered quantities are of interest. These changes are due to "in-flows and outflows", and relations can be established between changes within the considered fields, domains and periods and the "inflows and outflows". Considerations on the financial circumstances are for example conducted by establishing balances on the income and expenditure to obtain with these data information on the development of the financial situation of companies or persons. Many more examples of this kind could be cited that make clear the importance of balances for obtaining information in daily life.

We find balances on quantities like mass, momentum, energy etc. in almost every field of natural and engineering sciences. With these balances, basic equations are set up with the aid of existing conservation laws whose solution leads, in the presence of initial and boundary conditions, to the desired information on quantities. In order to obtain a definite information, the balance considerations have to be based not only on valid conservation laws (mass conservation, energy conservation, momentum conservation etc.) but also on definite specified domains. The field or the domain, on which balances are set up, has to be defined precisely to guarantee the unambiguity of the derived basic laws. A relation, that was derived by the consideration of a field is, in general, not applicable when domain modifications have taken place which were not included in the relations.

Fluid mechanics is based on the basic laws of mechanics and thermodynamics and moreover uses state equations in the derivations in order to establish relations between the state of a fluid. These state properties vary in the course of time or in space; however, the changes of state take place in accordance with the corresponding state equations while observing the conservation laws. For the derivation of the basic equations of fluid mechanics the following physical basic laws are followed:

- Mass conservation law (continuity equation)
- Momentum conservation law (equation of momentum) - energy conservation law (energy equation)
- Conservation for chemical species
- State equations

The above cited basic laws can now be applied to several "balance domains".

The size of the balance space is not important in general and it can include infinitesimal small balance domains (differential considerations) or finite volumes (integral considerations). Furthermore, the balance domains can lie in different coordinate systems and can carry out proper motions themselves

(Lagrangian and Euler's ways of consideration). In general, once selected balance domain is usually maintained, however, this is not necessary. Changes are admissible as long as they are known and thus can be included in the balance considerations.

Generally, in fluid mechanics, only integral considerations are made, i.e. these balances are set-up over different, favorable domains of interest. In the case of differential considerations, one finds in general attention only on balances with moving fluid elements (Lagrangian way of consideration) or space-fixed elements (Euler's way of consideration). Both have to be distinguished strictly and balances should always be set up separately for the Lagrangian and the Eulerian balance spaces.

Mixed balances lead to errors in general, however, transformations of final equations are possible. It is for example usual in fluid mechanics to transform the balance relations derived for a fluid element to space-fixed coordinate systems and thus to obtain balance relations for constant volumes. The connections between considerations in moving fluid elements and space-fixed coordinate systems are presented and the equations required for the transformation are derived. Particular attention is given here to the physical understanding of the principal connections, so that advantages and disadvantages of the different ways of consideration become clear. The advantages of the "Eulerian form" of the basic equations are brought out with view to the imposed boundary conditions for obtaining solutions. On the other hand the Lagrangian considerations allow the transfer of physical knowledge on the mechanics of moving bodies to fluid mechanics considera-tions.

When stating the basic equations in Lagrange variables, the following equations yield for a fluid element

- Mass Conservation : $\dfrac{\mathrm{d}(\delta m)_{\Re}}{dt} = 0$
- Newton's 2nd Law : $\dfrac{\mathrm{d}}{\mathrm{d}t}[(\delta m)_{\Re}(U_j)_{\Re}]$

$$= \sum (M_j)_{\Re} + (\Delta \dot{J}_j)_{\Re} + \sum (O_j)_{\Re}$$

- Energy Conservation : $\dfrac{\mathrm{d}}{\mathrm{d}t}(e)_{\Re} = \dfrac{\mathrm{d}}{\mathrm{d}t}(\dot{q})_{\Re} - P_{\Re}\dfrac{1}{\delta V_{\Re}}\dfrac{d(\delta V)_{\Re}}{dt} + \phi_{diss}$
- State Equation : $e_{\Re} = f(P_{\Re}, T_{\Re})\,\text{und}\,P_{\Re} = f(f(\rho_{\Re}, T_{\Re})$

The above summarizing presentations make it clear that generally in fluid mechanics considerations agree with principles that usually treated in thermodynamics, e.g. the energy equation (1st fundamental law of thermodynamics) and state equations of liquids and gases.

The above equations can also be expressed in field variables, such that the below cited set of differential equations for density ρ, pressure *P*, temperature *T*, internal energy e and three velocity components *Uj* (*j* = 1, 2, 3) are obtained:

- Mass conservation: $\frac{\partial \rho}{\partial t}+\frac{\partial(\rho U_i)}{\partial x_i}=0$

- Newton's 2nd law: $\rho\left[\frac{\partial U_j}{\partial t}+U_i\frac{\partial U_j}{\partial x_i}\right]=-\frac{\partial P}{\partial x_j}-\frac{\partial \tau_{ij}}{\partial x_i}+\rho g_i$

  with $\tau_{ij}=-\mu\left[\frac{\partial U_j}{\partial t}+\frac{\partial U_i}{\partial x_j}\right]+\frac{2}{3}\mu\delta_{ij}\frac{\partial U_k}{\partial x_k}$

- Energy conservation: $\rho\left[\frac{\partial e}{\partial t}+U_i\frac{\partial e}{\partial x_i}\right]=-\frac{\partial q_i}{\partial x_i}-P\frac{\partial U_j}{\partial x_j}-\tau_{ij}\frac{\partial U_j}{\partial x_i}$

  with $\dot{q}_i=-\lambda\frac{\partial T}{\partial x_i}$

- state equations: $e=f(P, \mathrm{T})$ and $P=f(\rho, T)$

Thus seven differential equations are available, when one inserts τ_{ij} and $\dot{q}_i$ in the previously mentioned equations, for altogether five unknowns. With this a closed system of differential equations which is given above, can be solved for specified initial and boundary conditions. It is therefore the respective initial and boundary conditions that define a given flow problem. The physical basic laws are identical for all flow problems. However, they comprise the conservation and state equations as well that are usually treated in thermodynamics.

THERMODYNAMIC CONSIDERATIONS

The thermodynamic state equations of fluids are often used in supplement for the solution of flow problems. However in the present text only "simple fluids", i.e. for homogenous liquids and gases for which the thermodynamic state can be expressed by a relation between pressure, temperature and density are considered. The statements are possible for substantial as well as for field quantities, i.e. it holds

$$P_{\Re}=f(T_{\Re}, \rho_{\Re}) \text{ or } P=f(T, \rho)$$

Thermodynamic state equations are known to be:

$$\frac{P_{\Re}}{\rho_{\Re}}=RT_{\Re} \text{ (thermodynamically ideal gases)}$$

$$\rho_{\Re}=\text{const (thermodynamically ideal liquids)}$$

If one defines with $\alpha_{\Re}=P_{\Re}, T_{\Re}, \rho_{\Re}, e_{\Re}$ and with $\alpha = P, T, \rho, e,...,$ the following relation holds, when the fluid element $\mathfrak{R}$ is located at the time t at position x_i:

$$\alpha_{\mathfrak{R}}(t)=\alpha(x_i,t) \rightsquigarrow \frac{d\alpha_{\mathfrak{R}}}{dt}=\frac{\partial\alpha}{\partial t}+U_i\frac{\partial\alpha}{\partial x_i}=\frac{D\alpha}{Dt}$$

The second part of equation states how temporal changes of substantial, thermodynamic quantities can be computed from the substantial derivative of corresponding field quantities.

In addition to the above introduced thermodynamic state properties $P_{\mathfrak{R}}$, $T_{\mathfrak{R}}$, $\rho_{\mathfrak{R}}$, $e_{\mathfrak{R}}$,..., other state properties can be defined whose introduction is of advantage in certain thermodynamic considerations. Some of them are:

- Specific volume: $u_{\mathfrak{R}}=\dfrac{1}{\rho_{\mathfrak{R}}}$
- Enthalpy: $h_{\mathfrak{R}}=e_{\mathfrak{R}}+P_{\mathfrak{R}}u_{\mathfrak{R}}$
- Free energy: $f_{\mathfrak{R}}=e_{\mathfrak{R}}-T_{\mathfrak{R}}s_{\mathfrak{R}}$ (Helmholtz potential)
- Free enthalpy : $g_{\mathfrak{R}}=h_{\mathfrak{R}}-T_{\mathfrak{R}}s_{\mathfrak{R}}$ (Gibb potential)

Accordingly, it is possible to apply certain mathematical operators in order to define "new" thermodynamic quantities. However, their introduction makes sense only when advantages result from the introduction into the thermodynamic considerations that are to be carried out with the new quantities.

In one of the above definitions for thermodynamic potentials, the entropy was used whose definition is given by a differential relation:

$$T_{\mathfrak{R}}ds_{\mathfrak{R}}=de_{\mathfrak{R}}+P_{\mathfrak{R}}du_{\mathfrak{R}} \quad \text{(Gibb relation)}$$

Integrating one obtains:

$$s_{\mathfrak{R}}=s(\mathfrak{R})_0+\int_{(e_{\mathfrak{R}})_0}^{e_{\mathfrak{R}}}\frac{1}{T_{\mathfrak{R}}}de_{\mathfrak{R}}+\int_{(u_{\mathfrak{R}})_0}^{u_{\mathfrak{R}}}\frac{P_{\mathfrak{R}}}{T_{\mathfrak{R}}}du_{\mathfrak{R}}$$

The above relations can be understood as identical definition of equations for the entropy s of a fluid element. When employing the relation one obtains with

$$\frac{ds_{\mathfrak{R}}}{dt}=\frac{Ds}{Dt} \quad \frac{de_{\mathfrak{R}}}{dt}=\frac{De}{dt} \quad \text{and} \quad \frac{du_{\mathfrak{R}}}{dt}=u_{\mathfrak{R}}\frac{\partial U_i}{\partial x_i}3$$

the following relation:

$$T_{\mathfrak{R}}\frac{ds_{\mathfrak{R}}}{dt}=\frac{de_{\mathfrak{R}}}{dt}+P_{\mathfrak{R}}\frac{du_{\mathfrak{R}}}{dt} \rightsquigarrow =T\frac{Ds}{Dt}=\frac{De}{dt}+P\frac{1}{\rho}\frac{\partial U_i}{\partial x_i}$$

or:

$$T\left(\frac{\partial s}{\partial t}+U_i\frac{\partial s}{\partial x_i}\right)=\left(\frac{\partial e}{\partial t}+U_i\frac{\partial e}{\partial x_i}\right)+\frac{P}{\rho}\left(\frac{\partial U_i}{\partial x_i}\right)$$

When one applies the mass conservation equation of the law of differential equations, it can be rearranged further:

$$\left(\frac{\partial U_i}{\partial x_i}\right) = -\frac{1}{\rho}\frac{D}{Dt}$$

of the mass conservation equation inserted in yields:

$$T\frac{Ds}{Dt} = \frac{De}{Dt} - \frac{P}{\rho^2}\frac{D\rho}{Dt}$$

From this relation further relations can be derived that are of importance for fluid mechanics considerations, e.g. for $s_{\Re}$ = const.:

$$\left(\frac{De}{Dt}\right)_s = \frac{P}{\rho^2}\frac{D\rho}{Dt} \rightsquigarrow \left(\frac{de_{\Re}}{dt}\right)_{s\Re} = \frac{P_{\Re}}{\rho_{\Re}^2} = \left(\frac{de_{\Re}}{d\rho_{\Re}}\right)_{s\Re}$$

For $\rho_{\Re}$ = const. or $u_{\Re}$ = const it holds:

$$T\left(\frac{Ds}{Dt}\right)_\rho = \left(\frac{De}{Dt}\right)_\rho \rightsquigarrow T_{\Re} = \left(\frac{de_{\Re}}{ds_{\Re}}\right)_{\rho\Re}$$

It holds further for $e_{\Re}$ = const.:

$$T\left(\frac{Ds}{Dt}\right)_e = -\frac{P}{\rho^2}\left(\frac{D\rho}{Dt}\right)_e \rightsquigarrow P_{\Re} = T_{\Re}\left(\frac{\partial s_{\Re}}{\partial \upsilon_{\Re}}\right)_{e\Re} = -T_{\Re}\rho_{\Re}^2\left(\frac{\partial s_{\Re}}{\partial \rho_{\Re}}\right)_{e\Re}$$

Further significant relations known from thermodynamics are needed in the following:

- Specific heat capacity of a fluid at constant volume

$$c_\upsilon = \left(\frac{\partial e_{\Re}}{\partial T_{\Re}}\right)_{\upsilon\Re} = T_{\Re}\left(\frac{\partial s_{\Re}}{\partial T_{\Re}}\right)_{\upsilon\Re}$$

- Specific heat capacity of a fluid at constant pressure

$$c_p = \left(\frac{\partial h_{\Re}}{\partial T_{\Re}}\right)_{P\Re} = T_{\Re}\left(\frac{\partial s_{\Re}}{\partial T_{\Re}}\right)_{P\Re}$$

where is $h_{\Re} = e_{\Re} + P_{\Re}\upsilon_{\Re}$

- Isothermal compressibility coeffcient

$$\alpha = -\frac{1}{\upsilon_{\Re}}\left(\frac{\partial \upsilon_{\Re}}{\partial P_{\Re}}\right)_{T\Re} = \frac{1}{\rho\Re}\left(\frac{\partial \rho_{\Re}}{\partial P_{\Re}}\right)_{T\Re}$$

- Thermal expansion coeffcient

$$\beta = \frac{1}{\upsilon_{\Re}}\left(\frac{\partial \upsilon_{\Re}}{\partial T_{\Re}}\right)_{P\Re} = -\frac{1}{\rho_{\Re}}\left(\frac{\partial \rho_{\Re}}{\partial T_{\Re}}\right)_{P\Re}$$

When one takes into consideration that the following relation holds

$$d_{\rho\Re} = \left(\frac{\partial \rho_{\Re}}{\partial T_{\Re}}\right)_{P\Re} dT_{\Re} + \left(\frac{\partial \rho_{\Re}}{\partial P_{\Re}}\right)_{T\Re} dP_{\Re}$$

the following relation can be formulated for all fluids:

$$\frac{1}{\rho_{\Re}} d\rho_{\Re} = \alpha dP_{\Re} = \beta dT_{\Re}$$

Or rearranged in terms of field variables:

$$\frac{1}{\rho}\frac{D\rho}{Dt} = \alpha \frac{DP}{Dt} - \beta \frac{DT}{Dt}$$

This relation allows the statement that all fluids of constant density, i.e. fluids having the property $\rho_{\Re}$ = const. or $(D\rho/Dt) = 0$, can be designated as incompressible. They react neither to pressure variations ($\alpha = 0$) nor to temperature variations ($\beta = 0$) for changes in volume or density.

For any fluids, the difference of the heat capacities results:

$$(c_p - c_\upsilon) = \frac{T_{\Re}}{\rho_{\Re}}\frac{\beta^2}{\alpha} = -\frac{T_{\Re}}{\rho_{\Re}} \cdot \beta \left(\frac{\partial P_{\Re}}{\partial T_{\Re}}\right)_{\rho\Re} = T_{\Re}\left(\frac{\partial P_{\Re}}{\partial T_{\Re}}\right)_{\rho\Re}\left(\frac{\partial \upsilon_{\Re}}{\partial T_{\Re}}\right)_{P\Re}$$

The above general relations can now be employed to derive the special relations that hold for the two thermodynamic ideal fluids that receive special attention in this manual, namely the ideal gas and the ideal fluid. For an ideal gas holds

$$\frac{P_{\Re}}{\rho_{\Re}} = RT_{\Re} \quad \text{consequently} \quad \frac{P}{\rho} = RT$$

and in addition $\left(\frac{\partial e_{\Re}}{\partial \upsilon_{\Re}}\right)_{T\Re} = \left(\frac{\partial e_{\Re}}{\partial P_{\Re}}\right)_{T\Re} = 0$ and c_υ = const, i.e. the internal energy of an ideal gas is a pure function of the temperature. For the isothermal compressibility coeffcient a and the thermal expansion coeffcient β yields

$$\alpha = \frac{1}{\rho_{\Re}}\left(\frac{\partial e_{\Re}}{\partial P_{\Re}}\right)_{T\Re} = \frac{1}{\rho_{\Re}}\frac{1}{RT_{\Re}} = \frac{1}{P_{\Re}}$$

$$\beta = -\frac{1}{\rho_{\Re}}\left(\frac{\partial \rho_{\Re}}{\partial T_{\Re}}\right)_{P\Re} = -\frac{1}{\rho_{\Re}}\left(\frac{P_{\Re}}{RT_{\Re}^2}\right) = \frac{1}{T_{\Re}}$$

and thus for the difference of the specific heat capacities:

$$c_p - c_\upsilon = \frac{T_{\Re}}{\rho_{\Re}}\frac{1}{T_{\Re}^2}P_{\Re} = \frac{P_{\Re}}{\rho_{\Re}T_{\Re}} = R$$

It can further be formulated for the change in density:

$$\frac{d\rho_{\Re}}{\rho_{\Re}} = \frac{dP_{\Re}}{P_{\Re}} - \frac{dT_{\Re}}{T_{\Re}}$$

As a further fluid of significance, we introduce the thermodynamic ideal liquid that distinguishes itself by $\alpha = 0$ and $\beta = 0$, i.e.

$$\frac{d\rho_{\Re}}{\rho_{\Re}}=0 \quad \text{(fluid of constant Density)}$$

For the difference of the heat capacities it can be computed:

$$c_p - c_\upsilon = -\frac{T_{\Re}}{\rho_{\Re}}\beta\left(\frac{\partial P_{\Re}}{\partial T_{\Re}}\right)_{\rho_{\Re}} \quad \text{bei } \beta = 0 \rightsquigarrow c_p = c_\upsilon.$$

When one employs the Gibb relation, $d\rho_{\Re} = d\upsilon_{\Re} = 0$ yields:

$$\left(\frac{\partial s_{\Re}}{\partial e_{\Re}}\right)_{\rho\Re} = \frac{1}{T_{\Re}}$$

Because $s_{\Re} \neq s_{\Re}\ (\rho_{\Re})$, the pressure in an ideal liquid is not taken into account as a thermodynamic quantity. It exists as a mechanical quantity, however, for an ideal fluid it is not part of a thermodynamic state equation.

A further physical property of a fluid, which is of significance when dealing with some of the flow problems presented in this book, is the velocity propagation of small pressure perturbations, the so-called sound velocity:

$$c^2 = \left(\frac{\partial P_{\Re}}{\partial \rho_{\Re}}\right)_{s\Re}$$

This quantity is defined as pressure modification due to the changes in density, the entropy being maintained constant, i.e. the propagation of small acoustic perturbations takes place isentropically.

When one takes into account the following relation for the cited sequence of partial derivations

$$1 = \left(\frac{\partial T_{\Re}}{\partial e_{\Re}}\right)_{\rho\Re}\left(\frac{\partial e_{\Re}}{\partial s_{\Re}}\right)_{\rho_{\Re}}\left(\frac{\partial s_{\Re}}{\partial T_{\Re}}\right)_{\rho_{\Re}}$$

and if one considers

$$\mathrm{T}_{\Re} = \left(\frac{\partial e_{\Re}}{\partial s_{\Re}}\right)_{\rho\Re} \quad c_\upsilon\left(\frac{\partial e_{\Re}}{\partial T_{\Re}}\right)_{\rho_{\Re}}$$

it holds:

$$c_\upsilon = T_{\Re}\left(\frac{\partial s_{\Re}}{\partial T_{\Re}}\right)_{\rho_{\Re}}$$

When taking into account the Maxwell relations

$$\left(\frac{\partial T_{\Re}}{\partial \rho_{\Re}}\right)_{s_{\Re}} = \frac{1}{\rho_{\Re}^2}\left(\frac{\partial P_{\Re}}{\partial s_{\Re}}\right)_{\rho_{\Re}}$$

$$\left(\frac{\partial \rho_{\Re}}{\partial s_{\Re}}\right)_{T\Re} = -\rho_{\Re}^2\left(\frac{\partial T_{\Re}}{\partial P_{\Re}}\right)_{\rho_{\Re}}$$

it can be formulated

$$\left(\frac{\partial s_{\Re}}{\partial T_{\Re}}\right)_{\rho_{\Re}}\left(\frac{\partial T_{\Re}}{\partial \rho_{\Re}}\right)_{\rho_{\Re}}\left(\frac{\partial \rho_{\Re}}{\partial s_{\Re}}\right)_{T_{\Re}} = -1$$

and it can also be written for the quantity c_υ

$$c_\upsilon = -T_{\Re}\left(\frac{\partial \rho_{\Re}}{\partial T_{\Re}}\right)_{s_{\Re}}\left(\frac{\partial s_{\Re}}{\partial \rho_{\Re}}\right)_{T_{\Re}}$$

Similarly it can be derived:

$$c_p = -T_{\Re}\left(\frac{\partial P_{\Re}}{\partial T_{\Re}}\right)_{s_{\Re}}\left(\frac{\partial s_{\Re}}{\partial P_{\Re}}\right)_{T_{\Re}}$$

For the relation of the heat capacities can be formulated

$$k = \frac{c_p}{c_\upsilon}\frac{\left(\frac{\partial P_{\Re}}{\partial T_{\Re}}\right)_{s_{\Re}}\left(\frac{\partial s_{\Re}}{\partial P_{\Re}}\right)_{T_{\Re}}}{\left(\frac{\partial \rho_{\Re}}{\partial T_{\Re}}\right)_{s_{\Re}}\left(\frac{\partial s_{\Re}}{\partial \rho_{\Re}}\right)_{T_{\Re}}} = \left(\frac{\partial P_{\Re}}{\partial \rho_{\Re}}\right)_{s_{\Re}}\left(\frac{\partial \rho_{\Re}}{\partial P_{\Re}}\right)_{T_{\Re}}$$

Under consideration of the definition equations for the sound velocity and for the isothermal compressibility coeffcient one obtains:

$$c^2 = \kappa\left(\frac{\partial P_{\Re}}{\partial \rho_{\Re}}\right)_{T_{\Re}} = \frac{\kappa}{\rho_{\Re}\alpha}$$

For the ideal gas with $\alpha = \frac{1}{P_{\Re}}$, considering the ideal gas equation, yields

$$c = \sqrt{\kappa R_{\Re} T_{\Re}}$$

For an ideal liquid with $\alpha \to 0$ holds:

$$c \to \infty$$

i.e. for a fluid with constant density an infinite large sound velocity results.

Chapter 3

Basics of Fluid Kinematics

GENERAL CONSIDERATIONS

The the most important basic knowledge of mathematics and physics with respect to fluid mechanics. This knowledge is needed to describe fluid flows or derive and construct basic equations of fluid mechanics in order to solve flow problems.

Here it is important to know that fluid mechanics is primarily interested in the velocity field $U_j(x_i, t)$ at initial and boundary conditions, and in the accompanying pressure field $P(x_i, t)$, i.e. fluid mechanics tries to describe flow processes in field variables.

This representation results in "Eulerian presentation" of fluid flows. This is best suited for the solution of flow problems and is thus applied in experimental, analytical and numerical fluid mechanics. The introduction of field quantities for the thermodynamic properties of a fluid, like e.g. the pressure $P(x_i, t)$, the temperature $T(x_i, t)$, the density $\rho(x_i, t)$, the internal energy $e(x_i, t)$ as well as for the molecular transport quantities, as the dynamic viscosity $\mu(x_i, t)$, the heat conductivity $\lambda(x_i, t)$ and the diffusion coefficients $D(x_i, t)$ so that a complete presentation of fluid mechanics is possible.

With the inclusion of diffusive transport quantities, i.e. the molecular heat trans-port $q_i(x_i, t)$, the molecular mass transport $m_i(x_i, t)$ the molecular momentum transport $t_{ij}(x_i, t)$, it is possible to formulate the conservation laws for mass, momentum and energy for general application.

The basic equations of fluid mechanics can thus be formulated locally, and hold for all flow problems in the same form. The differences in the solutions result from the different initial and boundary conditions that define the actual flow problems which enter into the solutions by the integration of the locally formulated basic quations.

Experience shows that the derivation of basic equations of fluid mechanics can be achieved in the easiest way if considerations are carried for fluid elements, i.e. by employing the "Lagrangian consideration" for the derivation of equations. The "Lagrange considerations" assumes that a fluid can be split up at a fixed time $t = 0$, in "marked elements" with the mass $\delta m_{\mathfrak{R}}$, the pressure $P_{\mathfrak{R}}$, the temperature $T_{\mathfrak{R}}$, the density $\rho_{\mathfrak{R}}$, the internal energy $e_{\mathfrak{R}}$ etc. The element with the index $\mathfrak{R}$ possesses also a velocity $(U_j)_{\mathfrak{R}}$, which is defined as Lagrangian velocity and which is always linked to the fluid element marked Basics of Fluid Kinematics once with as well as to all other quantities labelled with $\mathfrak{R}$.

In fluid mechanics, these are also designated as substantial quantities and always employed to derive the basic laws of fluid mechanics in an easily comprehensible way.

As the following considerations will show, the basic knowledge of mechanics gained in physics can be transferred in the most simple way into fluid mechanics by deriving the basic equations for fluid elements by introducing the Lagrange considerations.

SUBSTANTIAL DERIVATIVES

While one defines a generally as a substantial quantity, the derivation of fluid mechanics equations often require the total differential da to be employed.

$$d\alpha_{\mathfrak{R}} = \frac{\partial \alpha}{\partial t} dt + \frac{\partial \alpha}{\partial x_1}(dx_1)_{\mathfrak{R}} + \frac{\partial \alpha}{\partial x_2}(dx_2)_{\mathfrak{R}} + \frac{\partial \alpha}{\partial x_2}(dx_3)_{\mathfrak{R}}$$

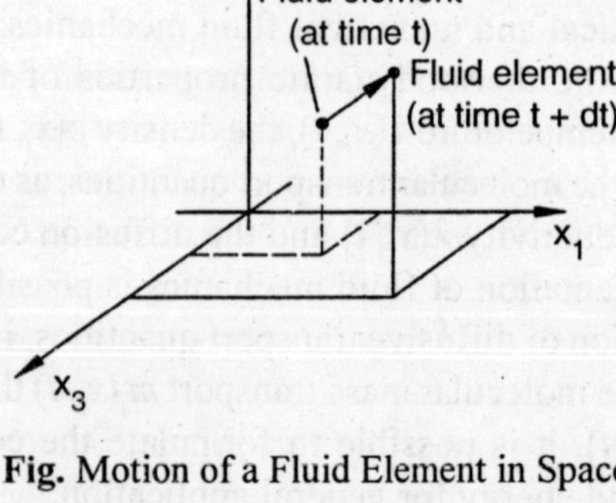

Fig. Motion of a Fluid Element in Space

The fluid element motion in space can be described as follows:

$$(dx_1) = (U_1)\,dt = U_1 dt$$
$$(dx_2) = (U_2)\,dt = U_2 dt$$
$$(dx_3) = (U_3)\,dt = U_3 dt$$

The transition of substational velocities $(U_i)_{\mathfrak{R}}$ to the field quantities U_i in equation is permissible, as at the time $t(x_i) = x_i(t)$ and thus $(U_i)\,(t) = U_i(x_i, t)$. Therefore generally it holds:

$$d\alpha_{\mathfrak{R}} = \frac{\partial \alpha}{\partial t} dt + \frac{\partial \alpha}{\partial x_1} U_1 dt \frac{\partial \alpha}{\partial x_2} U_2 dt + \frac{\partial \alpha}{\partial x_3} U_3 dt$$

and for the substantial time derivative with $\alpha_{\mathfrak{R}}(t) = \alpha(x_i, t)$ when $x_{\mathfrak{R}} = x_i$ at the time t it holds:

$$\frac{d\alpha_{\mathfrak{R}}}{dt} = \frac{\partial \alpha}{\partial t} + U_i \frac{\partial \alpha}{\partial x_i} =: \frac{D\alpha}{Dt}$$

where $(D\alpha/Dt)$ is the substantial time derivative of the field quantity $\alpha(x_i, t)$ with respect to time and the operator:

$$\frac{D}{Dt} := \frac{\partial}{\partial t} + U_i \frac{\partial}{\partial x_i}$$

indicates how the substantial derivative of a field variable is to be calculated. The operator D/Dt may only be applied to field quantities. When one applies D/Dt to the velocity field $U_j(x_i, t)$, the substantial acceleration results, i.e. the local acceleration which a fluid element experiences in a flow field at a point x_i at time t where $U_j(x_i, t)$ exists.

$$\frac{DU_j}{Dt} = \frac{\partial U_j}{\partial t} + U_i \frac{\partial U_j}{\partial x_i}$$

The substantial derivative plays an important role in the derivation of the momentum equation of fluid mechanics in Euler variables. In the acceleration term in Euler variables, four partial derivatives occur per momentum direction $j = 1, 2, 3$, one time derivative and three derivatives with respect to the space coordinates x_1, x_2 and x_3 i.e. the spatial derivatives $(\partial U_j/\partial x_i)$ multiplied by U_i occur in the substantial acceleration. These nonlinear terms in the resulting momentum equations lead to mathematical complications when flow problems are to be solved.

They prevent the application of the superposition principle of solutions, result in solution bifurcations, i.e. in multiple solutions for equal initial and boundary conditions and in coupled velocity fluctuations, e.g. in turbulent flows. The treatment of these nonlinear terms is considered in the presentations of this book. It is important that their significance is understood in detail as part of the acceleration term of fluid elements. It is important to realise that not only the temporal changes of the velocity field that lead to accelerations of fluid elements, but also the motion of a fluid element in a non-uniform velocity field.

MOTION OF FLUID ELEMENTS

Flow kinematics is a vast field and a comprehensive treatment, which is meant to give only an introduction into sub-domains of fluid mechanics, among others also into fluid kinematics.

To such an introduction belongs the treatment of path lines of particles, i.e. the computation of space curves along which marked fluid elements move in a fluid. Further, the computation of sweeping paths shall be treated, i.e. the "marked path" is to be computed which leaves a fixed injected tracer in a flow field. Both, the computation of path lines and of sweeping paths is of importance for the entire experimental flow mechanics, where it is often tried to get an insight in the process of flows by observations or also by quantitative measurements of the temporal changes of position of "'flow markers"'.

Path Lines of Fluid Elements

When one subdivides at the time $t = 0$ the entire domain of a flow field that is of interest into fluid elements and when one states the space coordinates of the mass centers of gravity of the each element in a coordinate system at the time $t = 0$, one achieves a marked fluid domain such that the position vector $\mathfrak{R}$:

$\{x_i\}_{\mathfrak{R},0} = \{x_i(t=0)\}\}_{\mathfrak{R}}$

is assigned to a fluid particle. Each of the moving fluid particles, resulting from the subdivision, marked and moving for $-\infty < t < +\infty$ is defined as a fluid element, that keeps its identity $0 \le t < \infty$ forever.

When kinematic considerations are carried out, only the motions of the in-dividual fluid elements are of interest. These considerations result for each fluid element in a separate and for the marked element characteristic path line. The computation of these path lines will be explained in the following. Here it is assumed that the flow field determining the fluid element motions is known.

As the velocity of a fluid element is only time-dependent, it follows from $d\{x_i\}_{\mathfrak{R}}/dt = \{U_i\}_{\mathfrak{R}}$, that as a path line of a fluid element the frequency locus:

$$t\{x_i(t)\}_{\mathfrak{R}} = \{x_i\}_{\mathfrak{R},0} + \int_0^t \{U_i(t')\}_{\mathfrak{R}}\,dt'$$

is to be understood. The position vector defined in this way for each moment in time t contains as a parameter the position vector of the particle defined at the time $t = 0$ i.e. $\mathfrak{R}$, i.e. $\{x_1\}_{\mathfrak{R},0}$. Now the identity $\{U_i\} = \{U_i\}$ can be introduced into the considerations, i.e. at a certain moment in time t holds:

$$\frac{d\{x_i\}_{\mathfrak{R}}}{dt} = \{U_i\}_{\mathfrak{R}} = \{U_i\}$$

The equals sign between the substantial velocity $\{U_i\}_{\mathfrak{R}}$ and $\{U_i\}$ existing at the moment in time t indicates that the identity $\{x_i\}_{\mathfrak{R}} = \{x_i\}$ which exists at time t justifies equating the substantial velocity $\{U_i\}_{\mathfrak{R}}$ with the field size $\{U_i\}$. For the components $\{x_i\}_{\mathfrak{R}}$ of the particle motion it holds therefore:

$$\frac{d\{x_i\}_{\mathfrak{R}}}{dt} = \{U_i\} \quad \text{or} \quad \frac{d\{x_i\}_{\mathfrak{R}}}{dt} = U_i$$

These differential equations have to be solved for $i = 1, 2, 3$ in order to determine the path lines of fluid elements. The differential quotient in the relation states that as a solution of the above differential equation the path line of a fluid element is obtained whose position was defined at the moment in time $t = 0$ with $\{x_i\}_{\Re, 0}$.

The general way of proceeding when defining path lines will be explained and made clear by the example stated below. The components of the flow velocity field shall be given:

$U_1 = x_1 (1 + t)$, $U_2 = -x_2$ und $U_3 = -x_3 t$

When one inserts these statements on the velocity field in the above indicated differential equations for the path lines of a fluid element, one obtains:

$$\frac{d\{x_1\}_{\Re}}{dt} = x_1(1+t),$$

$$\frac{d\{x_2\}_{\Re}}{dt} = -x_2,$$

and $$\frac{d\{x_3\}_{\Re}}{dt} = -x_3 t.$$

This law of differential equations can be solved now and results in the following solutions holding for the path lines of all fluid elements:

$$(x_1(t))_{\Re} = C_1 \exp\left[t + \frac{t^2}{2}\right]$$

$$(x_2(t))_{\Re} = C_2 \exp[-t]$$

$$(x_3(t))_{\Re} = C_3 \exp\left[-\frac{t^2}{2}\right].$$

When one considers a fluid element of interest which had the position coordinates (1, 1, 1)) at the time $t = 0$, then from the initial conditions for each of the equations in the introduced constants C_α result uniformly as:

$C_1 = C_2 = C_3 = 1$,

i.e. for the case considered here all integration constants are equal. For the path lines of this fluid element yields:

$$x_1 (t)_{\Re} = \exp\left[t + \frac{t^2}{2}\right],$$

$$x_2 (t)_{\Re} = \exp[-t]$$

$$x_3 (t) = \exp\left[-\frac{t^2}{2}\right].$$

This path line is presented spatially in Figure.

When one selects a particle whose position at time $t = 0$ showed different position coordinates, the integration constants C_1, C_2, C_3 change accordingly and a different path line results. Thus the path line is an "individual" property of a fluid element which in the resulting equations is determined by the flow field and the position of the fluid element at the time $t = 0$.

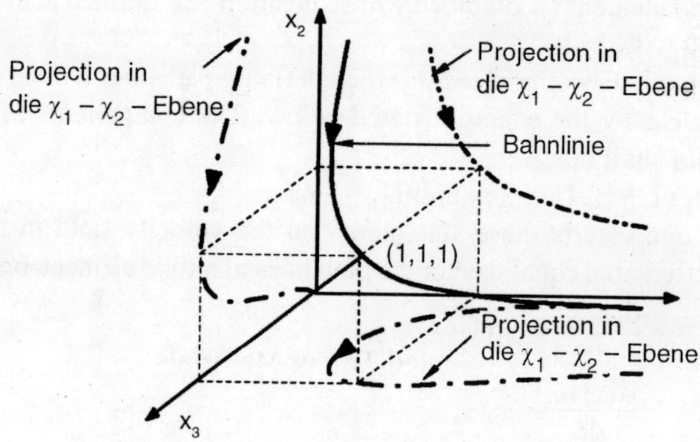

Fig. Spatial Path Line of the Considered Fluid Element

The general solution for the position coordinates (x_i) ,0, which a fluid element takes at the time $t = 0$ is obtained when one inserts these coordinates in the general solutions for the path line coordinates for the determination of the integration constants $C_\alpha(\alpha = 1, 2, 3)$ This results in:

$C_\alpha = (x_i), 0$

and thus in the general solutions for the path line coordinates:

$$(x_1(t)) = (x_{\Re,0})_1 \exp\left[t + \frac{t^2}{2}\right]$$

$$(x_2(t)) = (x_{\Re,0})_2 \exp[-t]$$

$$(x_3(t)) = (x_{\Re,0})_1 \exp\left[-\frac{t^2}{2}\right].$$

These yield the space curves, with the time t as parameter, which represent each curve points of the path lines of fluid elements.

For further explanation, the following two-dimensional velocity field is also considered:

$U_1 = x_1$, $U_2 = x_2(1 + 2t)$ und $U_3 = 0$.

With these data the following law of differential equations for the coordinates of the path lines of fluid elements can be formulated:

$$\frac{d(x_1)_\Re}{dt} = x_1, \frac{d(x_2)_\Re}{dt} = x_2(1+2t), \frac{d(x_3)_\Re}{dt} = 0$$

The solution of the third differential equation results in a constant that states in which plane the two-dimensional flow considerations are carried out. For the path line coordinates x_1 (t) and x_2 (t) it is computed:

$(x_1)_\Re = C_1 \exp[t]$, $(x_2)_\Re = C_2 \exp[t + t^2]$, $(x_3)_\Re = C_3$

When one computes the path line of the fluid element which at the time $t = 0$ took the coordinates (1,1,0), the result is:

$(x_1)_{\mathfrak{R}} = \exp[t], (x_2)_{\mathfrak{R}} = \exp[t+1], (x_3)_{\mathfrak{R}} = 0$

When one resolves the equation obtained for (x_1) with respect to time, it yields:

$t = \ln(x_1)_{\mathfrak{R}}$

When inserted in the solution for $(x_2)_{\mathfrak{R}}$ for two-dimensional path lines in the plane $x_1 - x_2$ – the following functional relation between $(x_1)_{\mathfrak{R}}$ and $(x_2)_{\mathfrak{R}}$ yields:

$(x_2)_{\mathfrak{R}} = (x_1)_{\mathfrak{R}}^{(1+\ln(x_1)_{\mathfrak{R}})}$

Sweeping Paths of Locally Injected Tracer Materials

It is usual in experimental fluid mechanics to gain qualitative insight in a flow process by injecting a continuous fluid tracer at a fixed position. This leads to a marked "fluid thread" which is carried with the flow and thus marks/traces the course of the flow.

When the exact course of the flow is of interest, quantitative evaluations of the location coordinates of sweeping paths of locally installed tracer materials are required. These evaluations can, based on the derivations stated below, be carried out with methods of flow kinematics.

Attention is drawn to the fact that this flow field is not source-free , thus it violates the requirements/demands of the continuity equation. This is, however, insignificant for the purely kinematic considerations mentioned here.

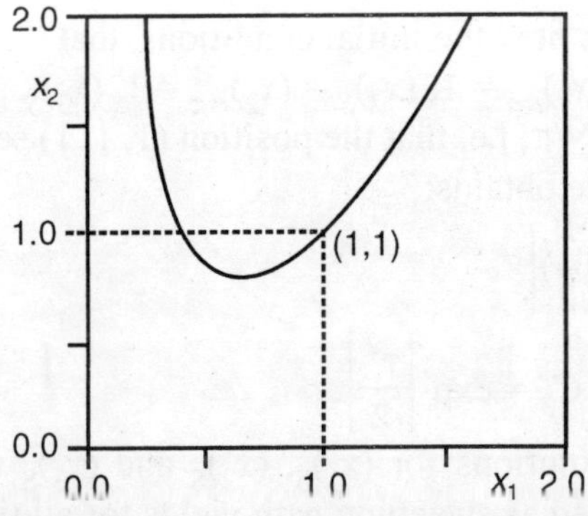

Fig. Path Line of the Flow Running Parallel to the Plane $x_1 - x_2$

A fluid particle marked with a tracer, e.g. an air particle or any other gas particle marked with smoke, or a water or fluid particle marked with colour, which at the time t is located at the position $\{x_i\} = \{x_i(t)\}_{\mathfrak{R}}$ must have passed the injection point for the tracer at a moment in time $(t - \tau)$, in order to be present as a marked particle at the point $\{x_i\}$ i.e. it holds :

$\{x_i(t)\}_{\mathfrak{R}} = \{x_i(t-\tau)\}_S$

Hence the way covered by a marked fluid element up to the time t can be computed as path line of the element that fulfills the condition i.e. a path line with the initial condition that for $t = \tau$ the fluid element held the position of the location coordinates of the injection point. The sweeping path thus is composed of the sum of the path lines of individual particles. For each individual marked particle of a sweeping path a parameter τ is introduced, which for $0 \le \tau \le t$ covers all parts of a sweeping path. It is therefore important to vary the parameter τ in the solution equations in order obtain the entire sweeping path.

The above short explanations shall be made clear again by way of an example, which is handled on the basis of the three-dimensional velocity field used above:

$U_1 = x_1 (1 + t),\ U_2 = -x_2$ and $U_3 = -x_3 t$

This velocity field yields the law of differential equations for the motion of a fluid element in space :

$$\frac{d(x_1)_S}{dt} = x_1(1+t), \quad \frac{d(x_2)_S}{dt} = x_1, \quad \frac{d(x_3)_S}{dt} = -x_3 t.$$

As a solution one obtains for the components$(x_1)_S$, $(x_2)_S$ and $(x_3)_S$ according to equation:

[1]The index s signifies that the location coordinate of the sweeping path is meant.

$$(x_1)_S = C_1 \exp\left[t(1+\frac{t}{2})\right],\ (x_2)_S = C_2 \exp\,[-t],\ (x_3)_S = C_3 \exp\left[-\frac{t^2}{2}\right].$$

When one inserts now the initial conditions, that

$$(x_1)_S = (x_1)_{t=\tau} = 1,\ (x_2)_S = (x_2)_{t=\tau} = 1,\ (x_3)_S = (x_3)_{t=\tau} = 1$$

was present for $t = \tau$, i.e. that the position (1, 1, 1) serves as an injection point of the tracer, one obtains:

$$C_1 = \exp\left[-\tau(1+\frac{\tau}{2})\right];$$

$$C_2 = \exp\,[\tau] \text{ and } C_3 = \exp\left[\frac{\tau^2}{2}\right].$$

Inserted in the solutions for $(x_1)_S$, $(x_2)_S$ and $(x_3)_S$ the equation of the frequency locus defined as sweeping path yields for all times:

$$(x_1)_S = \exp\left[t(1+\frac{t}{2}) - \tau(1+\frac{\tau}{2})\right],$$

$$(x_1)_S = \exp\,[-(t-\tau)],$$

$$(x_1)_S = \exp\left[1\frac{1}{2}(t^2 - \tau^2)\right]$$

When one wants to make visible the course of a sweeping path at a moment in time t (partly), one has to insert the value of t in the above equation in order to obtain in this way the equation of a space curve, with τ as a parameter. Here t is determined by the period of time $[\tau_1, \tau_2]$ of the tracer injection in (1, 1, 1) with $-\infty < \tau_1 < \tau_2 < t$. For $\tau_1 \rightarrow -\infty$, $\tau_2 = t$ and $t = 0$ yields:

$$(x_1)_S = \exp\left[-\tau(1+\frac{\tau}{2})\right],$$

$$(x_2)_S = \exp[\tau], \qquad -\infty < \tau < 0$$

$$(x_3)_S = \exp\left[\frac{\tau^2}{2}\right]$$

The course of this space curve is shown in 4.4. It indicates the sweeping path existing at the moment in time $\tau = 0$ (made visible from $\tau = -\infty$ to $\tau = 0$, the projections of the sweeping path into the main level of the Cartesian coordinate system are also introduced.

When one compares the equation for the sweeping path fixed by the space point (1, 1, 1) with the equations for the path line of a fluid element, stated for the same flow field, one realizes that path lines and sweeping paths are not identical for non-stationary flows. Only in the case of a stationary flow field path lines and sweeping paths are identical, as can be shown easily by the following considerations.

As a space curve is concerned here, the statement in x_1, x_2, x_3 – coordinates is appropriate. The definition x_S indicates that the location coordinates of a sweeping path are meant.

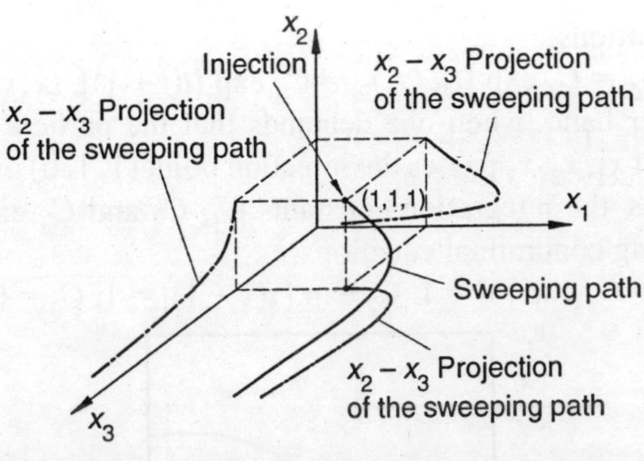

Fig. Sweeping Path for the Moment in Time $t = 0$. with Fluid Tracer Injections Between $\tau = -\infty$ and $\tau = 0$ at the Position (1, 1, 1)

Considering the stationary velocity field:

$U_1 = 2x_1$, $U_2 = -x_2$, $U_3 = -x_3$

one obtains for the path line of a fluid element the following differential equation:

$$\frac{d(x_1)_{\Re}}{dt} = 2x_1, \frac{d(x_2)_{\Re}}{dt} = -x_2, \frac{d(x_3)_{\Re}}{dt} = x_3$$

for $t = 0$ it shall be assumed that $(x_1)_{\Re} = (x_2)_{\Re} = (x_3)_{\Re} = 1$ so that in the solution

$$(x_1) = C_1 \exp[2t],\ (x_2)_{\Re} = C_2 \exp[-t],\ (x_3) = C_3 \exp[-t]$$

holds and thus the path line is stated as follows:

$$(x_1)_{\Re} = \exp[2t],\ (x_2)_{\Re} = \exp[-t],\ (x_3)_{\Re} = \exp[-t]$$

with $-\infty < t < \infty$. For the computation of the sweeping paths the solution can be employed again and C_1, C_2, C_3 can be computed such that it is claimed that at the time $t = \tau$ holds:

$$(x_1(t = \tau))_S = 1,\ (x_2(t = \tau))_S = 1,\ (x_3(t = \tau))_S = 1$$

Therefore it holds:

$$C_1 = \exp[-2\tau],\ C_2 = \exp[\tau],\ C_3 = \exp[\tau]$$

or as an equation for the sweeping path:

$$(x_1)_S = \exp[2(t - \tau)],\ (x_2)_S = \exp[-(t - \tau)],\ (x_3)_S = \exp[-(t - \tau)]$$

thus t being defined, and the range of values of t is defined by the period of time of the tracer injection. In the case that tracer substance is injected at all times, i.e. $-\infty < \tau < \infty$, equation yield the same curve. When the tracer injection is limited in time, one obtains as a visible sweeping path a corresponding part of the path line.

$$U_1 = x_1,\ U_2 = x_2(1 + 2t) \text{ and } U_3 = 0$$

which leads to the differential equations:

$$\frac{d(x_1)_{\Re}}{dt} = x_{\Re},\ \frac{d(x_2)_{\Re}}{dt} = x_{\Re}(1+2t),\ \frac{d(x_3)_{\Re}}{dt} = 0$$

with the solutions:

$$(x_1)_{\Re} = C_1 \exp[t],\ (x_2)_{\Re} = C_2 \exp[t(t + 1)],\ (x_3)_{\Re} = C_3$$

On the other hand, when one demands that the particle located at the time t in the point x_1, x_2, x_3 passes the injection point (1, 1, 0) of a tracer at the moment in time τ the integration constants C_1, C_2 and C_3 can be obtained from the following conditional equations:

$$C_1 \exp[\tau] = 1,\ C_2 \exp[\tau(\tau + 1)] = 1,\ C_3 = 0$$

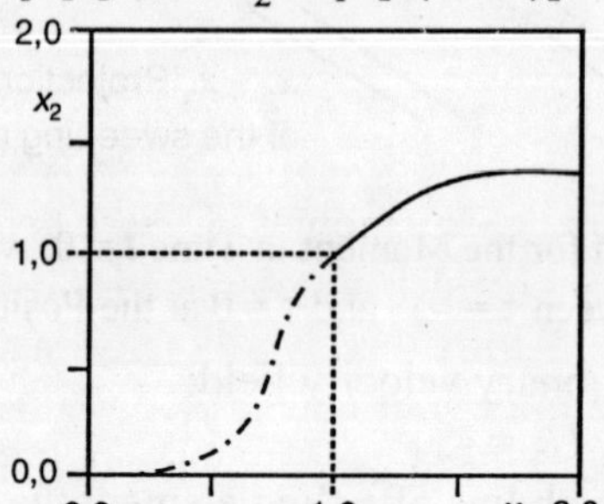

Fig. Sweeping Path in the Plane $x_1 - x_2$ (Fully Drawn Line Corresponds to $-\infty \leq \tau \leq 0$, Broken Line to $0 \leq \tau \leq \infty$)

These "constants" can now be inserted in equation again, where $C_3 = 0$

signifies that the sweeping path runs in the plane $x_1 - x_2$, i.e. in the plane $x_3 = 0$, and is described there by the following equations for the position coordinates $(x_1)_S$ and $(x_2)_S$

$$(x_1)_S = \exp\,[t - \tau],$$
$$(x_2)_S = \exp\,[t(t + 1) - \tau(\tau + 1)]$$

For the moment in time $t = 0$ the course of the sweeping path is:

$$(x_1)_S = \exp\,[-\tau],$$
$$(x_2)_S = \exp\,[-\tau(\tau + 1)]$$

in the domain $-\infty < \tau < \infty$.

From the equation for $(x_1)_S$ follows:

$$t = -\ln(x_1)_S$$

Inserted in the equation for $(x_2)_S$ at the time $t = 0$ the existing course of the sweeping path made visible by the injection of tracer material in $-\infty < \tau < \infty$ results in the plane $x_1 - x_2$

$$(x_2)_S = (x_1)_S^{(1-\ln(x_1)_S)} \quad \text{in the domain } 0 < (x1)_S < \infty$$

$$(x_2)_S = [(x_1)_S]^{(1-\ln(x_1)_S)}$$

KINEMATIC QUANTITIES OF FLOW FIELDS

Flow Lines of a Velocity Field

The considerations carried out in for fluid elements for the com-putation of path lines and sweepings paths have to be separated strictly from considerations for the determination of the flow lines of a flow field. Although for stationary flows flow lines, path lines and sweeping paths are identical, this does not justify to neglect the fundamental differences or even worse to assume that they do not exist. Only by a clear separation of the con-siderations it becomes generally comprehensible why under the steady-state condition of a flow field the above-stated identities of flow lines, path lines and sweeping paths exist.

When one considers the non-stationary flow field $U_j\,(x_i, t)$, flow lines can be defined for this field at any moment in time t so that space curves run parallel to the velocity vector which are defined as flow lines at each point. In the case that one considers initially a two-dimensional flow field, for which all velocity vectors lie parallel to the plane $x_1 - x_2$, then the above-indicated definition of the flow line leads to the following defining equation:

$$\frac{d(x_2)_\psi}{d(x_1)_\psi} = \frac{U_2}{U_1} \quad \text{for the time } t = \text{const}$$

The relation means that the gradient of the flow line is equal to the tangent of the angle formed by the velocity vector with the axis x_1 at the moment in time t. The index ψ taken up in the equation indicates that the stated coordinates $x_1 - x_2$ describe the flow line ψ.

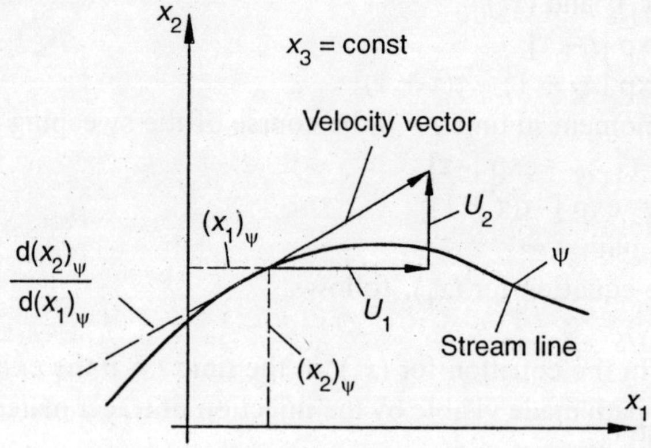

Fig. Sketch for Clarifying the Defining Equation for the Flow Line of a Flow Field

When one considers the defining equation for the flow line of a two dimensional flow field, it becomes comprehensible that in general for each moment in time t the quotient of U_2 and U_1 is a function of x_1 and x_2. The thus resulting differential equation has to be solved in order to conserve the equation for the flow line.

This will be explained on the basis of an example that starts from the following two-dimensional velocity field:

$U_1 = x_1 (1 + 2t);\quad U_2 = x_2 \quad \text{and} \quad U_3 = 0$

Introduced into the defining equation for the flow line one obtains:

$$\frac{d(x_2)_\psi}{d(x_1)_\psi} = \frac{U_2}{U_1} = \frac{x_2}{x_1(1+2t)}$$

or

$$\frac{d(x_2)_\psi}{x_2} = \frac{d(x_1)_\psi}{x_1} \frac{1}{2+2t}$$

By integration results:

$$\ln(x_2)_\psi = C + \frac{1}{1+2t} \ln(x_2)_\psi$$

Considering the flow line, passing at time $t = 0$ at the point (1, 1, 0) C = 0 results and thus:

$$(x_2)_\psi = [(x_1)_\psi]^{\frac{1}{1+2t}}$$

In the case that there is a three-dimensional velocity field, the above considerations, that were indicated for the projections of the flow lines into the planes $x_1 - x_2$., can be accepted in an analogous way. For the projections into the planes $x_1 - x_3$ – and $x_2 - x_3$. relations analogous to the defining equation result:

$$\frac{d(x_3)_\psi}{d(x_1)_\psi} = \frac{U_3}{U_1}$$

$$\frac{d(x_3)_\psi}{d(x_2)_\psi} = \frac{U_3}{U_2}$$

Thus the defining equations of the flow lines of a velocity field can be stated as follows:

$$\frac{d(x_1)_\psi}{U_1} = \frac{d(x_2)_\psi}{U_2}, \quad \frac{d(x_1)_\psi}{U_1} = \frac{d(x_3)_\psi}{U_3}, \quad \frac{d(x_2)_\psi}{U_2} = \frac{d(x_3)_\psi}{U_3}$$

or rewritten as:

$$\frac{d(x_1)_\psi}{U_1} = \frac{d(x_2)_\psi}{U_2} = \frac{d(x_3)_\psi}{U_3}$$

These differential relations for the flow line of a velocity field hold at each moment in time t. Their solution leads to a relation $(x_3)_\psi = \psi(x_1, x_2)$, which describes a space curve, the three-dimensional flow line.

Probably the most simple way to solve the law of differential equations is to seek a parameter solution $(x_1)_\psi = x_1\ (s)$, here s is a parameter, whose value at a certain reference point of the flow line is equal to zero and which takes increasing values along the flow line and in flow direction. When all values $-\infty < s < \infty$ are passed, through a presentation of the entire flow line is obtained.

By introducing s one obtains:

$$\frac{d(x_j)_\psi}{ds} = U_j\ (x_i, t) \qquad \text{for } t = \text{const and } j = s$$

a relation which represents for each coordinate $(x_j)_\psi$ a differential equation $j = 1, 2, 3$ describing the flow lines in space for t = const. If the flow line passing through the space point $[x_0]_j$ at time t is sought, $s = 0$, results from integrating the three differential equations, when $x_j(t) = x_{j,0}$ From this results the entire flow-line field as:

$$(x_j)_\psi = \psi_j\ (x_{0,j}\ , t, s)$$

In order to demonstrate the way of proceeding in determining three-dimensional flow-line fields, the following velocity field is to be considered again:

$U_1 = x_1(1 + t)$, $U_2 = -x_2$ and $U_3 = -x_3 t$

Thus a law/principle/theorem of differential equations for the flow lines of this velocity field results:

$$\frac{d(x_1)_\psi}{ds} = x_1(1+t) \quad \frac{d(x_2)_\psi}{ds} = -x_2 \quad \frac{d(x_3)_\psi}{ds} = -x_3 t$$

The integration of these equations yields:

$(x_1)_\psi = C_1 \exp [(1 + t)s]$,

$(x_2)_\psi = C_2 \exp [-s]$,

$(x_3)_\psi = C_3 \exp [-ts]$.

Searching for the flow line that passes through the point(1, 1, 1) one can chose this point as a point of reference and set $s = 0$ for $(x_i)_\psi = 1$ From this the following results for the integration constants:

$C_1 = C_2 = C_3 = 1$

One obtains thus:

$(x_1)_\psi = \exp [(1 + t)s]$,

$(x_2)_\psi = \exp [-s]$,

$(x_3)_\psi = \exp [-ts]$.

For the time $t = 0$ a flow line path results

$(x_1)_\psi = \exp [s]$, $(x_2)_\psi = \exp [-s]$, $(x_3)_\psi = 1$

So one obtains a flow line passing in the plane $x_3 = 1$ and described there by the relation

$$(x_2)_\psi = \frac{1}{(x_1)_\psi}$$

The entire flow line field is obtained when one introduces for s = 0 arbitrary position coordinates $(x_{0,i})$ so that for the position coordinates holds:

$(x_1)_\psi = (x_1)_{\psi,0} \exp [(1 + t)s]$,

$(x_2)_\psi = (x_2)_{\psi,0} \exp [-s]$,

$(x_3)_\psi = (x_3)_{\psi,0} \exp [-ts]$.

This set of equations indicates at each time t the flow lines passing through the point $\{x_j\}_{\psi,0}$ There the parameter s has the value $s = 0$. and for all other points of the flow line a value equal to zero. A clear assignment is thus occurring which guarantees that flow lines never intersect, as otherwise two velocities would exist simultaneously at this point of intersection. Thus occurring which guarantees that flow lines never intersect, as otherwise two velocities would exist simultaneously at this point of intersection. This is excluded because of the existence of clear velocity fields (except for stagnation points and singularities).

In the preceding section it was emphasized that in general flow lines, path lines and sweeping lines are not identical and the computed examples have made this clear have confirmed this. For stationary flows all three lines are identical, i.e.

- For stationary flow fields marked fluid elements move along flow lines, i.e. flow lines are equal to the path lines

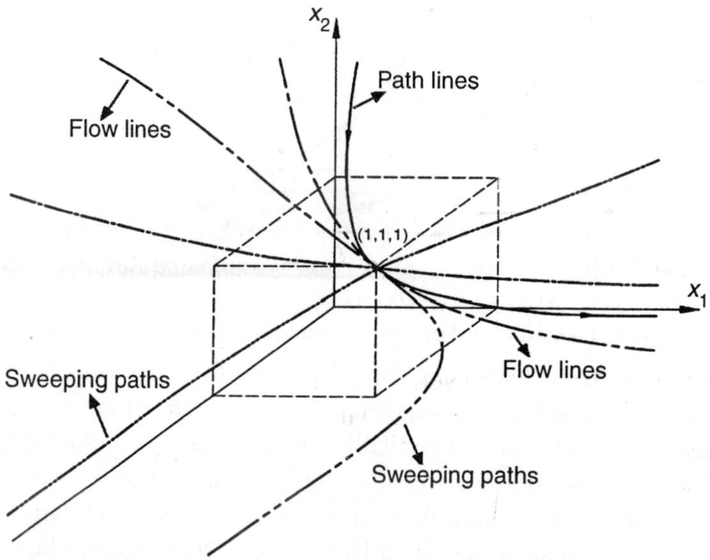

Fig. Comparison of Flow Lines, Path Lines and Sweeping Paths

- For stationary flow fields flow lines can be made visible by locally injected tracer particles, i.e. flow lines are equal to the sweeping paths

For non-stationary flows generally flow lines, path lines and sweeping paths are different space curves.

Stream function and Flow Lines of two-dimensional Flow Fields

For two-dimensional incompressible velocity fields $\{U_j\} = \{U_1, U_2, 0\}$ the stream function can be introduced as a field quantity. It is defined as follows:

$$U_1 = \frac{\partial_\psi}{\partial x_2} \text{ und } U_2 = \frac{\partial_\psi}{\partial x_1}$$

and is computed from the velocity field over the following line integrals:

$$\psi - \psi_0 = \int_{x_{2,0}}^{x_2} U_1 \mathrm{d}x_2 \quad \text{for } x_1 = \text{const}$$

$$\psi - \psi_0 = \int_{x_{1,0}}^{x_1} U_1 \mathrm{d}x_1 \quad \text{for } x_2 = \text{const}$$

The above equations show that the difference existing between two stream-

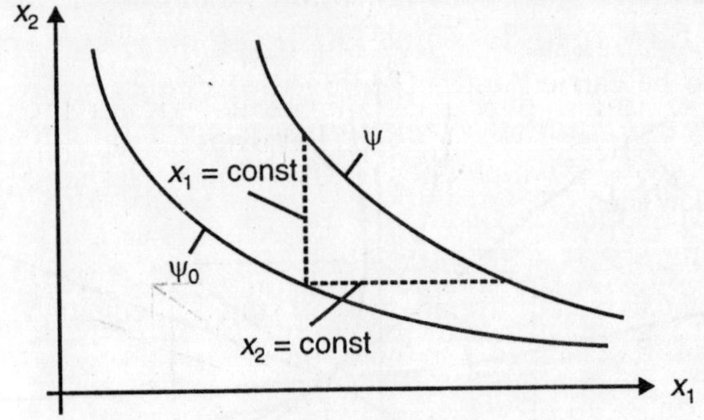

Fig. Explanation of the defining equations of stream function

function values is a measure for the volume flow that runs between two lines each with a constant stream-function value each, where a depth of the linear measure 1 is considered vertically to the plane $x_1 - x_2$. Accordingly the integrations along the line drawn and stated in the equations x_1 = const and x_2 = const yield identical volume-flow values, when the upper integration limits x_1 and x_2 are chosen such that the difference $\psi - \psi_0$ is equally large in both integrals.

For a fluid with ρ = const, i.e. for an ideal fluid, the identity of the integrals represents a mass-conservation relation. When one carries out the integrations stated in the equations along a line ψ = const i.e. for $d\psi = 0$, one obtains:

$$U_1 d(x_2)_\psi = U_2 d(x_1)\psi$$

or transcribed:

$$\frac{d(x_1)_\psi}{U_1} = \frac{d(x_2)_\psi}{U_2}$$

From the above indicated relation it can be said that the stream function by a comparison with the defination equation for the flow line stated in equation that the stream function defined for two-dimensional flow fields yields flow lines ψ for ψ = const.

While for the kinematic considerations is arbitrary mathematics set-ups for the velocity field could be used, the intro-duction of the stream function requires a reduction of the considerations to those velocity fields which have to fulfill physical conditions. Physically possible flow fields have to fulfill the mass-conservation law which can be formulated for ideal fluids (ρ = const) as follows:

$$\frac{\partial U_i}{\partial x_i} = \frac{\partial U_1}{\partial x_1} + \frac{\partial U_2}{\partial x_2} + \frac{\partial U_3}{\partial x_3} = 0$$

That a stream function fulfills the mass-conservation law for two-dimensional flows automatically, can easily be checked by inserting the definition in equation where the law/principle of Schwarz has to be applied.

When a flow field does not fulfill the mass-conservation law, the integrations to be carried out according to the equations result in solutions contradicting one another. This can be shown for the employed two-dimensional flow field which does not fulfill the mass-conservation law (i.e. the continuity equation):

$U_1 = x_1$; $U_2 = x_2(1 + 2t)$ and $U_3 = 0$

When one carries out the integration stated in equation in spite of this, from the defining equations for the stream function, i.e. from the relations results:

$$\frac{\partial_\psi}{\partial x_2} = x_1 \text{ and } \frac{\partial_\psi}{\partial x_1} = -x_2(1+2t)$$

By integration of these equations one obtains for the stream function:

$\psi = x_1\, x_2 + F(x_1, t) \qquad \psi = -x_1 x_2(1 + 2t) + G(x_2, t)$

or expressed otherwise:

$x_1 x_2 + F(x_1, t) \neq -x_1\, x_2(1 + 2t) + G(x_2, t)$

or $2(t + 1)x_1\, x_2 \neq G(x_2, t) - F(x_1, t)$

A comparison of the results of both the integrations, which yield the equations shows the contradiction resulting for the stream function.

The time dependence showing on the right side of the relation is missing on the left side. This results from the fact that the velocity field indicated according to although mathematically clearly defined, cannot exist physically; the velocity field does not fulfill the requirements determined by the mass-conservation law for ρ = const When one considers on the other hand the velocity field:

$U_1 = \exp[x_1(1 + t)]$; $U_2 = -x_2(1 + t) \exp[\text{x1}\ (1 + \text{t})]$; $U_3 = 0$ for which the relation is fulfilled, where it holds:

$$\frac{\partial U_1}{\partial x_1} = (1 + t) \exp[x_1(1 + t)] \qquad \text{and}$$

$$\frac{\partial U_2}{\partial x_2} = -(1 + t) \exp[x_1(1 + t)]$$

one obtains from the defining equations for the stream function :

$$\frac{\partial \psi}{\partial x_2} = \exp[x_1(1 + t)] \text{ and}$$

$$\frac{\partial \psi}{\partial x_1} = x_2(1 + t) \exp[x_1(1 + t)]$$

the following solution for the stream function ψ:

$\psi = x_2 \exp[x_1(1+t)] + C$

If one knows the value of the stream function for $x_{1,0}$ and $x_{2,0}$, it holds

$\psi_0 = x_{2,0} \exp[x_{1,0}(1+t)] + C$

$\psi - \psi_0 = x_2 \exp[x_1(1+t)] - x_{2,0} \exp[x_{1,0}(1+t)]$

The obtained result can also be computed for $x_1 = x_{1,0}$ = const. from the relation:

$\psi - \psi_0 = (x_2 - x_{2,0}) \exp[x_{1,0}(1+t)]$ for $x_1 = x_{1,0}$

This is indicated by the distribution $\psi(x_2)$ at the location $x_{1,0}$ an.

When one wants to determine the flow line path in the plane $x_1 - x_2$ – one has to consider y as a parameter in equation and derive the relation $x_2 - x_1$ – for this parameter from equation. Here ψ_0, x_1,0 and x_2, 0 are constants that can be chosen freely.

Divergence of a Flow Field

In this section mathematics operators shall be explained that are known from vector analysis and that can be applied to flow fields. Their derivations will also be repeated and considered as concerns their physical meaning.

Divergence of the stream field $\vec{U}$:

$$\frac{\partial U_i}{\partial x_i} = \frac{\partial U_1}{\partial x_1} + \frac{\partial U_2}{\partial x_2} + \frac{\partial U_3}{\partial x_3}$$

The above defined divergence of a flow field is to be considered as a scalar field which, in the presence of a steady velocity field, is defined at each point in space and can be computed from the velocity field that can be assumed to be known.

When one wants to perceive the physical meaning of the operator U_i applied to the velocity field $\frac{\partial}{\partial x_i}$ x_i the consideration of a fluid element, as stated in Fig.4.9, is recommended.

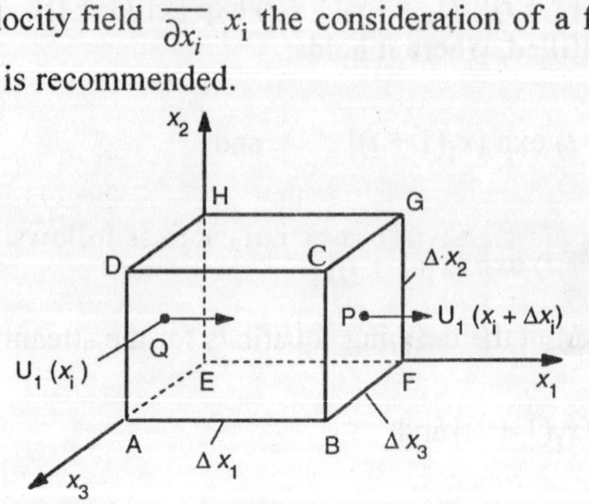

Fig. Fluid element for explaining the physical meaning of $\frac{\partial U_i}{\partial x_i}$

The edge lengths Δx_i were assumed to be very small, so that a velocity vector can be assigned to each surface such that this vector indicates with which velocity the considered surface of a fluid element moves. Accordingly, the surface AEHD moves in the direction x_1 with the velocity component of the velocity field present in point Q i.e. with $U_1\ (x_i) = U_1\ (x_1, x_2, x_3)$. In comparison, the surface BFGC moves with the velocity component $U_1(x_i + Dx_i, x_2, x_3)$.present in point *P*. This velocity component can be expressed by a Taylor series expansion as follows:

$$U_1(x_1 + Dx_1, x_2, x_3) = U_1(x_1) + \left(\frac{\partial U_1}{\partial x_1}\right)\Delta x_1 + \frac{1}{2}\left(\frac{\partial^2 U_1}{\partial x_1^2}\right)\Delta x_1^2 + \ldots$$

The difference velocity between the surfaces AEHD and BFGC can thus be computed as:

$$\Delta U_1(x_i, \Delta x_i) = \left(\frac{\partial U_1}{\partial x_1}\right)\Delta x_1 + \frac{1}{2}\left(\frac{\partial^2 U_1}{\partial x_1^2}\right)\Delta x_1^2 + \ldots$$

As a consequence of this velocity difference a volume increase or a volume decrease results, depending on which signs the derivations of the velocity fields, which can be stated as follows in a first approximation by multiplication with the surface $\Delta x_2\, \Delta x_3$, neglecting the terms of second and higher order in Δx_1:

$$\frac{d}{dt}(\delta V_1)_{\Re} = \Delta U_1(x_i)(\Delta x_2 \Delta x_3) = \frac{\partial U_1}{\partial x_1}\Delta x_1(\Delta x_2 \Delta x_3)$$

On the basis of simultaneously existing gradients of the velocity field in the directions x_2 – and x_3 in addition volume changes occur in the time unit, which again can be stated in a first approximation as follows:

$$\frac{d}{dt}(\delta V_2)_{\Re} = \frac{\partial U_2}{\partial x_2}\Delta x_2(\Delta x_1 \Delta x_3) \qquad \text{and}$$

$$\frac{d}{dt}(\delta V_3)_{\Re} = \frac{\partial U_3}{\partial x_3}\Delta x_3(\Delta x_1 \Delta x_2)$$

so that the entire volume change that can be expected in a flow field for a fluid element in the time unit can be indicated as follows:

$$\frac{d}{dt}(\delta V_3)_{\Re} = \sum_{\alpha=1}^{3}\frac{d}{dt}(\delta V_\alpha)_{\Re} = \frac{\partial U_i}{\partial x_i}(\delta V_{\Re})$$

or can be written as:

$$\frac{\partial U_i}{\partial x_i} = \frac{1}{\delta V_{\Re}}\frac{d}{dt}(\delta V_{\Re})$$

This relation makes emphasizes the physical significance of the divergence of a velocity field. In accordance the divergence of a velocity field states how big the volume change of a fluid element is that occurs per time and volume unit at a certain position in a flow field. At such locations of the flow field where the divergence of a velocity vector is equal to zero, there is no temporal volume change locally for a fluid element moving in the velocity field. When the divergence in sub-domains of the velocity field is computed negatively, a fluid element experiences volume decreases in these domains.

When one carries out considerations concerning the physical significance of the divergence of a velocity field at a stationary volume element of a fluid, inflows and outflows occur through the surfaces of the considered volume because of the existing velocity field, to an extent that the volume flowing in per time unit can be stated:

$$\dot{V}_{\text{inflow}} = U_i \Delta x_j \Delta_{xk} \qquad i = j, k$$

For the volume flowing out it can be computed:

$$\dot{V}_{\text{Out flow}} = \left[U_i + \frac{\partial U_i}{\partial x_i} \right] \Delta x_j \Delta x_k \qquad i = j, k$$

The relation makes clear that the sumation for $i = 1$ to 3 is stated in a sufficient6ly comprehensible way by the double index in $\partial U_i / \partial x_i$

The difference of inflows and outflows, considering DV = Dx_1, Dx_2 , Dx_3 can be computed as:

$$\Delta \dot{V} = \dot{V}_{\text{In flow}} - \dot{V}_{\text{Out flow}} = -\frac{\partial U_i}{\partial x_i} \Delta V$$

This relation makes clear that the presence of a positive divergence of the velocity field in volume is equal to a source, as more "'fluid volume"' is flowing out than flowing in. When, however, the divergence of a velocity field is negative, a sink occurs, as then the inflow in "'volume"' has to be larger than the outflow.

TRANSLATION, DEFORMATION AND ROTATION OF FLUID ELEMENTS

Analogous to considerations in solid-state mechanics, the deformations of fluid elements that occur due to existing velocity gradients are of interest in some flow-mechanics considerations.

When one includes the translatory motion and the rotation of a fluid element in the fluid deformations, the entire local state of motion and deformation can be stated by four "'geometrically easily separable"' sub-states. The pure translatory motion leads to a change of position of the fluid element marked to an extent that it holds:

$$d(x_j) = (U_j)_{\Re}\, dt = U_j dt$$

This relation expresses that the locally existing velocity field is responsible for the translatory motion of a fluid element, i.e. fluid elements move at each moment in time with the locally existing velocity vector.

In order to state or compute the rotation of a fluid element one has to describe both the angles $\Delta\Theta_1$ and $d\Delta\Theta_2$:

$$\Delta\Theta_1 = \left[\tan \frac{\frac{\partial U_2}{\partial x_1}(\Delta x_1)_{\Re}\,\Delta t}{(\Delta x_1)_{\Re}}\right] = \frac{\partial U_2}{\partial x_1}\Delta t$$

As a rotational speed of the fluid element the positive change of angle of the diagonal of the element occurring per time unit is defined:

Translation, deformation and rotation of fluid elements 107

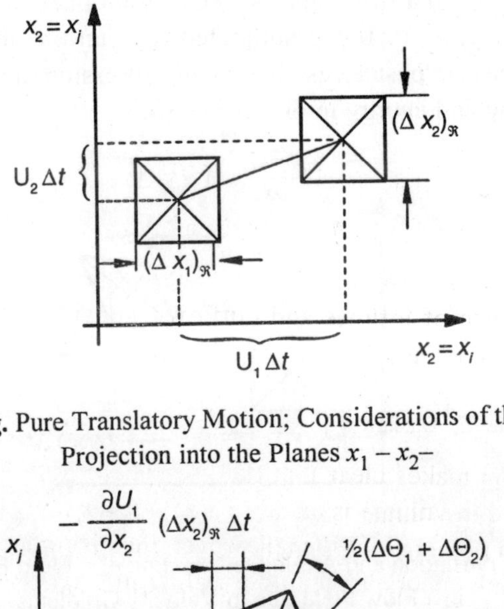

Fig. Pure Translatory Motion; Considerations of the Projection into the Planes $x_1 - x_2$–

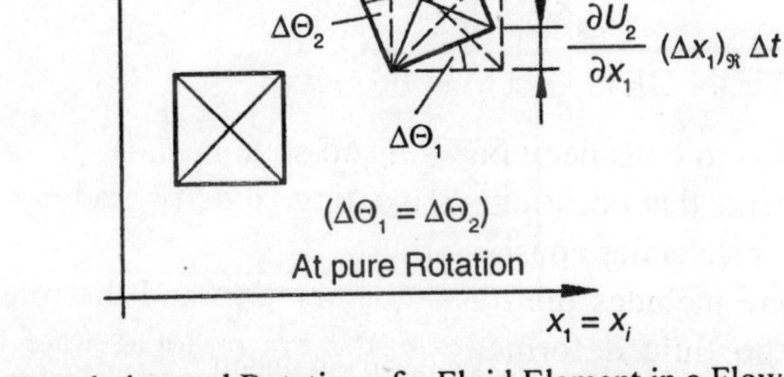

Fig. Translation and Rotation of a Fluid Element in a Flow Field

$$\dot{\Theta}_{21}^R = \left(\frac{d\Theta^R}{dt}\right) = \frac{1}{2}\left(\frac{d\Theta_1}{dt} + \frac{d\Theta_2}{dt}\right) = \frac{1}{2}\left(\frac{\partial U_2}{\partial x_1} - \frac{\partial U_1}{\partial x_2}\right)$$

Thus generally the components of the rotational speed vector $\{\omega_k\}$ for the locally occurring rotation of a fluid element per time unit can be stated as follows:

$$2\dot{\Theta}_{ij}^{R} = 2\omega_k = \frac{\partial U_i}{\partial x_j} = rot(\vec{U})$$

The quantity ω_k states the double rotational speed of the axis of the fluid element occurring in positive direction. The second diagonal of the considered fluid element rotates with the same angular speed. $\{\omega_k\}$ is an important kinematic quantity of the velocity field. It is defined as swirling/rotational power (Wirbelstrke) and is computed as half the rotation of the velocity field. Thus it is a field quantity of its own $\omega_k\ (x_i, t)$, for which holds:

$\omega_k(x_i, t) = 0$ when a flow field is free of rotation.

When $\omega_k(x_i, t) \neq 0$, flows subjected to rotations are present, whose swirling properties are best accessible when expressing the conservation laws for mass, impulse and energy in ω_k.

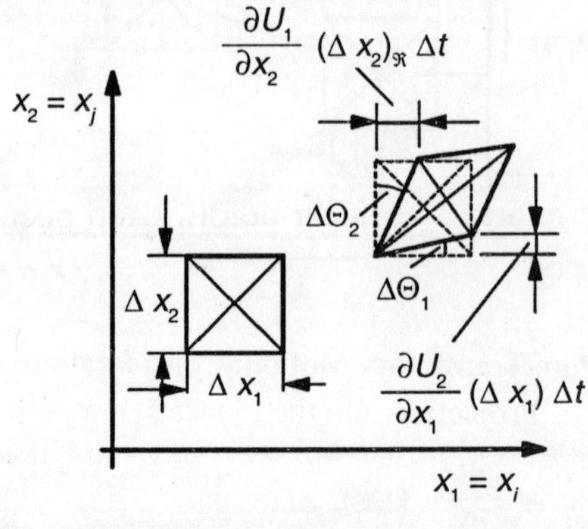

Fig. Translation and Angle Deformation of a Fluid Element in a Flow Field due to Velocity Gradients

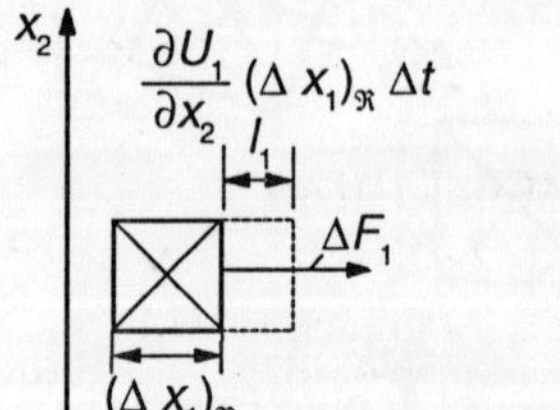

Fig. Elongation of a Volume Element due to Velocity Gradients in the Flow Field

When one considers next the angle deformation of a fluid element , one can see that for the angle deformation holds:

$$\dot{\Theta}_{21}^{D}=\frac{1}{2}\left(\frac{\mathrm{d}\Theta_1}{\mathrm{d}t}-\frac{\mathrm{d}\Theta_2}{\mathrm{d}t}\right) \quad \text{Deformation angular speed}$$

Thus for the angular deformation in the plane $x_1 - x_2$ holds generally

$$\dot{\Theta}_{21}^{D}=\frac{1}{2}\left(\frac{\partial U_2}{\partial x_1}+\frac{\partial U_1}{\partial x_2}\right)$$

Or generally:

$$\dot{\Theta}_{ij}^{D}=\frac{1}{2}\left(\frac{\partial U_j}{\partial x_i}+\frac{\partial U_i}{\partial x_j}\right) \qquad i\neq j$$

Analogous to considerations in solid-state mechanics the symmetry of the deformation tensor holds:

$$\dot{\Theta}_{ij}^{D}=\dot{\Theta}_{ij}^{D}$$

Finally one has to consider the dilatation of a fluid element which experiences strain rates due to the velocity gradient existing in a flow field. The linear deformation/change in length occurring due to an existing velocity gradient in direction x_1 can be stated as follows:

$$\frac{\mathrm{d}l}{\mathrm{d}t}=\lim_{\Delta t\to 0}\left(\frac{\partial U_1}{\partial x_1}\right)\frac{(\delta x_1)_{\Re}\Delta t}{\Delta t}=\left(\frac{\partial U_1}{\partial x_1}\right)(\delta x_1)_{\Re}$$

From this is computed the linear deformation occurring per length and time unit:

$$\frac{1}{(\Delta x_1)}\frac{\mathrm{d}(l_1)}{\mathrm{d}t}=\frac{\partial U_1}{\partial x_1}$$

When multiplying the linear deformation with the transverse area the apper-taining volume change results:

$$\frac{\mathrm{d}(\delta V_1)_{\Re}}{\mathrm{d}t}=\left(\frac{\partial U_1}{\partial x_1}\right)(\delta V_1)_{\Re}$$

Summed up over all three axis directions one obtains for the entire volume change per time unit:

$$\frac{1}{(\delta V)_{\Re}}\frac{\mathrm{d}(\delta V)_{\Re}}{\mathrm{d}t}=\frac{\partial U_i}{\partial x_i}$$

i.e. the divergence of the velocity field indicates how the volume of a fluid element at a point in space changes with time.

It is customary in literature to combine elongations of fluid elements and their angular deformations to a deformation tensor in such a way that it holds:

$$\varepsilon_{ij}=\frac{1}{2}\left(\frac{\partial U_j}{\partial x_i}+\frac{\partial U_i}{\partial x_j}\right) \quad \text{for } i\neq j$$

so that for the deformation tensor holds:

$$\{\varepsilon_{ij}\} = \begin{Bmatrix} \varepsilon_{11} & \varepsilon_{12} & \varepsilon_{13} \\ \varepsilon_{21} & \varepsilon_{22} & \varepsilon_{23} \\ \varepsilon_{31} & \varepsilon_{32} & \varepsilon_{33} \end{Bmatrix} \text{und } \varepsilon_{ij} = \varepsilon_{ij}$$

From the above considerations the following relation results:

$$\frac{\partial U_j}{\partial x_i} = \frac{1}{2}\left(\frac{\partial U_j}{\partial x_i} + \frac{\partial U_i}{\partial x_j}\right) + \frac{1}{2}\left(\frac{\partial U_j}{\partial x_i} + \frac{\partial U_i}{\partial x_j}\right)$$

i.e. it holds:

$$\frac{\partial U_j}{\partial x_i} = \varepsilon_{ij} + \frac{d\Theta_{ij}^R}{\partial x_i} = \varepsilon_{ij} + \frac{1}{2}\varepsilon_{ijk}\frac{\partial U_j}{\partial x_i}$$

The gradients existing in velocity fields are linked to deformations and rotations of fluid elements, the gradients allowing to state corresponding rates of deformation (normalized deformation per unit time) and rotational angle velocities. This has to be considered when employing the analogy between solid-state mechanics and fluid mechanics, in order to transfer considerations on deformations of elastic bodies, carried out in solid-state mechanics to rates of deformation of fluid elements occurring in fluid mechanics, due to existing gradients of the flow field.

Relative Motions

Considerations of the velocities at two points separated by (dx_i) result in the following relation:

$$U_j(x_i + \delta x_i, t) = U_j(x_i, t) + \frac{\partial U_j}{\partial x_i}\delta x_i + \ldots$$

or transcribed:

$$U_j(x_i + \delta x_i, t) = \underbrace{U_j(x_i, t)}_{\text{Translation}} + \overbrace{\frac{1}{2}\left(\frac{\partial U_j}{\partial x_i} - \frac{\partial U_i}{\partial x_j}\right)\delta x_i}^{\text{Rotation}} + \underbrace{\frac{1}{2}\left(\frac{\partial U_j}{\partial x_i} - \frac{\partial U_i}{\partial x_j}\right)\delta x_i}_{\text{Deformation}}$$

This relationship makes it clear that the velocity at the adjacent point $x_i + \delta x_i$ is composed of the translation by velocity at point $p(x_i)$, a rotational velocity around this point and a deformation velocity in this point. The components for $j = 1, 2, 3$ read:

$$j = 1: U_1(x_i + \delta x_i, t) = \left[U_1 + \frac{\partial U_1}{\partial x_1}\delta x_1 + \frac{1}{2}\left(\frac{\partial U_1}{\partial x_2} - \frac{\partial U_2}{\partial x_1}\right)\delta x_2\right.$$

$$+\frac{1}{2}\left(\frac{\partial U_1}{\partial x_3}-\frac{\partial U_3}{\partial x_1}\right)\delta x_3+\frac{1}{2}\left(\frac{\partial U_1}{\partial x_2}+\frac{\partial U_2}{\partial x_1}\right)\delta x_2+\frac{1}{2}\left(\frac{\partial U_1}{\partial x_3}+\frac{\partial U_3}{\partial x_1}\right)\delta x_3\Bigg]$$

$$j=2: U_2\,(x_i+\delta x_i, t)=\Bigg[U_2+\frac{\partial U_2}{\partial x_2}\delta x_2+\frac{1}{2}\left(\frac{\partial U_2}{\partial x_3}-\frac{\partial U_3}{\partial x_2}\right)\delta x_3$$

$$+\frac{1}{2}\left(\frac{\partial U_2}{\partial x_1}-\frac{\partial U_1}{\partial x_2}\right)\delta x_1+\frac{1}{2}\left(\frac{\partial U_2}{\partial x_3}+\frac{\partial U_3}{\partial x_2}\right)\delta x_3+\frac{1}{2}\left(\frac{\partial U_2}{\partial x_1}+\frac{\partial U_1}{\partial x_2}\right)\delta x_1\Bigg]$$

$$j=3: U_3\,(x_i+\delta x_i, t)=\Bigg[U_3+\frac{\partial U_3}{\partial x_3}\delta x_3+\frac{1}{2}\left(\frac{\partial U_3}{\partial x_1}-\frac{\partial U_1}{\partial x_3}\right)\delta x_1$$

$$+\frac{1}{2}\left(\frac{\partial U_3}{\partial x_2}-\frac{\partial U_2}{\partial x_3}\right)\delta x_2+\frac{1}{2}\left(\frac{\partial U_3}{\partial x_1}+\frac{\partial U_1}{\partial x_3}\right)\delta x_1+\frac{1}{2}\left(\frac{\partial U_3}{\partial x_1}+\frac{\partial U_2}{\partial x_3}\right)\delta x_2\Bigg]$$

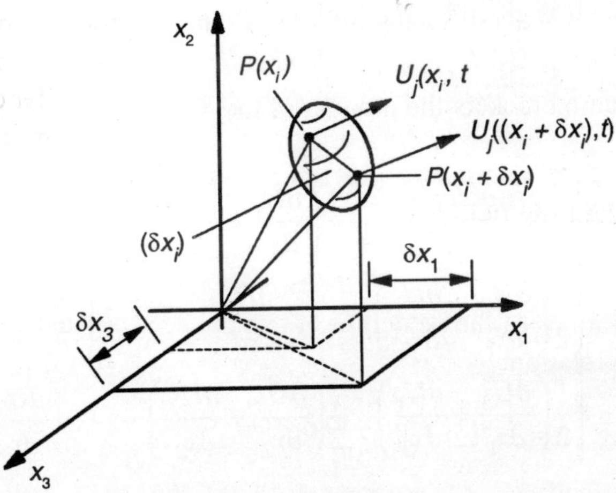

Fig. Relative motions in a fluid element

With the above equations the most general motions of fluid elements can now be described, i.e. the velocity in any point of a fluid element can be stated as the sum from the translation velocity of a reference point, a rotational motion around this point, i.e. the resulting speed of the rotation and a superimposed deformation velocity.

The different addends of the equations can be regrouped as follows

$$j=1: U_1(x_i)+\left[\frac{\partial U_1}{\partial x_1}\delta x_1+\frac{1}{2}\left(\frac{\partial U_1}{\partial x_2}+\frac{\partial U_2}{\partial x_1}\right)\delta x_1+\frac{1}{2}\left(\frac{\partial U_1}{\partial x_3}+\frac{\partial U_3}{\partial x_1}\right)\delta x_3\right]$$

$$+\left\{\frac{1}{2}\left(\frac{\partial U_1}{\partial x_2}-\frac{\partial U_2}{\partial x_1}\right)\delta x_1+\frac{1}{2}\left(\frac{\partial U_1}{\partial x_3}-\frac{\partial U_3}{\partial x_1}\right)\delta x_1\right\}$$

$$j=2:U_2(x_2)+\left[\frac{\partial U_2}{\partial x_2}\delta x_2+\frac{1}{2}\left(\frac{\partial U_2}{\partial x_1}+\frac{\partial U_1}{\partial x_2}\right)\delta x_1+\frac{1}{2}\left(\frac{\partial U_2}{\partial x_3}+\frac{\partial U_3}{\partial x_2}\right)\delta x_3\right]$$

$$+\left\{\frac{1}{2}\left(\frac{\partial U_2}{\partial x_1}-\frac{\partial U_1}{\partial x_2}\right)\delta x_1+\frac{1}{2}\left(\frac{\partial U_2}{\partial x_3}-\frac{\partial U_3}{\partial x_2}\right)\delta x_3\right\}$$

$$j=2:U_3(x_3)+\left[\frac{\partial U_3}{\partial x_3}\delta x_3+\frac{1}{2}\left(\frac{\partial U_3}{\partial x_2}+\frac{\partial U_2}{\partial x_3}\right)\delta x_2+\frac{1}{2}\left(\frac{\partial U_3}{\partial x_1}+\frac{\partial U_1}{\partial x_3}\right)\delta x_1\right]$$

$$+\left\{\frac{1}{2}\left(\frac{\partial U_3}{\partial x_2}-\frac{\partial U_2}{\partial x_3}\right)\delta x_2+\frac{1}{2}\left(\frac{\partial U_3}{\partial x_1}-\frac{\partial U_1}{\partial x_3}\right)\delta x_1\right\}$$

The expressions in front of the rectangular brackets represent the translation which is given by the following velocity vector:

$U_j(x_i, t) = \{U_1, U_2, U_3\}$

In the square brackets the product of the deformation tensor:

$$D_{ij}(x_i, t)=\begin{Bmatrix}\dfrac{\partial U_1}{\partial x_1} & \dfrac{1}{2}\left(\dfrac{\partial U_1}{\partial x_2}+\dfrac{\partial U_2}{\partial x_1}\right) & \dfrac{1}{2}\left(\dfrac{\partial U_1}{\partial x_3}+\dfrac{\partial U_3}{\partial x_1}\right)\\ \dfrac{1}{2}\left(\dfrac{\partial U_2}{\partial x_1}+\dfrac{\partial U_1}{\partial x_2}\right) & \dfrac{\partial U_2}{\partial x_2} & \dfrac{1}{2}\left(\dfrac{\partial U_2}{\partial x_3}+\dfrac{\partial U_3}{\partial x_2}\right)\\ \dfrac{1}{2}\left(\dfrac{\partial U_3}{\partial x_1}+\dfrac{\partial U_1}{\partial x_3}\right) & \dfrac{1}{2}\left(\dfrac{\partial U_3}{\partial x_2}+\dfrac{\partial U_2}{\partial x_3}\right) & \dfrac{\partial U_3}{\partial x_3}\end{Bmatrix}$$

and of the "'distance vector"'

$\{\delta x_i\} = \{\delta x_1, \delta x_2, \delta x_3\}$

is shown, while in the curved brackets the vector product of

$\{\delta x_i\} = \{\delta x_1, \delta x_2, \delta x_3\}$

and

$$2\omega_k=\left\{\left(\frac{\partial U_3}{\partial x_2}-\frac{\partial U_2}{\partial x_3}\right);\left(\frac{\partial U_1}{\partial x_3}-\frac{\partial U_3}{\partial x_1}\right);\left(\frac{\partial U_2}{\partial x_1}-\frac{\partial U_1}{\partial x_2}\right)\right\}$$

is shown. Thus it can be written:

$U_j(x + dx_i, t) = U_j(x_i, t) + D_{ij}(x_i, t)\, \delta x_i + \varepsilon_{ijk\omega k}\, (x_i, t)\, \delta x_i$

This relation again expresses the facts of the case that the total motion of a point $P'(x_i + dx_i)$ can be understood from the translation motion of the point $P(x_i)$., superimposed by deformation motions and rotational motions around P (x_i). The different parts translation, deformation and rotation can be taken from the sequence of a fluid element. For the two-dimensional case with $U_1 = u$ and $U_2 = v$ as well as $x_1 = x$ and $x_2 = y$.

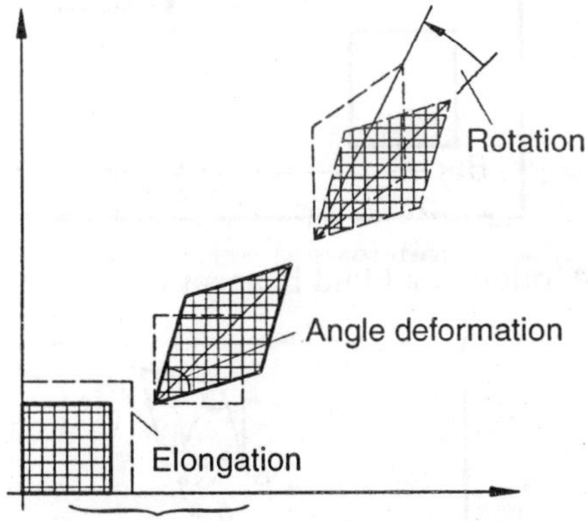

Fig. Illustration of the Translation Motion, the Deformation and the Rotation of a Fluid Element

The last named manner of motion requires:

$$\frac{\partial U2}{\partial x_1} = \frac{\partial U_1}{\partial x_2},$$

so that $\omega_3 = 0$.

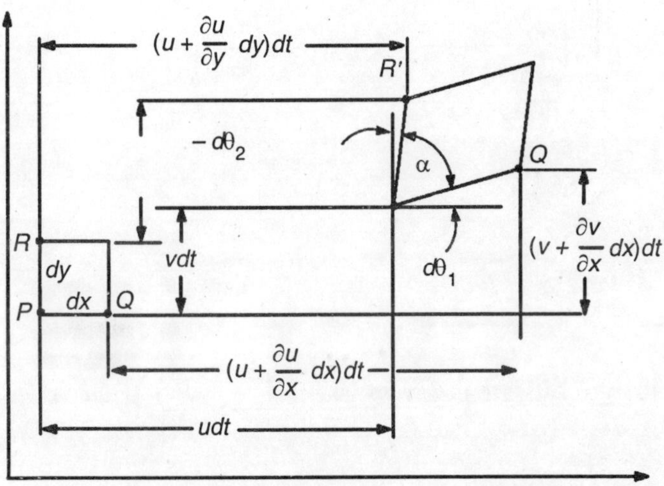

Fig. Translation, Deformation and Rotation of a Fluid Element Due to the Velocity Components u and v

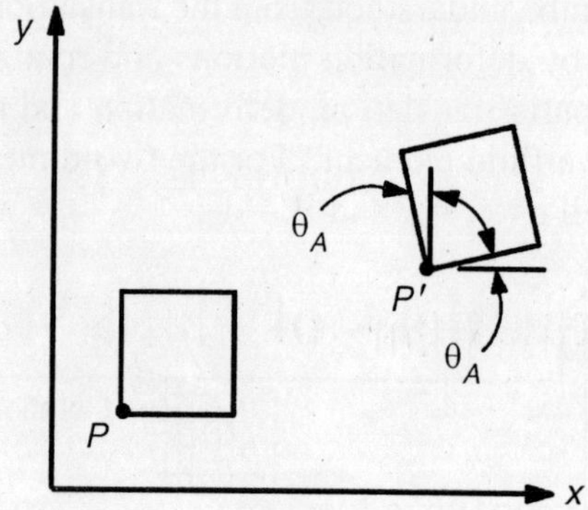

Fig. Translation Motion of a Fluid Element with Rotation Mit Rotation

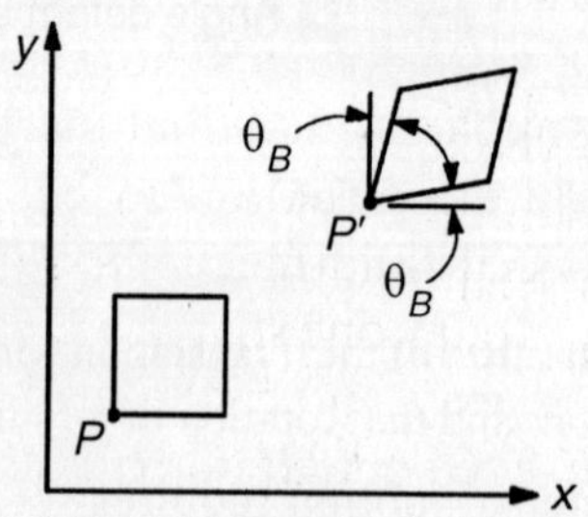

Fig. Translation Motion of a Fluid Element with Deformation

Chapter 4

Basic Equations of Fluid Mechanics

GENERAL CONSIDERATIONS

Fluid mechanics considerations are carried out in many domains, especially in the fields of engineering. Below there is a list which clearly indicates the far-reaching application of fluid-mechanics knowledge. The list lends itself to show the importance of fluid mechanics. While it was usual in the past to carry out special fluid-mechanics considerations on each of the below-listed domains, today one strives increasingly at developing and introducing generalized ways of consideration that are applicable unreservedly to all below-cited domains. This makes it necessary to derive the basic laws formulated for the application of solving flow problems so generally that they fulfill the requirements for broadest applicability indicated in the below-cited list. The objective of the derivations in this section is to formulate the conservation laws for mass, momentum, energy, chemical species etc. such that they can be applied to all the flow problems occurring in the following domains:

- Heat exchanger, cooling and drying technology
- Reaction technology and reactor layout
- Aerodynamics of vehicles and aero planes
- Semiconductor-crystal production, thin-film technology, vapour-phase separation processes
- Layout and optimization of pumps, valves and nozzles
- Usage of flow aggregates like bent pipes, junctions etc.
- Development of measuring instruments and production of sensors
- Ventilation, heating and air-conditioning techniques, function of labora-tory vents
- Problem solutions in roof ventilation and flows around buildings
- Production of electronic components, micro-systems analysis engineering
- Layout of stirrer systems, propellers and turbines
- Sub-domains of biomedicine and medical engineering
- Layout of baking ovens, melting furnaces as well as other fire places

- Development of engines, catalyzers and exhaust systems
- Combustion and explosion processes, energy generation, environmental
engineering
- Sprays, atomizing and coating technologies

Concerning the formulation of the basic equations of fluid mechanics it is easy to formulate the conservation equations for mass, momentum, energy and chemical species for a fluid element, i.e. to derive in the "Lagrange form" of the equations. In this way the derivations can be represented in an easily comprehensible way and it is possible to build the derivations upon the basic knowledge of physics. Derivations of the basic equations in the Lagrange form are usually followed by transformation considerations whose aim is to strive at local formulations of the conservation equations and to introduce field quantities into the mathematical representations, i.e. the "Euler form" of the conservation equations is sought for solutions of flow-mechanical problems. This requires to express temporal modifications of substantial quantities as temporal modifications of field quantities or to complement them in a deepening way.

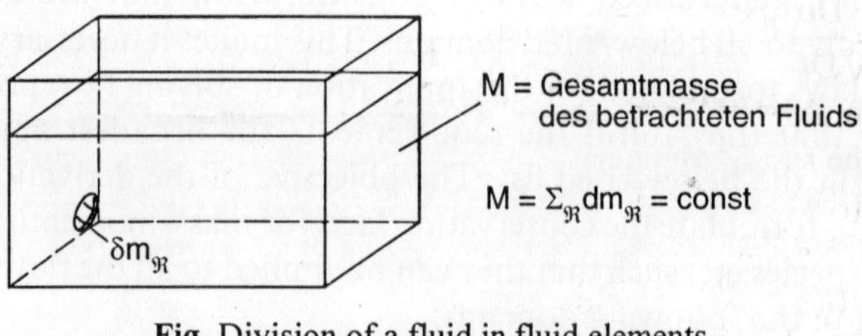

Fig. Division of a fluid in fluid elements

The considerations to be carried out apart from the assumption that at a certain point in time $t = 0$ the mass of a fluid is subdivided in fluid elements of the mass $\delta m_{\mathfrak{R}}$ i.e.

$$M = \sum_{\mathfrak{R}} \delta m_{\mathfrak{R}}$$

Here for each fluid element $\delta m_{\mathfrak{R}}$ is to be chosen large enough to make the assumption $\delta m_{\mathfrak{R}}$ = const possible with sufficient precision in spite of the molecular structure of the matter, it also allows to assign a arbitrary thermo-dynamical and fluid-mechanics properties $\alpha_{\mathfrak{R}}(x_{\mathfrak{R}}(t), t) = \alpha_{\mathfrak{R}}(t)$ for a fluid element and with a satisfactory precision for fluid-mechanics considerations.

The statement $\alpha_{\mathfrak{R}}(x_{\mathfrak{R}}, t)$, with $x_{\mathfrak{R}} = x_{\mathfrak{R}}(t)$, expresses that the thermodynamical or fluid-mechanics property, which is assigned to the considered fluid element and thus only represents a substantial quantity, is only a func-tion of time. This property of the element is changing with time, also due to the motion of the fluid element and for the description of this modification it is important that one follows the mass $\delta m_{\Re}$, i.e. knows $x_{\Re}(t)$ and also takes it into consideration as known. It is assumed that this motion of particles is constant and unequivocal, i.e. that the considered fluid element does not split up during the considerations of its motion. The fluid pertaining to the considered fluid element at the moment in time $t = 0$ remains also at all later moments in time. This signifies that it is not possible for two different fluid elements to take the same point in space at an arbitrary time: $x_{\Re}(t) = x_L(t)$ for $\Re \neq L$.

When a fluid element $\Re$ is at the position x_i, at the time t i.e. $x_i = (x_{\Re}(t))_i$ at the time t, then the substantial thermodynamic property or fluid property $\alpha_{\Re}(t)$ is equal to the field quantity α at the point x_i at time t:

$\alpha_{\Re}(t) = \alpha(x_i, t)$ when $(x_{\Re}(t))_i = x_i$ at t

For the temporal change of a quantity $\alpha_{\Re}(t)$ results :

$$\frac{d\alpha_{\Re}}{dt} = \frac{\partial\alpha}{\partial t} + \frac{\partial\alpha}{\partial x_i}\left(\frac{dx_i}{dt}\right)_{\Re}$$

With $(dx_i/dt)_{\Re} = (U_i)_{\Re} = U_i$ holds:

$$\frac{d\alpha_{\Re}}{dt} = \frac{D\alpha}{Dt} = \frac{\partial\alpha}{\partial t} + U_i\frac{\partial\alpha}{\partial x_i}$$

The operator $\alpha(x_i, t)$ applied to the field quantity $D/Dt = U_i\partial/\partial x_i$ is often defined as the substantial derivatve and will be applied in the subsequent derivations. Significance of individual terms are:

$\partial/\partial t = (\partial/\partial t)x_i$ = change with time at a fixed location, partial differentiation with respect to time

d/dt= total change with time (for a fluid element), total differentiation with respect to time

for e.g. for a fluid when $\alpha_{\Re} = \rho_{\Re}$ = const i.e. the density is constant, then it holds:

$$\frac{d\rho_{\Re}}{dt} = \frac{D\rho}{Dt} = 0 \text{ or } \frac{\partial\rho}{\partial t} = -U_i \cdot \frac{\partial\rho}{\partial x_i}$$

When at a certain point in space $\partial/\partial t\,(\alpha)_{xi} = 0$ indicate of stationary conditions, i.e. the field size $\alpha(x_i, t)$ is stationary and thus has no time dependency. On the other hand $d(\alpha_{\Re})/dt = D\alpha/Dt = 0$, is $\alpha_{\Re}(t) = \alpha(x_i, t)$ = const. i.e the field is independent of space and time.

MASS CONSERVATION (CONTINUITY EQUATION)

For fluid-mechanics considerations a “closed fluid system” can always be found, i.e. a system for whose total mass holds M = const. This is easily seen for a fluid mass, which is stored in a container. For fluid setups, control volumes can always be defined, within which the systemic total mass can be stated as constant. If necessary these control volumes can comprise the whole earth.

When one subdivides the fluid mass within the considered system in fluid elements with the sub-masses dm, then it holds true for the temporal change of the total mass:

$$0 = \frac{\mathrm{d}M}{\mathrm{d}t} = \frac{\mathrm{d}}{\mathrm{d}t}\sum_{\mathfrak{R}}(\delta m_{\mathfrak{R}}) = \sum_{\mathfrak{R}}\frac{\mathrm{d}}{\mathrm{d}t}(\delta m_{\mathfrak{R}})$$

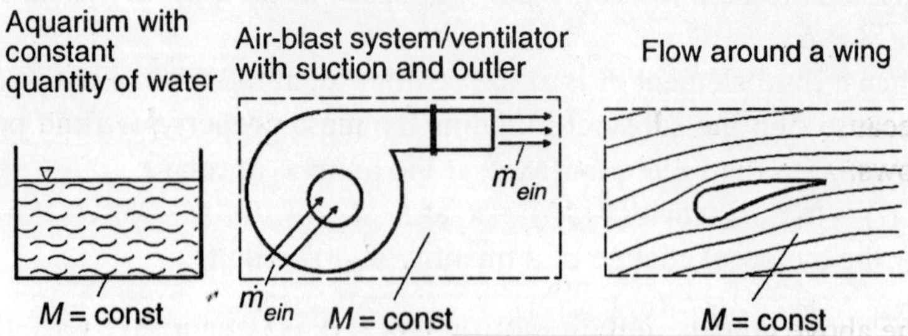

Fig. Different Fluid setups and aero foil within control volumes where M = const

This equation expresses that the mass conservation in the total fluid system is preserved when each individual fluid element conserves its mass $\delta m_{\mathfrak{R}}$. With this the balance equation for the mass conservation can be stated as follows in Lagrange notation:

$$\frac{\mathrm{d}(\delta m_{\mathfrak{R}})}{\mathrm{d}t} = 0$$

The basic molecular structure of matter and thermal motion connected with it indicates that to fulfil the above relation absolutely, it is necessary that $\delta m_{\mathfrak{R}} \to 0$ The consideration carried out in this book therefore requires that all the $\delta m_{\mathfrak{R}}$ are considered as finite but nevertheless as very small. A fluid element with position coordinates $(x_i)_{\mathfrak{R}}$ is shown.

The estimation of $\delta m_{\mathfrak{R}}$ is referred to considerations, where it is shown what dimensions a volume of an ideal gas has to have in order to define "sufficiently clearly" e.g. a density of the gas within the volume. The considerations carried out there would have to be repeated here in order to guarantee $\delta m_{\mathfrak{R}}$ = const, in spite of the molecular structure of the matter. With the choice of $\delta m_{\mathfrak{R}}$ const the conditions are set to carry out continuum-mechanics considerations for the motion of fluids, although the fluids show a molecular structure. In this context it is often referred to the continuum considerations in which usually fluid-mechanics considerations are carried out. Strictly speaking this means that the properties of the molecules, especially their transport properties, can only be introduced in fluid-mechanics considerations in integral form.

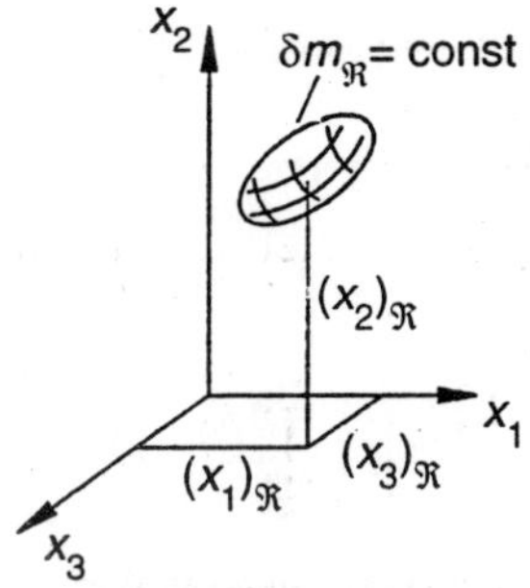

Fig. dm_B = Const, Condition for the Mass of a Fluid Element

Because of the above explanations the mass conservation can be stated as follows:

$$\frac{dM}{dt} = 0 \text{ and } \frac{d\delta m_{\mathfrak{R}}}{dt} = 0$$

The above considerations confirm that it is very simple to formulate the mass-conservation law in Lagrange variables. Working practically with the law of mass conservation, however, requires its representation in field quantities, i.e. the Lagrange form of mass-conservation law has to be brought into the Euler form.

Transformed into Euler variables (i.e. into field quantities) one obtains from equation for the mass conservation:

$$0 = \frac{d}{dt} = (\delta m_{\mathfrak{R}}) = \frac{d}{dt}(\rho_{\mathfrak{R}} \delta V_{\mathfrak{R}}) = \underbrace{\rho_{\mathfrak{R}} \frac{d(\delta / V_{\mathfrak{R}})}{dt}}_{I} + \underbrace{\delta V_{\mathfrak{R}} \frac{d(\rho_{\mathfrak{R}})}{dt}}_{II}.$$

For term I in equation using $\rho_{\mathfrak{R}} = \rho$ and $(x_{\mathfrak{R}}(t))_i = x_i$ at the time t: yields:

$$\rho_{\mathfrak{R}} \frac{d(\delta / V_{\mathfrak{R}})}{dt} = \rho \delta V_{\mathfrak{R}} \frac{\partial U_i}{\partial x_i}.$$

For term II one obtains:

$$\delta V_{\mathfrak{R}} \frac{d(\rho_{\mathfrak{R}})}{dt} = \delta V_{\mathfrak{R}} \left(\frac{\partial \rho}{\partial t} + U_i \frac{\partial \rho}{\partial x_i} \right).$$

With the above derivations the substantial derivation of the corresponding field quantity $d(\rho_{\mathfrak{R}})/dt$ could be applied for ρ, d. h. $D\rho/Dt$ in order to achieve the transition from the substantial quantity $\rho_{\mathfrak{R}}(t)$ to the field quantity $\rho(x_i, t)$. The same procedure is not possible with $d(\delta V_{\mathfrak{R}})/dt$ since there are no volume fields. From section 4, equation, concerning the temporal change of a fluid element is equal to the divergence of the velocity field is known, however:

$$\rho_{\mathfrak{R}} \frac{d(\delta V_{\mathfrak{R}})}{dt} = \delta V_{\mathfrak{R}} \frac{\partial U_i}{\partial x_i}.$$

When one inserts the equations one obtains:

$$\delta V_{\mathfrak{R}}\left[\frac{\partial\rho}{\partial t}+\rho\frac{\partial U_i}{\partial x_i}+U_i\frac{\partial\rho}{\partial x_i}\right]=0,$$

As $\delta V_{\mathfrak{R}} = 0$ it follows for the continuity equation in the most general form:

$$\frac{\partial\rho}{\partial t}+\rho\frac{\partial U_i}{\partial x_i}+U_i\frac{\partial\rho}{\partial x_i}=\frac{\partial\rho}{\partial t}+\frac{\partial(\rho U_i)}{\partial x_i}=0$$

The equation can, also be written as follows:

$$\frac{\partial\rho}{\partial t}+U_i\frac{\partial\rho}{\partial x_i}+\rho\frac{\partial U_i}{\partial x_i}=\frac{D\rho}{Dt}+\rho\frac{\partial U_i}{\partial x_i}=0.$$

This form of the continuity equation is not very useful for the solution of flow problems. However it is very well suited for presentation of the basic equations of fluid-mechanics in different ways, in order to bring out special physical facts. As an example the special form of the continuity equation for $\rho_{\mathfrak{R}}$ = const, i.e. $D\rho/Dt = 0$, from equation:

$$\frac{\partial U_i}{\partial x_i}=0,$$

i.e. the divergence of the velocity field is zero for fields of constant density. since the divergence of the velocity field is zero, the change in volume is also zero from equation, this can also be obtained from the equations.

$$\frac{1}{\rho}\frac{D\rho}{Dt}=\underbrace{\frac{1}{\rho}\left(\frac{\partial\rho}{\partial T}\right)_P}_{-\beta}\frac{DT}{Dt}+\underbrace{\frac{1}{\rho}\left(\frac{\partial\rho}{\partial P}\right)_T}_{+\alpha}\frac{DP}{Dt}=-\frac{\partial U_i}{\partial x_i}$$

This relation expresses for a fluid in the thermodynamic sense, $\rho = const$, i.e if the fluid density is constant, then it has to be thermodynamically incompressible.

$$\frac{1}{\rho}\frac{D\rho}{Dt}=-\frac{\partial U_i}{\partial x_i}=0=\alpha\frac{DP}{Dt}-\beta\frac{DT}{Dt}$$ with $\alpha = 0$ where a and β are:

$$\alpha=-\frac{1}{\upsilon}\left(\frac{\partial\upsilon}{\partial P}\right)_T=\frac{1}{\rho}\left(\frac{\partial\rho}{\partial P}\right)_T = \text{isothermal compressibility coefficient}$$

$$\beta=\frac{1}{\upsilon}\left(\frac{\partial\upsilon}{\partial T}\right)_P=-\frac{1}{\rho}\left(\frac{\partial\rho}{\partial T}\right)_P = \text{thermal expansion coefficient}$$

Thus the continuity equation holds in one of the two forms listed below:

$$\frac{\partial\rho}{\partial t}+\frac{\partial(\rho U_i)}{\partial x_i}=0 \qquad \text{(compressible flows)}$$

$$\frac{\partial U_i}{\partial x_i}=0 \qquad \text{(icompressible flows)}$$

SECOND NEWTONIAN LAW (MOMENTUM EQUATION)

The derivations of the equations of momentum for the three coordinate di-rections $j = 1, 2, 3$ in fluid mechanics, the second Newtonian law to a fluid element, i.e. the Lagrange formulation of the equation of momentum is chosen. For a fluid element it is stated that the time derivative of the momentum in the j-direction is equal to the sum of the external forces acting in this direction on the fluid element, plus the molecular-dependent input per unit time. The forces can be stated as inertia forces caused by gravitation forces (δM_j) and electromagnetic forces[2] as well as the surface forces caused by pressure (δO_j). After addition of a temporal change of momentum fed in by the molecular movement, the equation of motion can be formulated as follows:

$$\frac{d(\delta J_j)_{\Re}}{dt} = \underbrace{\sum(\delta M_j)_{\Re}}_{\text{Inertia forces}} + \underbrace{\sum(\delta O_j)_{\Re}}_{\text{surface forces}} + \underbrace{\left(\frac{d}{dt}(\delta J_M)_j\right)_{\Re}}_{\substack{\text{molecular-dependent}\\ \text{momentum input}}}$$

Here, as stated in Figure, $(\delta J_j)_{\Re} = \delta m (U_j)_{\Re}$

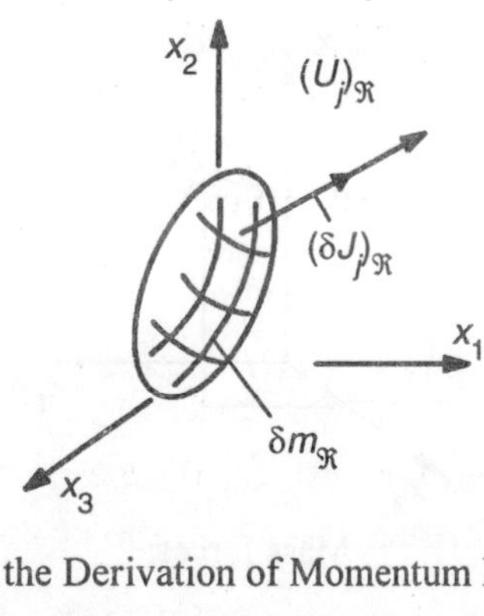

Fig. For the Derivation of Momentum Equations

The fluid element acts like a rigid body since it does not change its state of motion, i.e. its momentum, when no inertia or surfaces forces act on the fluid element and a molecular-dependent momentum input fails to appear. However, when forces are present, or when a momentum has input, the fluid element changes its momentum in accordance with the above-stated relation. It represents the Lagrange form of the equations of momentum ($j = 1,2,3$) of fluid mechanics.

In order to conserve the Euler form of the equation of momentum, it is important to express each of the terms contained in equation in field quantities. For the left side of the equation can be written as:

$$\frac{d(\delta J_j)_{\Re}}{dt} = \frac{d}{dt}[\delta m_{\Re}(U_j)_{\Re}] = \delta m_{\Re}\frac{d((U_j)_{\Re})}{dt} + (U_j)_{\Re}\frac{d(\delta m)_{\Re})}{dt}$$

Because of the universal mass conservation in formulation, the last term in equation is equal to zero:

$$\frac{d(\delta J_j)_{\Re}}{dt} = \delta m_{\Re} \frac{d((U_j)_{\Re})}{dt} = \delta m_{\Re}\left(\frac{\partial U_i}{\partial t} + U_i \frac{\partial U_i}{\partial x_i}\right)$$

This can be written as follows: $\delta m_{\Re} = \rho \delta V_{\Re}$ applying $\rho_{\Re} = \rho$ when: $(x_{\Re}(t))_i = x_i$ at the time t,

$$\frac{d(\delta J_j)_{\Re}}{dt} = \rho \delta V_{\Re}\left(\frac{\partial U_i}{\partial t} + U_i \frac{\partial U_i}{\partial x_i}\right)$$

Owing to the above derivations it is possible to state the left side of the equation of momentum in field quantities. For the right side the below-cited considerations can be carried out. The inertia forces can be represented in an easy way by the acceleration vector $\{g_j\}$, i.e. by its components g_j.

$$\sum_{\Re} (\delta M_j)_{\Re} = \text{Mass forces acting on a fluid element}$$

The mass forces acting on a fluid element can be stated by means of the

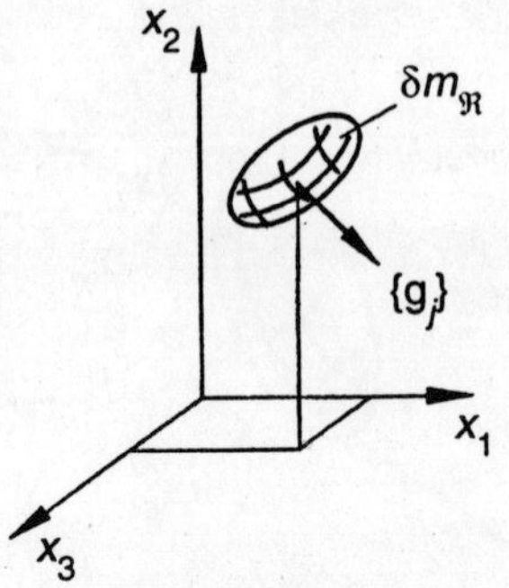

Mass forces:

Fig. Mass force on a fluid element

acceleration $\{g_j\} = \{g_1, g_2, g_3\}$ acting per measuring unit. The mass force acting on a fluid element in j-direction can be stated as follows:

$(\delta M_j) = (\delta m_{\Re}) g_j = \rho \delta V_{\Re} g_j$

Even when only gravitational acceleration is present, depending on the posi-tion of the coordinate system, several components of g_j may not be equal to zero

$$\sum_{\Re} (\delta O_j)_{\Re} = \text{Surface forces on a fluid element}$$

Fluids as they are treated in this book, i.e. fluids (e.g. water) and gases (e.g. air) are characterized by the way they can apply surface forces on a fluid element only by the molecular pressure. The pressure force acting on a fluid element is calculated as the difference of the forces acting on the areas that stand vertically on the considered axes. It holds since

The surface force resulting for the motion of the fluid element in *j*-direction is the sum of the attacking forces acting on *j*-planes.

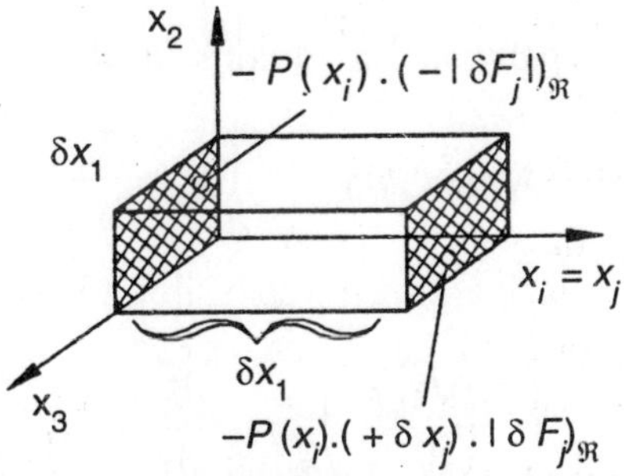

Surface forces:

Fig. Considerations Concerning Surface Force on a Fluid Element

$$(\delta O_j)\Re = -P(x_j)(-|\,\delta F_j\,|)_\Re - P(x_j + \delta x_j)\,(|\delta F_j|)$$

If one applies a Taylor series expansion for $P(x_i + \delta x_j)$ one obtains:

$$(\delta O_j)_\Re = +P\,(x_j)\,(|\delta F_j|)_\Re - \left[P(x_j) + \frac{\partial P}{\partial x_j}\delta x_j + \ldots\right](|\,\delta F_j\,|)_\Re$$

From this results for the surface force on a fluid element, neglecting all second and higher order terms:

$$(\delta O_j)_\Re = -\frac{\partial P}{\partial x_j}\delta V_\Re$$

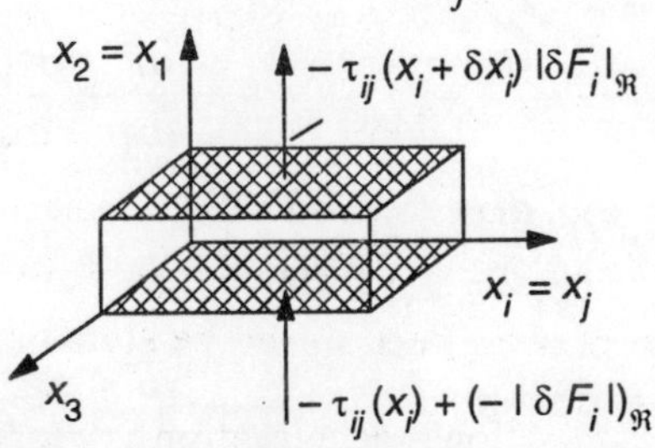

Molecular-dependent momentum input:

Fig. Considerations on the Molecular-dependent Momentum Input

$\left(\frac{\mathrm{d}}{\mathrm{d}t}(\delta J_M)_j\right)_\Re$ = molecular-dependent momentum input per unit time

When one defines the momentum *j*-fed into the direction *i*-per unit time and surface as t_{ij}, the input influencing the momentum *j*-of a fluid element is calculated as an input at the position x_i and as an output at the position $(x_i + \Delta x_i)$, i.e. it holds:

$$\left(\frac{\mathrm{d}(\delta J_M)_j}{\mathrm{d}t}\right)_j = -\tau_{ij}(x_i)(-|\delta F_j|)_{\Re} - \tau_{ij}(x_i + \Delta x_i)(|\delta F_i|)_{\Re}$$

By a Taylor series expansion one obtains for the term $\tau_{ij}(x_i + \delta x_i)$:

$$\left(\frac{\mathrm{d}(\delta J_M)_j}{\mathrm{d}t}\right)_{\Re} = +\tau_{ij}(x_i)(|\delta F_i|)_{\Re} - \tau_{ij}(x_i + \frac{\partial \tau_{ij}}{\partial x_i}\delta x_i ...](\delta F_i)_{\Re}$$

This results in:

$$\left(\frac{\mathrm{d}(\delta J_M)_j}{\mathrm{d}t}\right)_{\Re} = -\frac{\partial \tau_{ij}}{\partial x_i}\delta V_{\Re}$$

When one inserts all these derived relations after division by $\delta V_{\Re}$ the equation of momentum of fluid mechanics in direction j results, i.e. for $j = 1, 2, 3$ three equations result:

$$\rho\left(\frac{\partial U_j}{\partial t} + U_i\frac{\partial U_j}{\partial x_i}\right) = \rho g_j - \frac{\partial P}{\partial x_j} - \frac{\partial \tau_{ij}}{\partial x_{ij}}$$

As in this equation the volume of the fluid element $\delta V_{\Re}$ appearing in all terms was eliminated, the equation of momentum hold per unit volume. The momentum equation in the three coordinate directions results:

$$\rho\left[\frac{\partial U_1}{\partial t} + U_1\frac{\partial U_1}{\partial x_1} + U_2\frac{\partial U_1}{\partial x_2} + U_3\frac{\partial U_1}{\partial x_3}\right]$$

$$= -\frac{\partial P}{\partial x_1} - \frac{\partial \tau_{11}}{\partial x_1} - \frac{\partial \tau_{21}}{\partial x_2} - \frac{\partial \tau_{31}}{\partial x_3} + \rho g_1$$

$$\rho\left[\frac{\partial U_2}{\partial t} + U_1\frac{\partial U_2}{\partial x_1} + U_2\frac{\partial U_2}{\partial x_2} + U_3\frac{\partial U_2}{\partial x_3}\right]$$

$$= -\frac{\partial P}{\partial x_2} - \frac{\partial \tau_{12}}{\partial x_1} - \frac{\partial \tau_{22}}{\partial x_2} - \frac{\partial \tau_{32}}{\partial x_3} + \rho g_2$$

$$\rho\left[\frac{\partial U_3}{\partial t} + U_1\frac{\partial U_3}{\partial x_1} + U_2\frac{\partial U_3}{\partial x_2} + U_3\frac{\partial U_3}{\partial x_3}\right]$$

$$= -\frac{\partial P}{\partial x_3} - \frac{\partial \tau_{13}}{\partial x_1} - \frac{\partial \tau_{23}}{\partial x_2} - \frac{\partial \tau_{33}}{\partial x_3} + \rho g_3$$

From the fluid mechanics point of view only for ideal fluids $\tau_{ij} = 0$,

For fluids in general $\tau_{ij} \neq 0$ but for ideal fluids in terms of fluid mechanics

$\tau_{ij} = 0$. Hence the following forms of the momentum equations can be stated:

$$\rho\left(\frac{\partial U_j}{\partial t}+U_i\frac{\partial U_j}{\partial x_i}\right)=-\frac{\partial P}{\partial x_j}-\frac{\partial \tau_{ij}}{\partial x_i}+\rho g_j \quad \text{(viscous Fluide)}$$

$$\rho\left(\frac{\partial U_j}{\partial t}+U_i\frac{\partial U_j}{\partial x_i}\right)=-\frac{\partial P}{\partial x_j}+\rho g_j \quad \text{(ideal Fluide)}$$

THE NAVIER-STOKES EQUATIONS

In the above equation the molecular-dependent momentum input τ_{ij} occurring per surface and unit time, is an unknown term, i.e. it is introduced formally in derivations, without details being considered, as it can be formulated for several fluids. When one takes into consideration the symmetry of the term τ_{ij} i.e. $|\tau_{ij}| = |\tau_{ji}|$, holds, one notices/ascertains that there are the following unknowns in the equations:

U_1, U_2, U_3, P, τ_{11}, τ_{12}, τ_{13}, τ_{22}, τ_{23}, τ_{33} = 10 unknowns

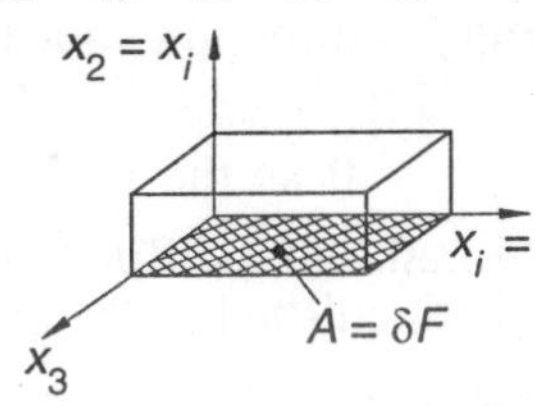

Fig. Momentum input due to flow through the plane δF_j

For these unknowns there are only four partial differential equations, the continuity equation and three equations of momentum, i.e. an incomplete equation system exists that does not permit the solution of flow problems. It is therefore necessary to state additional equations, i.e. to express the unknown terms τ_{ij} physically well-founded, as functions of $\partial U_j / \partial x_i$. This is done below for ideal gases, as their properties are known to a large extent from considerations in physics. From the below-stated derivations, relations for $\tau_{ij} = f(\partial U_j / \partial x_i)$ result, that are valid also for non-ideal gases and for a whole class of fluids whose molecular momentum-transport properties can be classified as "'Newtonian'". Thus the derived relations τ_{ij} are valid far beyond ideal gases and represent in this book the basic equations to describe the molecular-dependent momentum transport in Newtonian fluids.

When one considers a fluid element, one can see that the momentum *j* fed in direction *i* by a velocity field and can be stated as follows:

$$\dot{I}_{ij} = \rho \hat{U}_i \hat{U}_j \delta F_i$$

Assuming that the instantaneous velocity components are composed of the velocity of the fluid flow U_i and the share of the molecular motion u_i one obtains:

$$\rho\hat{U}_i\hat{U}_j = \rho(U_i + u_i)(U_j + u_j)$$
$$= \rho(U_iU_j + u_iU_j + u_jU_i + u_iu_j)$$

By time averaging, one obtains for the time-averaged total momentum change of the fluid element:

$$\rho\overline{\hat{U}_i\hat{U}_j} = \rho[\underbrace{\overline{U_iU_j}}_{\text{I}} + \underbrace{\overline{u_iU_j}}_{\text{II}} + \underbrace{\overline{u_jU_i}}_{\text{III}} + \underbrace{\overline{u_iu_j}}_{\text{IV}}]$$

The total momentum input consists of four terms that can be interpreted physically as follows:

Term I: Momentum input j in direction in i due to the velocity field of the Fluid.

Term II: Momentum input j in direction i due to the molecular motion in i-direction.

Term III: Momentum input j in direction i due to the molecular motion in j-direction.

Term IV: For $i \neq j$ is $u_i\, u_j = 0$ as the molecular motion in the three coordinate directions - are not correlated; for $i = j$ results the pressure.

The molecular motion is distinguished by the presence of free path lengths with finite dimensions, i.e. $l \neq 0$, the time averages $\overline{u_iU_j}$ and $\overline{u_jU_i}$ are unequal to zero. In order to calculate these contributions to τ_{ij} considerations on the molecular momentum transport, ideal gases are recommended. For the number of molecules moving in direction x_i and passing the plane A in Fig. 5.9 in the time Δt when $\delta x_1 = \delta x_2 = \delta x_3 = a$ it can be written:

$$z_i = \frac{1}{6}na^2u_i\Delta t$$

Where n is equal to the number of molecules per unit volume, a^2 the volume of area δF_i and u_i the mean velocity of the molecules in direction i. Connected to z_i, a mass transport through δF_i, can be stated as follows:

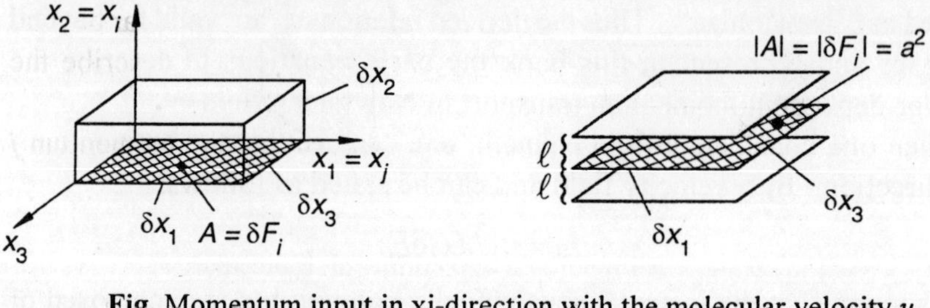

Fig. Momentum input in xi-direction with the molecular velocity u_i

$$mz_i = \frac{1}{6}\underbrace{(mn)}_{\rho}a^2 u_i \Delta t$$

where m represents the mass of a molecule and thus can be set $mn = \rho$.

When one considers now the auxiliary planes in Figure located to dF_i in the distance $\pm l$ and introduced for the derivations, in which the mean flow field owns the velocity components $U_j(x_i + l)$ and $U_j(x_i - l)$. In positive and negative i-directions, j-directional momentum input and discharge can be stated s follows:

$i_{ij}^{+} = + z_i m U_j (x_i - l)$ Momentum input over area δF_i

$i_{ij}^{-} = -z_i m U_j (x_i + l)$ Momentum discharge over area δF_i

Therefore the sum of the molecular-dependent input and discharge results:

$\Delta i_{ij} = z_i m[U_j (x_i - l) - U_j (x_i + l)]$

or with z_i inserted from equation :

$$\Delta i_{ij} = \frac{1}{6}\underbrace{(mn)}_{\rho}a^2 u_i \Delta t[U_j(x_i - l) - U_j(x_i + l)]$$

The momentum flow per unit area and time can be obtained by Taylor series expansion of velocity at around x_i by neglecting the higher order terms:

$$\tau_{ij}^{\mathrm{II}} = \frac{1}{a^2}\frac{\Delta i_{ij}}{\Delta t} = \frac{1}{6}\rho u_i \left[U_j(x_i) - \frac{\partial U_j}{\partial x_i} l - U_j(x_i) - \frac{\partial U_j}{\partial x_i} l \right],$$

so that for Term II in equation results:

$$\tau_{ij}^{\mathrm{II}} = -\frac{1}{3}\rho u_i l \frac{\partial U_j}{\partial x_i} = -\mu \frac{\partial U_j}{\partial x_i}.$$

Analogous to this it can be carried out considerations on τ_{ij}^{III} where for z_j can be written:

$$z_j = \frac{1}{6} n a^2 u_j \Delta t$$

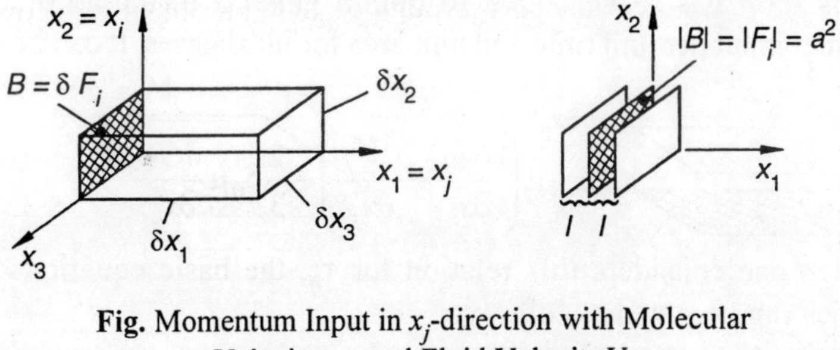

Fig. Momentum Input in x_j-direction with Molecular Velocity u_j -and Fluid VelocityU_i

In accordance with term III in equation a momentum input j-results which can be stated as follows:

$$i_{ij}^{+} = z_j m U_i(x_j - l)$$

$$i_{ij}^{-} = -z_j m U_i(x_j + l)$$

or

$$\Delta i_{ij} = z_j m[U_i(x_j - l) - U_i(x_j + l)]$$

Analogous to the derivations in equations yields:

$$i_{ij}^{\mathrm{III}} = -\underbrace{\frac{1}{3}(\rho u_j l)}_{\mu}\frac{\partial U_i}{\partial x_i} = -\mu\frac{\partial U_i}{\partial x_j}$$

For reasons of symmetry $\tau_{ij} = \tau_{ji}$, so that $u_i = u_j$ has to hold, i.e. the mean velocity field of the molecules is isotropic (no preferred velocity direction), so that it can be written:

$$\tau_{ij} = \tau_{ij}^{\mathrm{II}} + \tau_{ij}^{\mathrm{II}} = -\mu\left[\frac{\partial U_j}{\partial x_i} + \frac{\partial U_i}{\partial x_j}\right] \quad \text{for } i \neq j$$

This is the total momentum input τ_{ij} for $\rho = \text{const}$, i.e. when $d/dt(\delta V_{\Re}) = 0$, the thermodynamic state equation for an ideal fluid is assumed. For $\rho \neq$ const, an additional term needs to be added which is conditioned by the volume increase of a fluid element. For the volume increase of a fluid element at point x_i and time t,:

$$\frac{d(\delta V_{\Re})}{dt} = (\delta V_{\Re})\frac{\partial U_i}{\partial x_i}$$

For the corresponding surface increase holds:

$$\frac{d(\delta F_{\Re})}{dt} = \frac{2}{3}(\delta F_{\Re})\frac{\partial U_i}{\partial x_i}$$

With surface increase an increased momentum input results:

$$\tau_{ij} = +\mu\frac{2}{3}\delta_{ij}\frac{\partial U_k}{\partial x_k}$$

This term has to be added to obtain general formulas of the total momentum input per unit time and unit area for ideal gases. It can be stated as follows:

$$\tau_{ij} = -\mu\left(\frac{\partial U_j}{\partial x_i} + \frac{\partial U_j}{\partial x_j}\right) + \frac{2}{3}\delta_{ij}\mu\frac{\partial U_k}{\partial x_k}$$

When one considers this relation for τ_{ij}, the basic equations of fluid mechanics can be stated as follows:

Continuity equation:

$$\frac{\partial \rho}{\partial t}+\frac{\partial(\rho U_i)}{\partial x_i}=0$$

Momentum equations:

$$(j = 1, 2, 3)\ \left[\frac{\partial U_j}{\partial t}+U_i\frac{\partial U_j}{\partial x_i}\right]=-\frac{\partial P}{\partial x_j}-\frac{\partial \tau_{ij}}{\partial x_i}+\rho g_j$$

For Newtonian fluids

$$\tau_{ij}=-\mu\left[\frac{\partial U_j}{\partial x_i}+\frac{\partial U_i}{\partial x_j}\right]+\frac{2}{3}\delta_{ij}\mu\frac{\partial U_k}{\partial x_k}$$

With τ_{ij} from there exist equations for the 6 unknown terms τ_{ij} –in the momentum equations. The four equations, one continuity equation and three Navier-Stokes equations, contain five unknowns: P, ρ, U_j, so that an incomplete system of partial differential equations exists still. With the aid of the thermal energy equation and the thermodynamic state equation valid for the considered fluid, it is possible to obtain a complete system of partial differential equations that permits general solutions for flow problems, when initial and boundary conditions are present.

For ρ = const and μ = const, using $\frac{\partial^2 U_i}{\partial x_i \partial x_j}=\frac{\partial^2 U_i}{\partial x_j \partial x_j}=\frac{\partial}{\partial x_j}\left(\frac{\partial U_i}{\partial x_i}\right)=0$ stated:

Continuity equation: $\frac{\partial U_i}{\partial x_i}=0$

Navier-Stokes equations (j = 1, 2, 3):

$$\rho\left[\frac{\partial U_j}{\partial t}+U_i\frac{\partial U_j}{\partial x_i}\right]=\frac{\partial P}{\partial x_j}+\mu\frac{\partial^2 U_j}{\partial x_i^2}+\rho g_j$$

This system of equations comprises four equations for the four unknowns P, U_1, U_2, U_3. In principle, it can be solved for all flow problems to be investigated, when suitable initial and boundary conditions are given. For thermodynamic ideal fluids, i.e. ρ = const, a complete system of partial differential equations exists with the continuity equation and the momentum equations, which can be used for solutions of flow problems.

MECHANICAL ENERGY EQUATION

In many domains in which fluid-mechanic considerations are carried out, the mechanical energy equation is employed which can be derived from the momentum equation j -For this purpose one multiplies equation with U_j :

$$\rho\left[U_j\frac{\partial U_j}{\partial t}+U_iU_j\frac{\partial U_j}{\partial x_i}\right]=-U_j\frac{\partial P}{\partial x_j}-U_j\frac{\partial \tau_{ij}}{\partial x_i}+U_j\rho g_j$$

This equation can be transcribed as follows:

$$\rho\left[\frac{\partial}{\partial t}\left(\frac{1}{2}U_j^2\right)+U_i\frac{\partial}{\partial x_i}\frac{1}{2}U_j^2\right]$$

$$=-\frac{\partial(PU_j)}{\partial x_j}+P\frac{\partial U_j}{\partial x_j}-\frac{\partial(\tau_{ij}U_j)}{\partial x_i}+\tau_{ij}\frac{\partial U_j}{\partial x_i}+\rho g_jU_j$$

The equation states how the kinetic energy of a fluid element is changing at a location due to energy-production and dissipation terms that occur on the right side of the above equation. In order to discuss the significance of the different terms, the following modification of the last term is carried out, introducing a potential G from which the gravity is derived:

$$g_j=\frac{\partial G}{\partial x_j}\rightarrow\rho g_jU_j=-\rho\frac{\partial G}{\partial x_j}U_j$$

Thus, employing $\frac{\partial G}{\partial t}=0$:

$$\rho g_jU_j=-\rho\left[\frac{\partial G}{\partial t}+U_j\frac{\partial G}{\partial x_j}\right]=-\rho\frac{DG}{Dt}$$

The combined equations yield for the temporal change of the kinetic and potential energy of a fluid element:

$$\rho\frac{\mathrm{D}}{\mathrm{D}t}\left(\frac{1}{2}U_j^2+G\right)=\underbrace{-\frac{\partial(PU_j)}{\partial x_j}}_{\mathrm{I}}+\underbrace{P\frac{\partial U_j}{\partial x_j}}_{\mathrm{II}}\underbrace{-\frac{\partial(\tau_{ij}U_j)}{\partial x_i}}_{\mathrm{III}}+\underbrace{\tau_{ij}\frac{\partial U_j}{\partial x_i}}_{\mathrm{IV}}$$

Term I : This term describes the difference between input and discharge of pressure energy. Here it is referred to the considerations on ideal gases, in the framework of which was shown that P = $\frac{1}{3}\rho u^2$, i.e. expresses an energy per unit volume. Therefore the following can be said:

$(PU_j(x_i))$ = Input of pressure energy per unit area

$-(PU_j(x_i + Dx_i))$ = Discharge of pressure energy per unit area

Taylor series expansion and forming the difference yields for the energy per unit volume:

$$PU_j(x_i)-\left[PU_j(x_i)+\frac{\partial PU_j}{\partial x_j}+...\right]\simeq-\frac{(\partial PU)}{\partial x_j}$$

Term II: Considering:

$$\frac{\partial U_j}{\partial x_j} = \frac{1}{\delta V_{\Re}} \frac{\mathrm{d}(\delta V_{\Re})}{\mathrm{d}t}$$

the term:

$$\frac{\partial U_j}{\partial x_j} = \frac{P}{\delta V_{\Re}} \frac{\mathrm{d}(\delta V_{\Re})}{\mathrm{d}t}$$

proves to be the work done during expansion, occurring per unit volume.

Term III: When taking into consideration that t_{ij} represents the molecular-dependent momentum transport per unit area and unit time into a fluid element:

$-\frac{\partial(\tau_{ij}U_j)}{\partial x_j}$ = difference from the molecular-dependent input and discharge of the kinetic energy of the fluid.

Term IV: The term $\tau_{ij}\frac{\partial U_j}{\partial x_i}$ describes the dissipation of mechanical energy into heat. The above presentations show that the mechanical energy equation can be derived from the j momentum equation by multiplication with U_j. It is to make sure that no independent equation and should be employed along with the momentum equation for the solution of fluid-mechanical problems.

A special form of the mechanical energy equation is the Bernoulli equation which can be derived from the general form of the mechanical energy equation:

$$\rho\frac{\mathrm{D}}{\mathrm{D}t}\left(\frac{1}{2}U_j^2 + G\right) = -\frac{\partial P}{\partial x_j}U_j - \frac{\partial \tau_{ij}}{\partial x_i}U_j$$

For $\tau_{ij} = 0$ and $\frac{\partial P}{\partial t} = 0$, as well as ρ = const. Holds:

$$\rho\frac{\mathrm{D}}{\mathrm{D}t}\left[\frac{1}{2}U_j^2 + G\right] = -\rho\left[\frac{\partial \frac{P}{\rho}}{\partial t} + U_j\frac{\partial\left(\frac{P}{\rho}\right)}{\partial x_j}\right] = \frac{\mathrm{D}\frac{\mathrm{P}}{\rho}}{\mathrm{D}t}$$

$$\rho\frac{\mathrm{D}}{\mathrm{D}t}\left[\frac{1}{2}U_j^2 + \frac{P}{\rho} G\right] = 0 \rightsquigarrow \frac{1}{2}U_j^2 + \frac{P}{\rho} + G = \text{const.}$$

This form of mechanical energy equation can be employed in many engineering applications for an imputation estimate of flow processes.

THERMAL ENERGY EQUATION

The mechanical energy equation is derivable from the momentum equation, so that both equations have to be considered as not being independent from the another. From the derivations resulted the equation in the following form:

$$\rho\left[\frac{\mathrm{D}}{\mathrm{D}t}\left(\frac{1}{2}U_j^2\right)\right]=-\frac{\partial(PU_j)}{\partial x_j}+P\frac{\partial U_j}{\partial x_j}-\frac{\partial(\tau_{ij}U_j)}{\partial x_i}+\tau_{ij}\frac{\partial_{ij}U_j}{\partial x_i}+\rho g_j U_j$$

When one sets up the equation for the total energy balance, the consideration stated below results, which starts from the entire internal kinetic and potential energy of a fluid element and considers its evolution as a function of time:

$$\frac{\mathrm{d}}{\mathrm{d}t}\left(\delta m_{\mathfrak{R}}\left[\frac{1}{2}U_j^2+e+G\right]\right)=\delta m_{\mathfrak{R}}\frac{\mathrm{d}}{\mathrm{d}t}[\ldots]+[\ldots]\frac{\mathrm{d}\delta m_{\mathfrak{R}}}{\mathrm{d}t}$$

For the temporal change of the total energy of a fluid element results with $\delta m_{\mathfrak{R}}$ = const, d.h. $\frac{\mathrm{d}}{\mathrm{d}t}(\delta m_{\mathfrak{R}})=0$:

$$\frac{\mathrm{d}}{\mathrm{d}t}\left(\delta m_{\mathfrak{R}}\left[\frac{1}{2}U_j^2+e+G\right]\right)=\delta m_{\mathfrak{R}}\frac{\mathrm{D}}{\mathrm{D}t}\left(\frac{1}{2}U_j^2+e+G\right)$$

This is the total energy change which has to be considered concerning the derivation of the total energy equation.

The change of the total energy of the fluid element can emanate from the heat conduction, which yields the following inputs minus the discharges of heat:

$-\frac{\partial \dot{q}_i}{\partial x_i}\delta V_{\mathfrak{R}}$ = Energy input per unit time by heat conduction can originate from the convective transport of pressure energy and by the input of kinetic energy due to molecular transport into the fluid element:

$-\frac{\partial \dot{q}_i}{\partial x_i}(\tau_{ij}U_j)\delta V_{\mathfrak{R}}$ = molecular-dependent input of kinetic energy

The following total energy balance is thus resulting:

$$\rho\delta V_{\mathfrak{R}}\frac{\mathrm{D}}{\mathrm{D}t}\left[\frac{1}{2}U_j^2+e+G\right]=-\frac{\partial\dot{q}_i}{\partial x_i}\delta V_{\mathfrak{R}}-\frac{\partial(PU_j)}{\partial x_j}\delta V_{\mathfrak{R}}-\frac{\partial(\tau_{ij}U_j)}{\partial x_i}\delta V_{\mathfrak{R}},$$

as $\delta V_{\mathfrak{R}}\neq 0$ follows:

$$\rho\frac{\mathrm{D}}{\mathrm{D}t}\left[e+\left(\frac{1}{2}U_j^2+G\right)\right]=-\frac{\partial\dot{q}_i}{\partial x_i}-\frac{\partial(PU_j)}{\partial x_j}-\frac{\partial(\tau_{ij}U_j)}{\partial x_i}.$$

When one deducts from this the mechanically derived parts, i.e. by subtracting the equation from the equation for the mechanical energy, given here once again:

$$\rho\frac{\mathrm{D}}{\mathrm{D}t}\left[\frac{1}{2}U_j^2+G\right]=-\frac{\partial(PU_j)}{\partial x_j}+P\frac{\partial U_j}{\partial x_j}-\frac{\partial(\tau_{ij}U_j)}{\partial x_i}+\tau_{ij}\frac{\partial U_j}{\partial x_i}$$

one obtains the thermal energy equation:

$$\underbrace{\rho\frac{De}{Dt}}_{I}=-\underbrace{\frac{\partial\dot{q}_i}{\partial x_i}}_{II}-\underbrace{P\frac{\partial U_j}{\partial x_j}}_{III}-\underbrace{\tau_{ij}\frac{\partial U_j}{\partial x_i}}_{IV}$$

Term I: Temporal change of the internal energy of a fluid per unit volume.

Term II: Heat supply per time and unit area.

Term III : Work done per unit volume and unit time.

Term IV : Irreversible transfer of mechanical energy into heat, per unit volume and unit time.

Considering the energy equation from technical thermodynamics

$$dq_{\Re}=de_{\Re}+P_{\Re}d\upsilon_{\Re}-dl_{diss}$$

and the sign convention is usual in technical thermodynamics, that the energy to be dissipated by a fluid element has to be regarded as negative, it results:

$$\frac{de_{\Re}}{dt}=\frac{De}{Dt};\quad \frac{dq_{\Re}}{dt}=-\frac{1}{\rho}\frac{\partial\dot{q}_i}{\partial x_i};\quad P_{\Re}\frac{d\upsilon_{\Re}}{dt}=\frac{1}{\rho}P\frac{\partial U_j}{\partial x_j}$$

and $\frac{dl_{diss}}{dt}=\frac{1}{\rho}\tau_{ij}\frac{\partial U_j}{\partial x_i}$

The above derivations thus lead to the form of energy equation used in thermodynamics.

Different forms of the thermal energy equation can be derived from the relation, this is advantageous for most of the fluid-mechanical computations, to substitute the internal energy (e) by pressure and temperature, the following relations being employed.

Generally it can be written for thermodynamically simple systems (fluids):

$$\left(\frac{\partial e}{\partial\upsilon}\right)_T d\upsilon+\left(\frac{\partial e}{\partial T}\right)_\upsilon dT=\left(\frac{\partial e}{\partial\upsilon}\right)_T d\upsilon+c_\upsilon\, dT$$

Considering the Maxwell relations of thermodynamics it can be written:

$$\left(\frac{\partial e}{\partial\upsilon}\right)_T=-P+T\left(\frac{\partial P}{\partial T}\right)_\upsilon$$

so that it holds:

$$\rho\frac{De}{Dt}=\left[-P+T\left(\frac{\partial P}{\partial T}\right)_\rho\right]\frac{\partial U_i}{\partial x_i}+\rho c_\upsilon\frac{DT}{Dt}$$

The thermal energy equation thus can be written:

$$\rho c_\upsilon\frac{DT}{Dt}=\frac{\partial\dot{q}_i}{\partial x_i}-T\left(\frac{\partial P}{\partial T}\right)_\rho\frac{\partial U_i}{\partial x_i}-\tau_{ij}\frac{\partial U_j}{\partial x_i}$$

For an Ideal gas, as $\left(\frac{\partial P}{\partial T}\right)_\rho = \frac{P}{T}$ and $\dot{q}_i = -\lambda \frac{\partial T}{\partial x_i}$ yields

$$\rho c_\upsilon \frac{\mathrm{D}T}{\mathrm{D}t} = \lambda \frac{\partial^2 T}{\partial x_i^2} - P\frac{\partial U_i}{\partial x_i} - \tau_{ij}\frac{\partial U_j}{\partial x_i}$$

For an ideal fluid as $\frac{\partial U_i}{\partial x_i} = 0$ and $c_\upsilon = c_p$ therefore:

$$\rho c_p \frac{\mathrm{D}T}{\mathrm{D}t} = \lambda \frac{\partial^2 T}{\partial x_i^2} - \tau_{ij}\frac{\partial U_j}{\partial x_i}$$

Therefore the equation for the change of the total energy can be derived by addition of the equations for the mechanical and thermal energies:

Equation for mechanical energy:

$$\rho \frac{\mathrm{D}}{\mathrm{D}t}\left(\frac{1}{2}U_j^2 + G\right) = -\frac{\partial}{\partial x_j}(PU_j) + P\frac{\partial U_j}{\partial x_j} - \frac{\partial}{\partial x_i}(\tau_{ij}U_j) + \tau_{ij}\frac{\partial U_j}{\partial x_i}$$

Equation for thermal energy:

$$\rho \frac{\mathrm{D}e}{\mathrm{D}t} = -\frac{\partial \dot{q}_i}{\partial x_i} - P\frac{\partial U_i}{\partial x_i} - \tau_{ij}\frac{\partial U_i}{\partial x_i}$$

Equation for the total energy:

$$\rho \frac{\mathrm{D}}{\mathrm{D}t}\left(\frac{1}{2}U_j^2 + G + e\right) = -\frac{\partial \dot{q}_i}{\partial x_i} - \frac{\partial}{\partial x_j}(PU_j) - \frac{\partial}{\partial x_i}(\tau_{ij}U_j)$$

$$= \lambda \frac{\partial^2 T}{\partial x_i^2} - \frac{\partial}{\partial x_j}(PU_j) - \frac{\partial}{\partial x_i}(\tau_{ij}U_j)$$

From this final relation the Bernoulli equation can be derived which is often used for fluid-mechanical considerations:

Ideal Fluid: (ρ = const): no heat conduction and viscous dissipation

$$\rho \frac{\mathrm{D}}{\mathrm{D}t}\left(\frac{1}{2}U_j^2 + G\right) = U_j\frac{\partial P}{\partial x_j} = -U_i\frac{\partial P}{\partial x_i}$$

For a steady flow:

$$\rho \frac{\mathrm{D}}{\mathrm{D}t}\left[\overbrace{\rho \frac{\partial}{\partial t}\left(\frac{1}{2}U_j^2 + G\right)}^{=0} + U_i\frac{\partial}{\partial x_i}\left(\frac{1}{2}U_j^2 + G\right)\right] = -U_i\frac{\partial P}{\partial x_i}$$

$$\frac{\partial}{\partial x_i}\left(\frac{1}{2}U_j^2 + G + \frac{P}{\rho}\right) = 0$$

or after integration:

$$\frac{1}{2}U_j^2 + G + \frac{P}{\rho} = \text{const.}$$

Ideal Gas: $P/\rho = RT$, no heat conduction and neglecting viscous dissipation as well as of the potential energy

$$\rho\frac{\mathrm{D}}{\mathrm{D}t}\left(\frac{1}{2}U_j^2 + e\right) = -\frac{\partial}{\partial x_j}(PU_j) = -\frac{\partial}{\partial x_i}(PU_j)$$

For steady flow:

$$\rho\frac{\partial}{\partial x_i}\left(\frac{1}{2}U_j^2 + e\right) = -\frac{\partial}{\partial x_j}(PU_i) = -P\frac{\partial U_i}{\partial x_i} - U_i\frac{\partial P}{\partial x_i}$$

From the continuity equation follows for steady flows:

$$\rho\frac{\partial U_i}{\partial x_i} - U_i\frac{\partial \rho}{\partial x_i}$$

Inserted consideration of $e = c_\upsilon T$

$$\rho\frac{\partial}{\partial x_i}\left(\frac{1}{2}U_j^2 + e\right) = \rho\frac{\partial}{\partial x_i}\left(\frac{1}{2}U_j^2\right) + \rho c_\upsilon\frac{\partial T}{\partial x_i} = \frac{P}{\rho}\frac{\partial \rho}{\partial x_i} - \frac{\partial P}{\partial x_i}$$

$$\frac{\partial}{\partial x_i}\left(\frac{1}{2}U_j^2\right) = \frac{\partial \rho}{\partial x_i} - c_\upsilon\frac{\partial T}{\partial x_i}$$

Introducing:

$$\frac{\partial T}{\partial x_i} = \frac{P}{R\rho^2}\frac{\partial \rho}{\partial x_i} + \frac{1}{R\rho}\frac{\partial P}{\partial x_i}$$

$$\frac{\partial}{\partial x_i}\left(\frac{1}{2}U_j^2\right) = \frac{\partial}{\partial x_i}\left(\frac{1}{2}U_j^2\right) = \frac{P}{\rho^2}\frac{\partial \rho}{\partial x_i}\left(1 + \frac{c_\upsilon}{R}\right) - \frac{1}{\rho}\frac{\partial P}{\partial x_i}\left(1 \pm \frac{c_\upsilon}{R}\right)$$

$$= -\frac{\kappa}{\kappa - 1}\frac{\partial}{\partial x_i}\left(\frac{P}{\rho}\right)$$

$$\frac{\partial}{\partial x_i}\left[\frac{1}{2}U_j^2 + \frac{\kappa}{\kappa - 1}\left(\frac{P}{\rho}\right)\right] = 0 \Rightarrow \frac{1}{2}U_j^2 + \frac{\kappa}{\kappa - 1}\left(\frac{P}{\rho}\right) = \text{const}$$

BASIC EQUATIONS IN DIFFERENT COORDINATE SYSTEMS

Continuity Equation

The derivations carried out for the continuity equation in Cartesian coordinates result in:

$$\frac{\partial \rho}{\partial t} + \frac{\partial(\rho U_i)}{\partial x_i} = 0 \quad \text{or}$$

$$\frac{\partial \rho}{\partial t}+\frac{\partial(\rho U_1)}{\partial x_1}+\frac{\partial(\rho U_2)}{\partial x_2}+\frac{\partial(\rho U_3)}{\partial x_3}=0$$

For r = const: $\frac{\partial U_1}{\partial x_1}+\frac{\partial U_2}{\partial x_2}+\frac{\partial U_3}{\partial x_3}=0$

In cylindrical coordinates (r, φ, z) with (U_r, U_j, U_z) results the following equation:

$$\frac{\partial \rho}{\partial t}+\frac{\partial(\rho U_r)}{\partial r}+\frac{1}{r}\frac{\partial(\rho U_\varphi)}{\partial \varphi}+\frac{\partial(\rho U_z)}{\partial z}+\frac{\rho U_r}{r}=0$$

and for ρ = const equation reduces to:

$$\frac{\partial \rho}{\partial t}+\frac{\partial U_r}{\partial r}+\frac{1}{r}\frac{\partial U_\varphi}{\partial \varphi}+\frac{\partial U_z}{\partial z}+\frac{U_r}{r}=0$$

In spherical coordinates (r, θ, ϕ) the continuity equation can be stated as shown below for (U_r, U_θ, U_ϕ):

$$\frac{\partial \rho}{\partial t}+\frac{1}{r^2}\frac{\partial}{\partial r}(\rho r^2 U_r)+\frac{1}{r\sin\theta}\frac{\partial}{\partial \theta}(\rho U_\theta \sin\theta)+\frac{1}{r\sin\theta}\frac{\partial}{\partial \phi}(\rho U_\phi)=0$$

Here the below-mentioned coordinates in the derivations of the relations were employed.

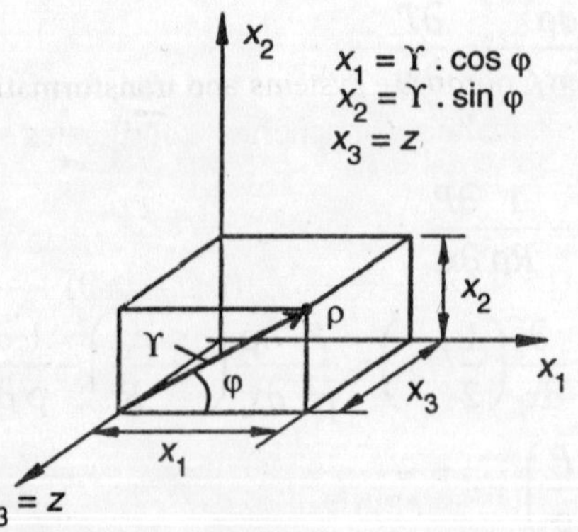

Cylindrical coordinates

Fig. Coordinate Systems and Transformation Equations for Cylindrical Coordinates

The use of cylindrical coordinates in the derivations of the basic equations leads to the metric coefficients introduced for the transformation of the equations

$h_r = 1$; $h_\varphi = r$; $h_z = 1$

for the general continuity equation:

$$\frac{\partial(\rho U_i)}{\partial x_i}=\frac{1}{r}\frac{\partial}{\partial r}(r\rho U_r)+\frac{1}{r}\frac{\partial}{\partial \varphi}(\rho U_\varphi)+\frac{\partial(\rho U_z)}{\partial z},$$

or for the continuity equation with ρ = const:

$$\frac{\partial U_r}{\partial r}+\frac{1}{r}\frac{\partial U_\varphi}{\partial \varphi}+\frac{\partial U_z}{\partial z}\frac{U_r}{r}=0$$

Analogous to the above derivations of the continuity equation in cylindrical coordinates one obtains for spherical coordinates:

$h_r = 1; h_\upsilon = r; h_\varphi = r \sin \theta$

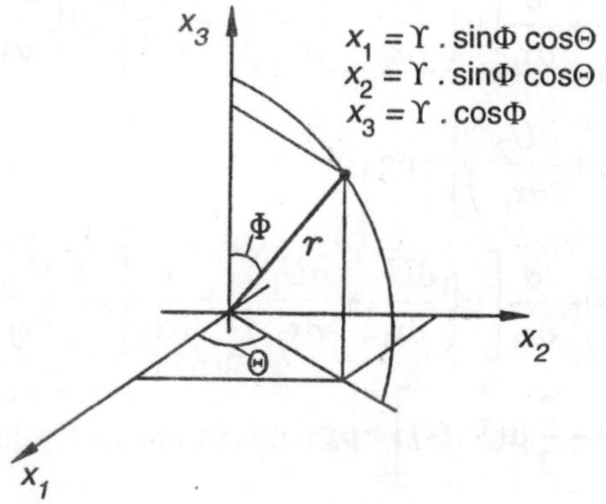

Spherical coordinates

Fig. Coordinate systems and transformation equations for spherical coordinates

and thus for

$$\frac{\partial(\rho U_i)}{\partial x_i}=\frac{1}{r^2}\frac{\partial}{\partial r}(\rho r^2 U_r)+\frac{1}{r\sin\theta}\frac{\partial}{\partial\theta}(\rho U_\theta\sin\theta)+\frac{1}{r\sin\theta\, r\sin\theta\,\partial\phi}(\rho U_\phi)$$

the continuity equation in spherical coordinates with ρ = const results:

$$\frac{\partial(\rho U_i)}{\partial x_i}=\frac{1}{r^2}\frac{\partial}{\partial r}(r^2 U_r)+\frac{1}{r\sin\theta}\frac{\partial}{\partial\theta}(U_\theta\sin\theta)+\frac{1}{r\sin\theta}\frac{\partial}{\partial\phi}(U_\phi)=0$$

Analogous to the transformation of the continuity equation in cylindrical and spherical coordinate systems, the different terms of the Navier-Stokes equations can also be transferred, which can be stated in cartesian coordinates as follows for Newtonian fluids:

$$\rho=\frac{DU_j}{Dt}=\rho\left[\frac{\partial U_j}{\partial t}+U_i\frac{\partial U_j}{\partial x_i}\right]$$

$$=\frac{\partial P}{\partial x_j}+\frac{\partial}{\partial x_i}\left[\mu\left(\frac{\partial U_j}{\partial x_i}+\frac{\partial U_i}{\partial x_j}\right)-\frac{2}{3}\mu\delta_{ij}\frac{\partial U_\kappa}{\partial x_\kappa}\right]+\rho g_j$$

Written out for j = 1, 2, 3:

$$\rho\frac{DU_1}{Dt}=-\frac{\partial P}{\partial x_1}+\frac{\partial}{\partial x_1}\left[2\mu\frac{\partial U_1}{\partial x_1}-\frac{2}{3}\mu(\nabla.\vec{U})\right]$$

$$+\frac{\partial}{\partial x_2}\left[\mu\left(\frac{\partial U_1}{\partial x_2}+\frac{\partial U_2}{\partial x_1}\right)\right]+\frac{\partial}{\partial x_3}\left[\mu\left(\frac{\partial U_3}{\partial x_3}+\frac{\partial U_1}{\partial x_1}\right)\right]+\rho g_1$$

$$\rho\frac{DU_2}{Dt}=-\frac{\partial P}{\partial x_2}+\frac{\partial}{\partial x_1}\left[\mu\left(\frac{\partial U_2}{\partial x_1}+\frac{\partial U_1}{\partial x_2}\right)\right]+\frac{\partial}{\partial x_2}\left[2\mu\left(\frac{\partial U_2}{\partial x_2}-\frac{2}{3}\mu(\nabla.\vec{U})\right)\right]$$

$$+\frac{\partial}{\partial x_3}\left[\mu\left(\frac{\partial U_3}{\partial x_3}+\frac{\partial U_2}{\partial x_1}\right)\right]+\rho g_2$$

$$\rho\frac{DU_3}{Dt}=-\frac{\partial P}{\partial x_3}+\frac{\partial}{\partial x_1}\left[\mu\left(\frac{\partial U_3}{\partial x_1}+\frac{\partial U_1}{\partial x_3}\right)\right]+\frac{\partial}{\partial x_2}\left[\mu\left(\frac{\partial U_3}{\partial x_2}+\frac{\partial U_2}{\partial x_3}\right)\right]$$

$$+\frac{\partial}{\partial x_3}\left[2\mu\frac{\partial U_3}{\partial x_3}-\frac{2}{3}\mu(\nabla.\vec{U})\right]+\rho g_3$$

where it holds $\nabla.\vec{U}=\frac{\partial U_\kappa}{\partial x_\kappa}$

- Momentum Equations in Cartesian Coordinates – Momentum equations with τ_{ij} –terms:

x_1 –Component:

$$\rho\left(\frac{\partial U_1}{\partial t}+U_1\frac{\partial U_1}{\partial x_1}+U_2\frac{\partial U_1}{\partial x_2}+U_3\frac{\partial U_1}{\partial x_3}\right)$$
$$=-\frac{\partial P}{\partial x_1}-\left(\frac{\partial \tau_{11}}{\partial x_1}+\frac{\partial \tau_{21}}{\partial x_2}+\frac{\partial \tau_{31}}{\partial x_3}\right)+\rho g_1$$

x_2 –Component:

$$\rho\left(\frac{\partial U_2}{\partial t}+U_1\frac{\partial U_2}{\partial x_1}+U_2\frac{\partial U_2}{\partial x_2}+U_3\frac{\partial U_2}{\partial x_3}\right)=-\frac{\partial P}{\partial x_2}$$
$$-\left(\frac{\partial \tau_{12}}{\partial x_1}+\frac{\partial \tau_{22}}{\partial x_2}+\frac{\partial \tau_{32}}{\partial x_3}\right)+\rho g_2$$

x_3 –Component:

$$\rho\left(\frac{\partial U_3}{\partial t}+U_1\frac{\partial U_3}{\partial x_1}+U_2\frac{\partial U_3}{\partial x_2}+U_3\frac{\partial U_3}{\partial x_3}\right)=-\frac{\partial P}{\partial x_3}$$
$$-\left(\frac{\partial \tau_{13}}{\partial x_1}+\frac{\partial \tau_{23}}{\partial x_2}+\frac{\partial \tau_{33}}{\partial x_3}\right)+\rho g_3$$

- Navier-Stokes equations for r and μ equally constant:

x_1 –Component:

$$\rho\left(\frac{\partial U_1}{\partial t}+U_1\frac{\partial U_1}{\partial x_1}+U_2\frac{\partial U_1}{\partial x_2}+U_3\frac{\partial U_1}{\partial x_3}\right)$$

$$=-\frac{\partial P}{\partial x_1}+\mu\left(\frac{\partial^2 U_1}{\partial x_1^2}+\frac{\partial^2 U_1}{\partial x_2^2}+\frac{\partial^2 U_1}{\partial x_3^2}\right)+\rho g_1$$

x_2 –Component:

$$\rho\left(\frac{\partial U_2}{\partial t}+U_2\frac{\partial U_2}{\partial x_1}+U_2\frac{\partial U_2}{\partial x_2}+U_3\frac{\partial U_2}{\partial x_3}\right)=-\frac{\partial P}{\partial x_2}$$

$$+\mu\left(\frac{\partial^2 U_2}{\partial x_1^2}+\frac{\partial^2 U_2}{\partial x_2^2}+\frac{\partial^2 U_2}{\partial x_3^2}\right)+\rho g_2$$

x_3 –Component:

$$\rho\left(\frac{\partial U_3}{\partial t}+U_1\frac{\partial U_3}{\partial x_1}+U_2\frac{\partial U_3}{\partial x_2}+U_3\frac{\partial U_3}{\partial x_3}\right)=-\frac{\partial P}{\partial x_3}$$

$$+\mu\left(\frac{\partial^2 U_3}{\partial x_1^2}+\frac{\partial^2 U_3}{\partial x_2^2}+\frac{\partial^2 U_3}{\partial x_3^2}\right)+\rho g_3$$

- Momentum Equations in Cylindrical Coordinates – Momentum equations with τ_{ij} –terms:

r –Component:

$$\rho\left(\frac{\partial U_r}{\partial t}+U_r\frac{\partial U_r}{\partial r}+\frac{U_\varphi}{r}\frac{\partial U_r}{\partial \varphi}-\frac{U_\varphi^2}{r}+U_z\frac{\partial U_r}{\partial z}\right)=-\frac{\partial P}{\partial r}$$

$$-\left(\frac{1}{r}\frac{\partial}{\partial r}(rr_{rr})+\frac{1}{r}\frac{\partial r_{r\varphi}}{\partial \varphi}-\frac{r_{\varphi\varphi}}{r}-\frac{r_{\varphi\varphi}}{r}+\frac{\partial r_{rz}}{\partial z}\right)+\rho g_r$$

φ–Component:

$$\rho\left(\frac{\partial U_\varphi}{\partial t}+U_r\frac{\partial U_\varphi}{\partial r}+\frac{U_\varphi}{r}\frac{\partial U_\varphi}{\partial \varphi}+\frac{U_r U_\varphi}{r}+U_z\frac{\partial U_\varphi}{\partial z}\right)$$

$$=-\frac{1}{r}\frac{\partial P}{\partial \varphi}-\left(\frac{1}{r^2}\frac{\partial}{\partial r}(r^2 r_{r\varphi})+\frac{1}{r}\frac{\partial r_{\varphi\varphi}}{\partial \varphi}+\frac{\partial r_{\varphi z}}{\partial z}\right)+\rho g_\varphi$$

z–Component:

$$\rho\left(\frac{\partial U_z}{\partial t}+U_r\frac{\partial U_z}{\partial r}+\frac{U_\varphi}{r}\frac{\partial U_z}{\partial \varphi}+U_z\frac{\partial U_z}{\partial z}\right)$$

$$= -\frac{\partial P}{\partial z} - \left(\frac{1}{r}\frac{\partial}{\partial r}(r r_{rz}) + \frac{1}{r}\frac{\partial r_{\varphi z}}{\partial \varphi} + \frac{\partial r_{zz}}{\partial z}\right) + \rho g_z$$

— Navier-Stokes equations for ρ and μ equally constant:

ρ–Component:

$$\rho\left(\frac{\partial U_r}{\partial t} + U_r\frac{\partial U_r}{\partial r} + \frac{U_\varphi}{r}\frac{\partial U_r}{\partial \varphi} - \frac{U_\varphi^2}{r} + U_z\frac{\partial U_r}{\partial_z}\right)$$

$$= -\frac{\partial P}{\partial r} + \mu\left[\frac{\partial}{\partial r}\left(\frac{1}{r}\frac{\partial}{\partial r}(rU_r)\right) + \frac{1}{r^2} + \frac{\partial^2 U_r}{\partial \varphi^2} - \frac{2}{r^2}\frac{\partial U_\varphi}{\partial \varphi} + \frac{\partial^2 U_r}{\partial z^2}\right] + \rho g_r$$

φ–Component:

$$\rho\left(\frac{\partial U_\varphi}{\partial t} + U_r\frac{\partial U_\varphi}{\partial r} + \frac{U_\varphi}{r}\frac{\partial U_\varphi}{\partial \varphi} + \frac{U_r U_\varphi}{r} + U_z\frac{\partial U_z}{\partial_z}\right)$$

$$-\frac{1}{r}\frac{\partial p}{\partial \varphi} + \mu\left[\frac{\partial}{\partial r}\left(\frac{1}{r}\frac{\partial}{\partial r}(rU_\varphi)\right) + \frac{1}{r^2}\frac{\partial^2 U_\varphi}{\partial \varphi^2} + \frac{2}{r^2}\frac{\partial U_r}{\partial \varphi} + \frac{\partial^2 U_\varphi}{\partial z^2}\right] + \rho g_\varphi$$

z–Component:

$$\rho\left(\frac{\partial U_z}{\partial t} + U_r\frac{\partial U_z}{\partial r} + \frac{U_\varphi}{r}\frac{\partial U_z}{\partial \varphi} + U_z\frac{\partial U_z}{\partial_z}\right)$$

$$= -\frac{\partial P}{\partial z} + \mu\left[\frac{1}{r}\frac{\partial}{\partial r}\left(r\frac{\partial U_z}{\partial r}\right) + \frac{1}{r^2}\frac{\partial^2 U_z}{\partial \varphi^2} + \frac{\partial^2 U_z}{\partial z^2}\right] + \rho g_z$$

- Momentum Equations in Spherical Coodinates

— Momentum equations with τ_{ij} –terms

$$\rho\left(\frac{\partial U_r}{\partial t} + U_r\frac{\partial U_r}{\partial r} + \frac{U_\theta}{r}\frac{\partial U_r}{\partial \theta} + \frac{U_\varphi}{r\sin\theta}\frac{\partial U_r}{\partial \phi} - \frac{U_\theta^2 + U_\phi^2}{r}\right)$$

$$= -\frac{\partial P}{\partial r} - \left(\frac{1}{r^2}\frac{\partial}{\partial r}(r^2\tau_{rr}) + \frac{1}{r\sin\theta}\frac{\partial}{\partial \theta}(\tau_{r\theta}\sin\theta)\right.$$

$$\left. + \frac{1}{r\sin\theta}\frac{\partial \tau_{r\phi}}{\partial \phi} - \frac{\tau_{\theta\theta} + \tau_{\theta\theta}}{r}\right) + \rho g_r$$

θ–Component:

$$\rho\left(\frac{\partial U_\theta}{\partial t} + U_r\frac{\partial U_\theta}{\partial r} + \frac{U_\theta}{r}\frac{\partial U_\theta}{\partial \theta} + \frac{U_\varphi}{r\sin\theta}\frac{\partial U_\theta}{\partial \phi} + \frac{U_r + U_\theta}{r} - \frac{U_\phi^2\cot\theta}{r}\right)$$

$$= -\frac{1}{r}\frac{\partial p}{\partial \theta} - \left(\frac{1}{r^2}\frac{\partial}{\partial r}(r^2\tau_{r\theta}) + \frac{1}{r\sin\theta}\frac{\partial}{\partial \theta}(\tau_{\theta\theta}\sin\theta)\right.$$

$$+\frac{1}{r\sin\theta}\frac{\partial\tau_{\theta\phi}}{\partial\phi}+\frac{\tau_{r\theta}}{r}-\frac{\cot\theta}{r}\tau_{\theta\theta}\Bigg)+\rho g_\theta$$

ϕ–Component:

$$\rho\left(\frac{\partial U_\phi}{\partial t}+U_r\frac{\partial U_\phi}{\partial r}+\frac{U_\theta}{r}\frac{\partial U_\phi}{\partial\theta}+\frac{U_\phi}{r\sin\theta}\frac{\partial U_\phi}{\partial\phi}+\frac{U_\phi+U_r}{r}+\frac{U_\theta U_\phi}{r}\cot\theta\right)$$

$$=-\frac{1}{r\sin\theta}\frac{\partial p}{\partial\phi}-\Bigg(\frac{1}{r^2}\frac{\partial}{\partial r}(r^2\tau_{r\phi})+\frac{1}{r}\frac{\partial\tau_{\theta\phi}}{\partial\theta}$$

$$+\frac{1}{r\sin\theta}\frac{\partial\tau_{\phi\phi}}{\partial\phi}+\frac{\tau_{r\phi}}{r}+\frac{2\cot\theta}{r}\tau_{\theta\phi}\Bigg)+\rho g_\phi$$

Navier-Stokes Equations for ρ and μ Equally Constant r–Component:

$$\rho\left(\frac{\partial U_r}{\partial t}+U_r\frac{\partial U_r}{\partial r}+\frac{U_\theta}{r}\frac{\partial U_r}{\partial\theta}+\frac{U_\phi}{r\sin\theta}\frac{\partial U_r}{\partial\phi}-\frac{U_\theta^2+U_\phi^2}{r}\right)$$

$$=-\frac{\partial P}{r^2\sin\theta}+\mu\Bigg(\nabla^2U_r-\frac{2}{r^2}U_r-\frac{2}{r^2}\frac{\partial U_\theta}{\partial\theta}-\frac{2}{r^2}U_\theta\cot\theta$$

$$-\frac{2}{r^2\sin\theta}\frac{\partial U_\phi}{\partial\phi}\Bigg)+\rho g_r$$

θ–Component:

$$\rho\left(\frac{\partial U_\theta}{\partial t}+U_r\frac{\partial U_\theta}{\partial r}+\frac{U_\theta}{r}\frac{\partial U_\theta}{\partial\theta}+\frac{U_\phi}{r\sin\theta}\frac{\partial U_\theta}{\partial\phi}+\frac{U_r+U_\theta}{r}\frac{U_\phi^2\cot\theta}{r}\right)$$

$$=-\frac{1}{r}\frac{\partial P}{\partial\theta}+\mu\Bigg(\nabla^2U_\theta+\frac{2}{r^2}\frac{\partial U_r}{\partial\theta}$$

$$-\frac{U_\theta}{r^2\sin^2\theta}-\frac{2\cos\theta\partial U_\phi}{r^2\sin^2\theta}\frac{\partial U_\phi}{\partial_\phi}\Bigg)+\rho g_\theta$$

ϕ–Component:

$$\rho\left(\frac{\partial U_\phi}{\partial t}+U_r\frac{\partial U_\phi}{\partial r}+\frac{U_\phi}{r}\frac{\partial U_\phi}{\partial\theta}+\frac{U_\phi}{r\sin\theta}\frac{\partial U_\phi}{\partial\phi}+\frac{U_\phi+U_r}{r}+\frac{U_\theta U_\phi}{r}\cot\theta\right)$$

$$=-\frac{1}{r\sin\theta}\frac{\partial P}{\partial\phi}+\mu\Bigg(\nabla^2U_\phi-\frac{U_\phi}{r^2\sin^2\theta}$$

$$+\frac{2}{r^2\sin^2\theta}\frac{\partial U_r}{r^2\sin^2\theta}\frac{\partial U_r}{\partial\phi}\Bigg)+\rho g_\phi$$

In these equations:

$$\nabla^2 = \frac{1}{r^2}\frac{\partial}{\partial r}\left(r^2\frac{\partial}{\partial r}\right) + \frac{1}{r^2\sin\theta}\frac{\partial}{\partial\theta}\left(\sin\theta\frac{\partial}{\partial\theta}\right) + \frac{1}{r^2\sin^2\theta}\left(\frac{\partial^2}{\partial\phi^2}\right)$$

- Components of the Molecular Momentum Transport Tensor in Cartesian Coordinates:

$$\tau_{11} = -\mu\left[2\frac{\partial U_1}{\partial x_1} - \frac{2}{3}(\nabla.\vec{U})\right];$$

$$\tau_{22} = -\mu\left[2\frac{\partial U_2}{\partial x_2} - \frac{2}{3}(\nabla.\vec{U})\right];$$

$$\tau_{33} = -\mu\left[2\frac{\partial U_3}{\partial x_3} - \frac{2}{3}(\nabla.\vec{U})\right]$$

$$\tau_{12} = \tau_{21} = -\mu\left[\frac{\partial U_1}{\partial x_2} + \frac{\partial U_2}{\partial x_1}\right];$$

$$\tau_{23} = \tau_{32} = -\mu\left[\frac{\partial U_2}{\partial x_3} + \frac{\partial U_3}{\partial x_2}\right];$$

$$\tau_{31} = \tau_{13} = -\mu\left[\frac{\partial U_3}{\partial x_1} + \frac{\partial U_1}{\partial x_3}\right].$$

above being employed

$$(\nabla.\vec{U}) = \frac{\partial U_1}{\partial x_1} + \frac{\partial U_2}{\partial x_2} + \frac{\partial U_3}{\partial x_3} = \frac{\partial U_\kappa}{\partial x_\kappa}$$

- Components of the Molecular Transport Tensor in Cylindrical Coordinates:

$$\tau_{rr} = -\mu\left[2\frac{\partial U_r}{\partial r} - \frac{2}{3}(\nabla.\vec{U})\right];$$

$$\tau_{\varphi\varphi} = -\mu\left[2\left(\frac{1}{r}\frac{\partial U_\varphi}{\partial\varphi} + \frac{U_r}{r}\right) - \frac{2}{3}(\nabla.\vec{U})\right];$$

$$\tau_{zz} = -\mu\left[2\frac{\partial U_z}{\partial x_z} - \frac{2}{3}(\nabla.\vec{U})\right]$$

$$\tau_{r\varphi} = \tau_{\varphi r} - \mu\left[r\frac{\partial}{\partial r}\left(\frac{U_\varphi}{r}\right) + \frac{1}{r}\frac{\partial U_r}{\partial\varphi}\right];$$

$$\tau_{\varphi z} = \tau_{z\varphi} - \mu\left[\frac{\partial U_\varphi}{\partial z} + \frac{1}{r}\frac{\partial U_z}{\partial\varphi}\right];$$

$$\tau_{zr} = \tau_{rz} = -\mu\left[\frac{\partial U_z}{\partial r} + \frac{\partial U_r}{\partial z}\right].$$

above being employed

$$(\nabla.\vec{U}) = \frac{1}{r}\frac{\partial}{\partial r}(rU_r) + \frac{1}{r}\frac{\partial U_\varphi}{\partial \varphi} + \frac{\partial U_z}{\partial z}$$

Components of the Molecular Momentum Transport Tensor in Spherical Coordinates:

$$\tau_{rr} = -\mu\left[2\frac{\partial U_r}{\partial r} - \frac{2}{3}(\nabla.U)\right]$$

$$\tau_{\theta\theta} = -\mu\left[2\left(\frac{1}{r}\frac{\partial U_\theta}{\partial \theta} + \frac{U_r}{r}\right) - \frac{2}{3}(\nabla.U)\right]$$

$$\tau_{\phi\phi} = -\mu\left[2\left(\frac{1}{r\sin\theta}\frac{\partial U_\phi}{\partial \phi} + \frac{U_r}{r} + \frac{U_\theta \cot\theta}{r}\right) - \frac{2}{3}(\nabla.U)\right]$$

$$\tau_{r\theta} = \tau_{\theta r} - \mu\left[r\frac{\partial}{\partial r}\left(\frac{U_\theta}{r}\right) + \frac{1}{r}\frac{\partial U_r}{\partial \theta}\right]$$

$$\tau_{\theta\phi} = \tau_{\phi\theta} - \mu\left[\frac{\sin\theta}{r}\frac{\partial}{\partial \theta}\left(\frac{U_\phi}{\sin\theta}\right) + \frac{1}{r\sin\theta}\frac{\partial U_\theta}{\partial \phi}\right]$$

$$\tau_{\phi r} = \tau_{r\phi} - \mu\left[\frac{1}{r\sin\theta}\frac{\partial U_r}{\partial \phi} + r\frac{\partial}{\partial r}\left(\frac{U_\phi}{r}\right)\right]$$

$$(\nabla.U) = \frac{1}{r^2}\frac{\partial}{\partial r}(r^2 U_r) + \frac{1}{r\sin\theta}\frac{\partial}{\partial \theta}(U_\theta \sin\theta) + \frac{1}{r\sin\theta}\frac{\partial U_\phi}{\partial_\phi}$$

Dissipation Function $\tau_{ij}\frac{\partial U_j}{\partial x_i} = \mu\Phi_\mu$:

Cylindrical coordinates:

$$\Phi_\mu = 2\left[\left(\frac{\partial U_1}{\partial x_1}\right)^2 + \left(\frac{\partial U_2}{\partial x_2}\right)^2 + \left(\frac{\partial U_3}{\partial x_3}\right)^2\right] + \left[\frac{\partial U_2}{\partial x_1} + \frac{\partial U_1}{\partial x_2}\right]^2$$

$$+\left[\frac{\partial U_2}{\partial x_2} + \frac{\partial U_2}{\partial x_3}\right]^2 + \left[\frac{\partial U_1}{\partial x_3} + \frac{\partial U_3}{\partial x_1}\right]^2 + \frac{2}{3}\left[\frac{\partial U_1}{\partial x_1} + \frac{\partial U_2}{\partial x_2} + \frac{\partial U_3}{\partial x_3}\right]^2$$

Cylindrical coordinates:

$$\Phi_\mu = 2\left[\left(\frac{\partial U_r}{\partial r}\right)^2 + \left(\frac{1}{r}\frac{\partial U_\varphi}{\partial \varphi} + \frac{U_r}{r}\right)^2 + \left(\frac{\partial U_z}{\partial x_z}\right)^2\right]$$

$$+\left[r\frac{\partial}{\partial r}\left(\frac{U_\varphi}{r}\right)+\frac{1}{r}\frac{\partial U_z}{\partial \varphi}\right]^2+\left[\frac{1}{r}\frac{\partial U_z}{\partial \varphi}+\frac{\partial U_\varphi}{\partial x_z}\right]^2+\left[\frac{\partial U_r}{\partial x_z}+\frac{\partial U_z}{\partial r}\right]^2$$

$$-\frac{2}{3}\left[\frac{1}{r}\frac{\partial}{\partial r}(rU_r)+\frac{1}{r}\frac{\partial U_\varphi}{\partial_\varphi}+\frac{\partial U_z}{\partial x_z}\right]^2$$

Spherical coordinates:

$$\Phi_\mu = 2\left[\left(\frac{\partial U_r}{\partial r}\right)^2+\left(\frac{1}{r}\frac{\partial U_\theta}{\partial \theta}+\frac{U_r}{r}\right)^2+\left(\frac{1}{r\sin\theta}\frac{\partial U_\varphi}{\partial \varphi}+\frac{U_r}{r}+\frac{U_\theta \cot\theta}{r}\right)^2\right]$$

$$+\left[r\frac{\partial}{\partial r}\left(\frac{U_\theta}{r}\right)+\frac{1}{r}\frac{\partial U_r}{\partial \theta}\right]^2+\left[\frac{\sin\theta}{r}\frac{\partial}{\partial \theta}\left(\frac{U_\phi}{\sin\theta}\right)+\frac{1}{r\sin\theta}\frac{\partial U_\theta}{\partial \phi}\right]^2$$

$$+\left[\frac{1}{r\sin\theta}\frac{\partial U_r}{\partial \phi}+r\frac{\partial}{\partial r}\left(\frac{U_\phi}{r}\right)\right]^2$$

$$-\frac{2}{3}\left[\frac{1}{r^2}\frac{\partial}{\partial r}(r^2U_r)+\frac{1}{r\sin\theta}\frac{\partial}{\partial \theta}(U_\theta \sin\theta)+\frac{1}{r\sin\theta}\frac{\partial U_\phi}{\partial \phi}\right]^2$$

The above equations can be solved in connection with the initial and boundary conditions describing the actual flow problems.

SPECIAL FORMS OF THE BASIC EQUATIONS

Due to the multitude of fluid-mechanical considerations special forms have crystallized from the equations. These are the vortex-power equation, the Bernoulli equation and the Crocco equation, that have already been treated before. The derivations will moreover treat the Kelvin theorem as a basis for explanations of its physical significance. The objective of the considerations is to bring out clearly the prerequisites under which the special forms of the basic equations are valid. Only the simplified treatments of flow problems that are sought with the special forms of the equations lead to secure results.

Transport Equation for Vortex Power

The vortex power ω_i is a property of the flow field which can be employed advantageously in considerations of rotating fluid motions. It can be computed from the velocity field as follows:

$$\vec{\omega}_\kappa = \nabla \times \vec{U} = -\varepsilon_{ijk}\frac{\partial U_j}{\partial x_i} = \left(\frac{\partial U_j}{\partial x_i}-\frac{\partial U_i}{\partial x_j}\right)$$

For a fluid with the properties $\rho = \text{const}$ and $\mu = \text{const}$, the Navier-Stokes equation can be written in the following way:

$$\rho\left[\frac{\partial U_j}{\partial t}+U_i\frac{\partial U_j}{\partial x_i}\right]=-\frac{\partial P}{\partial x_j}+\mu\frac{\partial^2 U_j}{\partial x_i^2}+\rho g_j$$

or in vector form:

$$\left[\frac{\partial \vec{U}}{\partial t}+(\vec{U}.\nabla)\vec{U}\right]=-\frac{1}{\rho}\nabla P+\nu\nabla^2\vec{U}+\rho\vec{g}$$

Considering that this vector form of the Navier-Stokes equation can also be written as:

$$\frac{\partial \vec{U}}{\partial t}+\nabla\left(\frac{1}{2}\vec{U}.\vec{U}\right)\vec{U}\times(\nabla\times\vec{U})=-\frac{1}{\rho}\nabla P+\nu\nabla^2\vec{U}+\vec{g}$$

When one applies the operator $\nabla\times(\ldots)$. to each of the terms appearing in the above equation one obtains:

$$\frac{\partial\vec{\omega}}{\partial t}-\nabla\times(\vec{U}\times\vec{\omega})=\nu\nabla^2\vec{\omega}$$

Making use of the relation valid for vectors:

$$\nabla\times(\vec{U}\times\vec{\omega})=\vec{U}(\nabla.\vec{\omega})-\vec{\omega}(\nabla.\vec{U})-(\vec{U}.\nabla)\vec{\omega}+(\vec{\omega}.\nabla)\vec{U}$$

where $\nabla.\vec{\omega}=0$ as the divergence of the rotation of each vector is equal to zero, and where at the same time ρ = const $\nabla.\vec{U}=0$ holds owing to the continuity equation. When one introduces all this into the above equations, the transport equation for the vortex power reads:

$$\frac{\partial\vec{\omega}}{\partial t}+(\vec{U}\nabla)\vec{\omega}=(\vec{\omega}\nabla)\vec{U}+\nu\nabla^2\vec{\omega}$$

or in tensor notation:

$$\frac{D\omega_j}{Dt}=\frac{\partial\omega_j}{\partial t}+U_i\frac{\partial\omega_j}{\partial x_i}=\omega_j\frac{\partial U_i}{\partial x_j}+\nu\frac{\partial^2\omega_j}{\partial x_i^2}$$

The equation does not contain the pressure term, from this it is apparent that the vortex-power field can be determined without knowledge of the pressure distribution. To be able to compute the pressure, one forms the divergence of the Navier-Stokes equation and obtains for $g_j = 0$:

$$\frac{\partial^2}{\partial x_i^2}\left(\frac{P}{\rho}\right)\omega_j^2+U_j\frac{\partial^2 U_j}{\partial x_i^2}-\frac{1}{2}\frac{\partial^2 U_j^2}{\partial x_i^2}$$

Thus yields the Poisson equation for the computation of the pressure. For two-dimensional flows, for which the vortex-power vector stands vertical on the flow plane, $(\vec{\omega}.\nabla)\vec{U}=0$. The transport equation for the vortex power therefore reads:

$$\frac{\partial \omega_j}{\partial t} + U_i \frac{\partial \omega_j}{\partial x_i} = \nu \frac{\partial^2 \omega_j}{\partial x_i^2}$$

The Bernoulli Equation

The general momentum equation is transferred into the Euler equations, by assuming an ideal fluid to derive the Bernoulli equation, i.e:

$$\rho \frac{\mathrm{D}U_j}{\mathrm{D}t} = \rho \left[\frac{\partial U_j}{\partial t} + U_j \frac{\partial U_j}{\partial x_i} \right] = -\frac{\partial P}{\partial x_j} + \rho g_j$$

Multiplying this equation by U_j, one obtains the mechanical energy equation valid for dissipation-free fluid flows:

$$\rho \frac{\mathrm{D}}{\mathrm{D}t}\left(\frac{1}{2}U_j^2\right) = \rho \frac{\partial}{\partial t}\left(\frac{1}{2}U_j^2\right) + U_i \frac{\partial}{\partial x_i}\left(\frac{1}{2}U_j^2\right) = U_j \frac{\partial P}{\partial x_j} + \partial g_j U_j.$$

When one introduces the potential field G for the presentation of g_j so as:

$$g_j = -\frac{\partial G}{\partial x_j}$$

the last term of the equation reads:

$$\rho g_j U_j = -\rho U_j \frac{\partial G}{\partial x_j} = -\rho \frac{\mathrm{D}G}{\mathrm{D}t} + \rho \frac{\partial G}{\partial t},$$

one obtains for $\frac{\partial G}{\partial t} = 0$

$$\rho \frac{\mathrm{D}}{\mathrm{D}t}\left[\left(\frac{1}{2}U_j^2\right) + G\right] = -U_j \frac{\partial P}{\partial x_j}.$$

Considering that it holds:

$$\rho \frac{\mathrm{D}}{\mathrm{D}t}\left(\frac{P}{\rho}\right) = \rho \frac{\partial}{\partial t}\left(\frac{P}{\rho}\right) \rho U_j \frac{\partial}{\partial x_j}\left(\frac{P}{\rho}\right)$$

and that moreover the following conversions are possible:

$$\rho \frac{\partial}{\partial t}\left(\frac{P}{\rho}\right) = \frac{\partial P}{\partial t} - \frac{P}{\rho}\frac{\partial \rho}{\partial t},$$

$$\frac{\mathrm{D}}{\mathrm{D}t} \rho U_j \frac{\partial}{\partial x_j}\left(\frac{P}{\rho}\right) = U_j \frac{\partial P}{\partial x_j} - \frac{P}{\rho} U_j \frac{\partial \rho}{\partial x_j},$$

then it can be written:

$$\rho \frac{\mathrm{D}}{\mathrm{D}t}\left(\frac{P}{\rho}\right) = \frac{\partial P}{\partial t} + U_j \frac{\partial P}{\partial x_j} - \frac{P}{\rho}\frac{\mathrm{D}\rho}{\mathrm{D}t}$$

From the equations :

$$\rho\frac{D}{Dt}\left[\left(\frac{1}{2}U_j^2\right)+G\right]=-U_j\frac{\partial P}{\partial x_j}=-\rho\frac{D}{Dt}\left(\frac{P}{\rho}\right)+\frac{\partial P}{\partial t}-\frac{P}{\rho}\frac{D\rho}{Dt},$$

or after conversion of some terms:

$$\rho\frac{D}{Dt}\left[\left(\frac{1}{2}U_j^2\right)+\frac{P}{\rho}+G\right]=\frac{\partial P}{\partial t}-\frac{P}{\rho}\frac{D\rho}{Dt}$$

or

$$\rho\frac{D}{Dt}\left[\left(\frac{1}{2}U_j^2\right)+\frac{P}{\rho}+G\right]=\frac{\partial P}{\partial t}+\frac{\partial U_j}{\partial x_j}P.$$

For stationary pressure fields $\frac{\partial P}{\partial t}=0$, and for ρ = const, the Bernoulli equation can be stated as follows:

$$\frac{1}{2}U_j^2+\frac{P}{\rho}+G=\frac{1}{2}U_j^2+\frac{P}{\rho}-x_j g_j=\text{const}$$

The above derivations make clear under which conditions the well-known Bernoulli equation holds.

From the above derivations, general form of the mechanical energy equation by including dissipative flow field can be written in other form:

$$\rho\frac{D}{Dt}\left[\frac{1}{2}U_j^2+\frac{P}{\rho}+G\right]=\frac{\partial P}{\partial t}+P\frac{\partial U_j}{\partial x_j}+\frac{\partial}{\partial x_i}(\tau_{ij}U_j)-\tau_{ij}\frac{\partial U_j}{\partial x_i}$$

left side of this form of the mechanical energy contains all terms of the Bernoulli equation.

Crocco Equation

The Crocco equation is a special form of the momentum equation which shows in an impressive manner how purely fluid-mechanical considerations can be supplemented by thermodynamical insights. The Crocco equation connects the vortex power of a flow field to the entropy of the considered fluid. It can be shown from this equation that isotropic flows are free from rotation and vice versa under certain conditions. So when one recognizes a flow field to be isentrope, the simplified rotation-free flow fields considerations can be applied.

For the derivation of the Crocco equation one starts from the Navier-Stokes equation, as it is stated in equation supplemented by $\nu = 0$, i.e. one introduces an ideal fluid into the considerations, to neglect inertia forces.

$$\frac{\partial\vec{U}}{\partial t}+\nabla\left(\frac{1}{2}\vec{U}.\vec{U}\right)-\vec{U}\times(\nabla\times\vec{U})=-\frac{1}{\rho}\nabla P$$

It was shown that it holds:

$$T_{\Re}.ds_{\Re} = de_{\Re} + P_{\Re}.d\upsilon_{\Re} = de_{\Re} + P_{\Re} d\left(\frac{1}{\rho_{\Re}}\right)$$

With $e_{\Re} = h_n - P_{\Re}/\rho_{\Re}$ the following relation holds:

$$\mathrm{d}h_{\Re} - \mathrm{d}\frac{P_{\Re}}{\rho_{\Re}} = -P_{\Re}\mathrm{d}\frac{1}{\rho_{\Re}} + T_{\Re}\mathrm{d}s_{\Re}$$

because of $\mathrm{d}\left(\frac{P_{\Re}}{\rho_{\Re}}\right) = P_{\Re}.\mathrm{d}\left(\frac{1}{\rho_{\Re}}\right) + \frac{1}{\rho_{\Re}}.\mathrm{d}P_{\Re}$

$$-\frac{1}{\rho_{\Re}}\mathrm{d}P_{\Re} = T_{\Re}\,\mathrm{d}s_{\Re} - \mathrm{d}h_{\Re}$$

This relation can also be written:

$$\frac{1}{\rho}\nabla P = T\nabla s - \nabla h$$

equation is inserted in equation to yield:

$$\frac{\partial \vec{U}}{\partial t} + \nabla\left(\frac{1}{2}\vec{U}.\vec{U}\right) - \vec{U}\times(\nabla\times\vec{U}) = T\nabla s - \nabla h$$

For stationary adiabatic processes the thermal energy equation in the following form:

$$\rho\frac{\mathrm{D}h}{\mathrm{D}t} = \frac{\mathrm{D}P}{\mathrm{D}t}$$

From the momentum equation it follows further:

$$\rho\frac{\mathrm{D}}{\mathrm{D}t}\left(\frac{1}{2}\vec{U}\vec{U}\right) = -\vec{U}\nabla P$$

Thus:

$$\rho\frac{\mathrm{D}}{\mathrm{D}t}\left(h + \frac{1}{2}\vec{U}\vec{U}\right) = \frac{\mathrm{D}P}{\mathrm{D}t} - \vec{U}\nabla P$$

$$\rho\frac{\mathrm{D}}{\mathrm{D}t}\left(h + \frac{1}{2}\vec{U}\vec{U}\right) = \frac{\mathrm{D}P}{\mathrm{D}t}$$

equation is inserted in under stationary flow conditions to yield:

$$\vec{U}\times\vec{\omega} + T\nabla s = \nabla\left(h + \frac{1}{2}\vec{U}\vec{U}\right)$$

If flow is considered along a flow line, then $\nabla(h + 1/\vec{U}.\vec{U})$ is a vector vertical to the considered flow line. $\vec{U}\times\vec{\omega}$ is also a vector and lies also vertical to the flow line. Thus $T\nabla s$ as well lies vertical to the fluid motion along a flow line, therefore it can be stated:

$$U_n\omega_n + T\frac{\mathrm{d}s}{\mathrm{d}n} = \frac{\mathrm{d}}{\mathrm{d}n}\left(h + \frac{1}{2}\vec{U}\vec{U}\right)$$

when $\left(h+\frac{1}{2}\vec{U}\vec{U}\right)$ is constant along a flow field, then $\frac{d}{dn}\left(h+\frac{1}{2}\vec{U}\vec{U}\right)$ thus:

$$U_n\omega_n + T\frac{ds}{dn} = 0$$

If $\omega_n = 0$ then $ds/dn = 0$, thus rotation-free flows are isentropic and vice versa. If the flow is assumed to be stationary and in the absence of viscosity,the inertial forces turn out to be zero.

Further Forms of the Energy Equation

The close connection between fluid mechanics and thermodynamics becomes clear from different forms of the energy equation, The notation are as follows:

Symbols	*Explanation*	*Dimension*
c_p	Heat capacity at constant pressure, per mass unit	$L^2/(Tt^2)$
$c\upsilon$	Heat capacity at constant volume, per mass unit	$L^2/(Tt^2)$
$\dot{e}_{total}$	Total energy of the fluid, per unit mass	ML^2/t^2
e	Internal energy, per unit mass	ML^2/t^2
g, g_i	External mass acceleration	L/t^2
G	Potential energy, potential of G	ML^2/t^2
h	Enthalpy	ML^2/t^2
P	Pressure field	$M/(Lt^2)$
q, q_i	Heat flow per unit area	M/t^3
T	Absolute temperature	T
$\vec{U}$, U_i	Velocity field	L/t
V	Volume	L^3
x_i	Cartesian coordinates	L
β	Thermal expansion coefficient	$1/T$
ρ	Fluid density field	M/L^3
τ, τ_{ij}	Molecular momentum transport	$M/(Lt^2)$

Mass conservation (Continuity Equation)

Equations in vector and tensor notation	*Notes/observations*
$\frac{D\rho}{Dt} = -\rho(\nabla.\vec{U})$	for $\frac{D\rho}{Dt} = 0; (\nabla.u) = 0$
$\frac{D\rho}{Dt} = -\rho\frac{\partial U_i}{\partial x_i}$	or $\frac{\partial U_i}{\partial x_i} = 0$

Equation of Motion (Momentum Equation)

Special Form	*Equations in vector and Tensor Notation*	*Notes/observations*
Imposed convection	$\rho\frac{D\vec{U}}{Dt} = -\nabla P - [\nabla.\tau] + \rho\vec{g}$ $\rho\frac{DU_j}{Dt} = -\frac{\partial P}{\partial x_j} - \frac{\partial \tau_{ij}}{\partial x_i} + \rho g_j$	For $\tau = 0$ one obtains Euler equations
Free convection	$\rho\frac{D\vec{U}}{Dt} = -[\nabla.\tau] - \rho\beta\vec{g}\nabla T$ $\rho\frac{DU_j}{Dt} = -\frac{\partial \tau_{ij}}{\partial x_i} - \rho\beta g_j \nabla T$	This equation comprises approximation by Boussinesque assumptions

Energy Equations

Special Form	*Equations in vector and tensor notation*	*Notes/ observations*
Written for $e_{\text{total}} = e + \frac{1}{2}\vec{U}^2 + G$	$\rho\frac{De_{total}}{Dt} = (\nabla.\vec{q}) - (\nabla.\rho\vec{U}) - (\nabla.[\tau.\vec{U}])$ $\rho\frac{De_{total}}{Dt} = -\frac{\partial \dot{q}_i}{\partial x_i} - \frac{\partial (PU_i)}{\partial x_i} - \frac{\partial (\tau_{ij}U_j)}{\partial x_i}$	Exact only for G time independent
$e + \frac{1}{2}\vec{U}^2$	$\rho\frac{D(e+\frac{1}{2}\vec{U}^2)}{Dt} = (\nabla.\vec{q}) - (\nabla.\rho\vec{U}) - (\nabla.[\tau.\vec{U}]) + \rho(\vec{U}.\vec{g})$ $\rho\frac{D(e+\frac{1}{2}U_i^2)}{Dt} = -\frac{\partial \dot{q}_i}{\partial x_i} - \frac{\partial (PU_i)}{\partial x_i} - \frac{\partial (\tau_{ij}U_j)}{\partial x_i} + \rho U_i g_i$	

Special Form	*Equations in vector and tensor notation*	*Notes/observations*
	$\rho\frac{D\frac{1}{2}\vec{U}_i^2}{Dt} = -(\vec{U}.\nabla P)$	

	$-(\vec{U}.[\nabla.\tau])+\rho(\vec{U}.\vec{g})$	
$\frac{1}{2}\vec{U}^2$	$\rho\frac{D\frac{1}{2}\vec{U}_i^2}{Dt}=-U_i\frac{\partial P}{\partial x_i}-U_i\frac{\partial \tau_{ij}}{\partial x_i}+\rho U_i g_i$	
e	$\rho\frac{De}{Dt}=-\nabla.\vec{q})-P(\nabla.\vec{U})-(\tau:\nabla\vec{U})$ $\rho\frac{De}{Dt}=\frac{\partial \dot{q}_i}{\partial x_i}-P\frac{\partial U_i}{\partial x_i}-\tau_{ij}\frac{\partial U_i}{\partial x_j}$	the P Term containing is zero for $\frac{D\rho^{\cdot}}{Dt}=0$
h	$\rho\frac{Dh}{Dt}=-(\nabla.\vec{q})-P(\tau:\nabla\vec{U})+\frac{DP}{Dt}$ $\rho\frac{Dh}{Dt}=-\frac{\partial \dot{q}_i}{\partial x_i}-\tau_{ij}\frac{\partial U_i}{\partial x_j}+\frac{DP}{Dt}$	
Written $c\upsilon$ and T	$\rho c_\upsilon\frac{DT}{Dt}=-(\nabla.\vec{q})-T\left(\frac{\partial P}{\partial T}\right)_\rho$ $(\nabla.\vec{U})+\frac{DP}{Dt}-(\tau:\nabla\vec{U})$ $\rho c_\upsilon\frac{DT}{Dt}=-\frac{\partial \dot{q}_i}{\partial x_i}-T\left(\frac{\partial P}{\partial T}\right)_\rho\left(\frac{\partial P_i}{\partial x_i}\right)$ $-\tau_{ij}\frac{\partial U_i}{\partial x_j}$	For and ideal Gas $\left(\frac{\partial P}{\partial T}\right)_\rho=\frac{P}{T}$
Written four c_p and T	$\rho c_p\frac{DT}{Dt}=-(\nabla.\vec{q})+\left(\frac{\partial \ln V}{\partial \ln T}\right)_\rho\frac{DP}{Dt}$ $\rho c_p\frac{DT}{Dt}=-\frac{\partial \dot{q}_i}{\partial x_i}+\left(\frac{\partial \ln V}{\partial \ln T}\right)_\rho\frac{DP}{Dt}$ $-\tau_{ij}\frac{\partial U_i}{\partial x_j}$	For an ideal Gas $\left(\frac{\partial \ln V}{\partial \ln T}\right)_\rho=1$

TRANSPORT EQUATION FOR CHEMICAL SPECIES

In many domains of engineering science investigations of fluids with chemical reactions are required which make it necessary to enlarge the considerations carried out to-date. It is necessary to state the basic equations of fluid mechanics for the different chemical components:

- Local modification of the mass of the $\frac{\partial \rho A}{\partial t}\rho V_{\Re}$ chemical component A.

- Modification of the mass of component A $-\frac{\partial}{\partial x_i}\rho A(U_A)_i \delta V_{\Re}$ by inflow and outflow
- Production of the chemical component A $r_A \delta V_{\Re}$ by chemical reaction

Thus yields a mass balance:

$$\frac{\partial \rho_A}{\partial t}\delta V_{\Re} = -\frac{\partial}{\partial x_i}[\rho_A (U_A)_i]\delta V_{\Re} + r_A \delta V_{\Re}$$

and the equation for the mass conservation for the chemical component A of a fluid results as:

$$\frac{\partial \rho_A}{\partial t} + \frac{\partial}{\partial x_i}[\rho_A (U_A)_i] = r_A$$

For a chemical component B, as a consequence of equal considerations:

$$\frac{\partial \rho_B}{\partial t} + \frac{\partial}{\partial x_i}[\rho_B (U_B)_i] = r_B$$

The addition of these equations yields:

$$\frac{\partial \rho}{\partial t} + \frac{\partial (\rho U_i)}{\partial x_i} = 0$$

i.e. the total mass conservation equation for a mixture of different components is equal to the continuity equation for a fluid which consists of one chemical component only. By considering Fick's law of diffusion it can be stated:

$$\frac{\partial \rho_A}{\partial t} + \frac{\partial}{\partial x_i}(\rho_A U_i) = \frac{\partial}{\partial x_i}\left[\rho D_{AB}\frac{\partial (C_A/C)}{\partial x_i}\right] + r_A$$

For ρ = const and D_{AB} = const, one obtains:

$$\frac{\partial \rho_A}{\partial t} + \rho_A \overbrace{\frac{\partial U_i}{\partial x_i}}^{=0} + U_i \frac{\partial \rho_A}{\partial x_i} = D_{AB}\frac{\partial^2 \rho_A}{\partial x_i} + r_A$$

or expressed in terms of concentration, C_A

$$\frac{DC_A}{Dt} = \left[\frac{\partial C_A}{\partial t} + U_i \frac{\partial C_A}{\partial x_i}\right] = D_{AB}\frac{\partial^2 C_A}{\partial x_i^2} + R_A$$

with $r_A = A \cdot R_A$.

Chapter 5

Gas Dynamics

Gas dynmaics is a sub-domain of fluid mechanics which deals with the motion of gases at high velocities. The study of gases in motion. In general, matter exists in any of three states: solid, liquid, or gas. Liquids are incompressible under normal conditions; water is a typical example. In contrast, gases are compressible fluids; that is, their density varies depending on the pressure and temperature. The air surrounding a high-speed aircraft is an example.

Gas dynamics can be treated in a variety of ways. One such way deals with gases as a continuum. The structure of gases on the particle level is called rarefied gas dynamics. Gases in motion are subject to certain fundamental laws. These are the laws of the conservation of mass, momentum, and energy. In the case of the dynamics of incompressible fluids, it is usually sufficient to satisfy only the laws of conservation of mass and momentum. This distinction constitutes the fundamental difference between high-speed gas dynamics and hydrodynamics. If irreversibilities are involved, a fourth equation called the entropy balance equation may be considered. Whereas mass, momentum, and energy are conserved, the entropy is not. Real problems are irreversible; that is, losses such as friction are involved. However, as a first approximation such effects are generally not considered.

To understand why this is important, we must consider the structure of a gas. Air, the medium of greatest interest to us, is a mixture of gases. It is roughly 79% nitrogen, a gas composed of diatomic molecules, 20% Oxygen, another diatomic gas, and 1% of other gases such as Argon, Carbon di-oxide etc. which do not significantly alter the properties of air.

At any temperature above absolute zero, the molecules of a gas are in constant motion. The speed of this "random thermal motion" varies from molecule to molecule, and instant to instant. The variation occurs through collisions between molecules. At any given instant, one expects to find molecules traveling at many different speeds and in all directions. If one finds the mean velocity of all these molecules, one comes up with the mean velocity of the gas as a whole: if it is flowing, there is a finite mean velocity. If it is just contained in a stagnant chamber, the mean may be close to zero. However, if one finds the mean speed, i.e., ignore the directions, and just see how fast things are moving, one gets a value which is proportional to the temperature. In fact, what we call "temperature" is a measure of the kinetic energy of the molecules. Thus, the greater the temperature, the greater the speed of random thermal motion of molecules.

Pressure is the force, per unit area, felt on a surface due to the "momentum flux" through the surface. If our surface is an imaginary one in a gas, then the "momentum flux" is the rate at which molecules are carrying "momentum", i.e., their own mass times their velocity, across the surface. If it is a solid surface where we feel the pressure, this is due to the momentum transferred to the surface, when the molecules collide with the surface and change direction. Thus pressure is related to how fast the molecules are moving (the temperature), how massive the molecules are (the molecular weight), and how many molecules there are, per unit volume (number density, related to density).

The temperature and the type of gas determine the "speed of sound". The speed of sound is the speed at which the smallest imaginable disturbances in pressure, travel through an undisturbed medium. It is a property of the medium. For disturbances to propagate, molecules must collide with each other, transferring momentum and kinetic energy. Thus, the speed of sound is roughly equal to the mean speed of the molecules! In fact, the relation is:

Speed of Sound = square root of (γ RT) where γ called the "ratio of specific heats" of the gas, and R is the "gas constant" of the gas. The gas constant is obtained by dividing the Universal Gas Constant (8314 in SI units) by the molecular weight in kilograms per kilogram-mole. Typically, in air at 0 deg. Celsius (273.2 deg. K), the speed of sound is roughly 340 meters per second.

Gravitational forces and their influence on flows can be neglected here for the most part. Considerations of the pressure differences to be expected by gravitational forces in gas flows show that this is justified:

$$\Delta P = -\rho g_j x_j = \rho g \Delta z,$$

which can be determined for an ideal gas ($P = \rho RT$) with the following relation:

$$\frac{\Delta P}{P} = g\frac{\Delta z}{P} \approx 9{,}81\frac{\Delta z}{287T}\frac{\Delta z}{287T}\left[\frac{m}{s^2}\frac{ms^2K}{m^2K}\right].$$

When inserting $T \approx 293$ K it can be seen that the relative pressure changes that depend on gravitation assume values around 1% only when vertical displacements of about 100 m occur. As gas-dynamic considerations are usually restricted to installations of much smaller dimensions, it is justified to simplify the flow-mechanical basic equations for considerations in gas dynamics by neglecting the gravitational forces. For many flow-mechanical considerations it is permissible to regard gaseous fluids also as incompressible when the occurring fluid velocities are small as compared to the sound velocity of the fluid. This can be explained for a stagnation point flow by the following consideration:

$$P_S = P_\infty + g\frac{\Delta_\infty^2}{2} \text{ with } \rho = \text{const}$$

When the compressibility of a fluid is to be considered, e.g. for an ideal gas at high flow velocities, for the stagnation pressure under adiabatic conditions holds, where in the framework of the stream tube theory gasdynamic considerations were carried out already:

$$P_S = P_\infty\left(1+\frac{k-1}{2k}\frac{\rho_\infty}{P_\infty}U_\infty^2\right)^{\frac{k}{k-1}}$$

or written as series expansion:

$$P_S = P_\infty\left[1+\frac{\rho_\infty}{2P_\infty}U_\infty^2+\frac{1}{2k}\left(\frac{\rho_\infty U_\infty^2}{P_\infty}\right)^2+\cdots\right]$$

From this follows that for k ≈ 1, 4 and $T \approx 293$, there have to exist velocities in the range of 70 m/s before stagnation pressures of incompressible and compressible fluids differentiate by about 2% As gas dynamics is the science of compressible fluid flows, it thus treats flows at high velocities where the fluid compressibility has to be included into the considerations. Such flows are characterized by the Mach number *Ma*,, so that the compressibility of a fluid is to be considered only onwards from:

$$Ma = \frac{U}{c} = \frac{U}{\sqrt{kRT}} \geq 0{,}2$$

As the sound velocity $C\ (T)$ is given by the temperature field and for ≈ 300K, velocities of $C \approx 350$m/s exist, gas-dynamic aspects have to be considered with fluid flows only at high Reynolds numbers. Therefore it is permissible to neglect the viscosity term in the momentum equation. Thus the basic equations of flow mechanics for gas-dynamic considerations can be stated as shown below. In these equations the relations holding for ideal gases were introduced for the internal energy: $e = c_a T$ and c_0 = const:

Continuity equation

$$\frac{\partial\rho}{\partial t}+\frac{\partial(\rho U_i)}{\partial x_i} = 0,$$

Momentum equation (j = 1, 2, 3)

$$\rho\left[\frac{\partial U_j}{\partial t}+U_i\frac{\partial U_j}{\partial x_i}\right] = -\frac{\partial P}{\partial x_i},$$

Energy equation

$$\rho c_v\left[\frac{\partial T}{\partial t}+U_i\frac{\partial T}{\partial x_i}\right] = -P\frac{\partial U_j}{\partial x_i},$$

where the energy equation) is given for adiabatic fluid flows. Together with the state equation for ideal gases, a closed system of differential equations exists which can be solved for given boundary conditions. Gas-dynamics problems are thus solvable in principle.

The possible solutions require special considerations, however, the appearance of high flow velocities is linked to specific phenomena which differentiate gas dynamics sharply from other sub-domains of flow mechanics.

As the following considerations will show, the presence of high Mach numbers, $Ma = U/c$, leads to the emergence of "discontinuity surfaces" (compression shocks) in which the pressure (and other flow quantities) experience a jump. This makes a special procedure necessary when solving flow problems, as the employment of the differential form of the basic equations usually requires that the quantities describing a flow are steady in the flow area.

There is also the fact that when treating fluid flows at high Mach numbers, processes occur that are linked to different time scales. These are the time scales of the diffusion Δt_{Diff}, the convection Δt_{Conv} and the sound propagation Δt_{Sound}:

$$\Delta t_{\text{Diff}} = \Delta t_{\text{conv}} = \frac{L_c}{U_c}; \Delta t_{\text{Sound}} = \frac{L_c}{c}.$$

For Δt_{Conv} Δt_{Diff} it results:

$$\text{Re} = \frac{\Delta t_{\text{Diff}}}{\Delta t_{\text{Conv}}} = \frac{U_c \cdot L_c}{\nu} \gg 1$$

i.e. during the time in which a flow covers a certain distance, the molecular transport at high flow velocities manages only to overcome a negligible distance, i.e. at high Re-numbers the formation of thin boundary layers comes about. The latter are neglected in the introducing considerations presented here. From the point of view of characteristic times it results analogously for the Mach number:

$$Ma = \frac{\Delta t_{\text{Conv}}}{\Delta t_{\text{Sound}}} = \frac{U_c}{c},$$

i.e. the Mach number shows how quickly flows proceed in fluids, relative to the characteristic velocity of the propagation of small disturbances, i.e. the sound velocity c. c depends on the thermo-dynamic change of state and can be stated as follows:

$$k = c_p/c_v = \text{Relation of the heat capacities}$$

$$c = \sqrt{kRT} \quad R = \text{Specific gas constant}$$

$$T = \text{Absolute temperature}$$

This context between the sound velocity and the thermodynamical state quantities pressure and density can be presented as follows: Considered is the propagation of a small (i.e. isentropic) disturbance at the velocity c in a fluid at rest.

This is a non-stationary process which by changing the reference system (the observer moves together with the flow) can be modified into a stationary problem, Now the momentum equation can be employed as a balance of forces at a control volume around the disturbance:

$$F\,[P - (P + dP\,)] = \rho Fc[(c + dU) - c]$$

The equation gives the following relation:

$$-dP = \rho c dU$$

For the mass conservation it can be stated:

$$\rho Fc = (r + dr)(c + dU)F$$

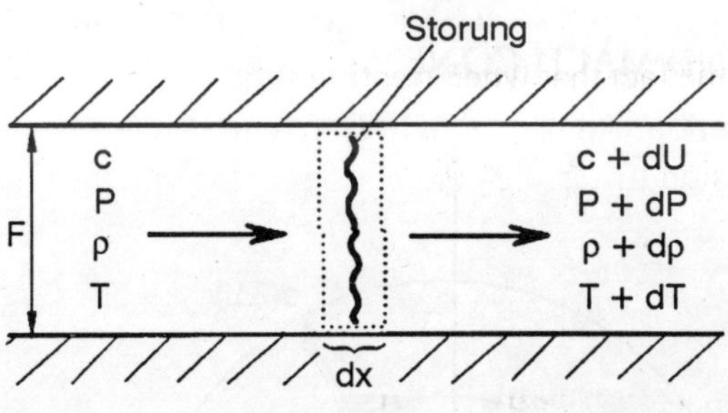

Fig. Propagation of a Disturbance in a Compressible Fluid

so that it holds:

$$d\rho = -\rho\frac{dU}{c}$$

Thus it also holds:

$$\frac{dP}{d\rho} = c^2 = \left(\frac{\partial P}{\partial \rho}\right)_{\text{ad}}$$

as no heat exchange is included in the present considerations.

The sound velocity is therefore a local quantity, i.e. it depends on the locally existing pressure changes. With the local value $c(x_i, t)$ the local Mach number can be computed at each point of a flow field $U_j(x_i, t)$, so that the corresponding Mach number field can be assigned to the flow field, i.e. $Ma(x_i,$

t.) This local *Ma*-number expresses essentially how quickly at each point of the flow field disturbances propagate relative to the existing flow velocity.

From a historic point of view it is interesting that Newton was the first scientist to compute the sound velocity for gases, although on the assumption of an isothermal process in which no temperature changes occur due to the sound propagation. He thus obtained in his considerations:

$$c_{\text{Newton}} = \sqrt{\frac{P}{\rho}} = \sqrt{RT} < c$$

Only a full century later Marquis de Laplace corrected the result of Newton's computations by recognizing that the temperature fluctuations produced by sound disturbances and also the temperature gradients connected with them are very small. Laplace recognized that it is not possible to radiate the heat . The $\sqrt{k}$ -correction of Newton's formula introduced by Laplace led to the correct propagation velocity of sound waves in ideal gases:

$$c = \sqrt{kRT}$$

Attention is drawn once again to the fact that via this formula a sound velocity field also $T(x_i, t)$ is assigned to each temperature field $c(x_i, t)$ of an ideal gas.

MACH LINES AND MACH CONE

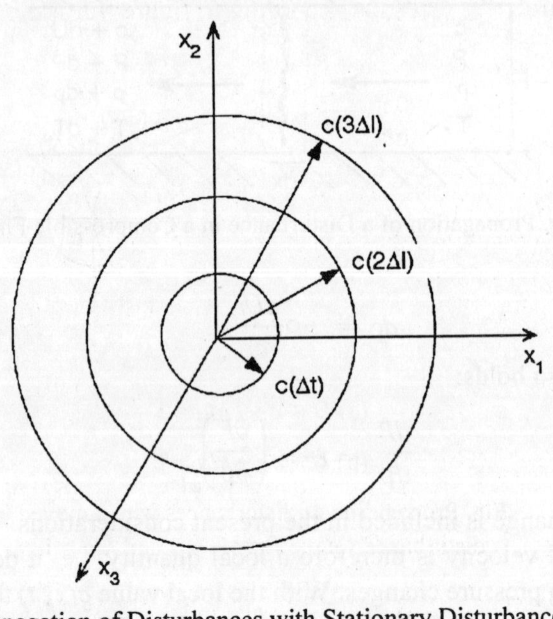

Fig. Propagation of Disturbances with Stationary Disturbance Source

When considering a disturbance emanating from a point source in the origin of a coordinate system, same will propagate radially at the velocity c if the point source does not carry out any motion, i.e. the surfaces of disturbance of the same phase represent spherical surfaces when the propagation takes place in a field of constant temperature.

When there is on the other hand a temperature field with variations of temperature, these variations are reflected as deformations of the spherical surfaces. The propagation takes place more rapidly in the direction of high temperatures, as is predicted in equation. Possible temperature distributions impair thus the symmetry of the propagation of sound waves.

When one extends now the considerations of the propagation of disturbances to moving disturbance sources of small dimensions, for U < c, i.e. $Ma < 1$ and U > c, i.e. $Ma > 1$ By moving the disturbance source at a velocity smaller than the propagation velocity of the disturbances as diagrammed in figure a propagation image results which does not show any more the "symmetric propagation of the disturbances" will be obtained in the propagation direction of the disturbance source, a concentration of the emitted disturbance waves takes place, i.e. an observer standing in positive x_1-direction will register a frequency increase of the disturbance as compared to the starting frequency of the disturbance source. In the opposite direction on the other hand a frequency decrease takes place toward the emitted disturbance.

When one computes this frequency change for the frequency increase in positive x_1-direction, it results according to figure for $Ma < 1$:

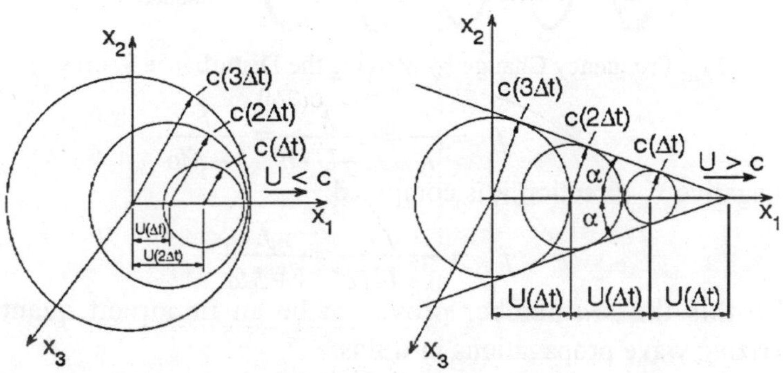

(a) $U < c \rightsquigarrow Ma < 1$

(b) $U < c \rightsquigarrow Ma > 1$

Fig. Propagation of disturbances with a moved disturbance source for $Ma < 1$ and $Ma > 1$

$\lambda' = \dfrac{c - U_i \cdot \ell_i}{f}$ or with $U_i \cdot \ell_i = U$ where $\lambda' = \dfrac{(c-U)}{f}$ and ℓ_i states the unit vector in the direction of propagation (f = frequency of the disturbance)

Thus it holds for f'.

$$f' = \frac{f}{1 - U_i \cdot \ell_i / c} = \frac{f}{1 - (U/c)}$$

c ist die Ausbreitungsgeschwindingkeit deremittierten welle.

$$\lambda' = \frac{[c - U_i \cdot \ell_i]}{f}$$

$$f' = \frac{f}{\left[1 - \frac{1}{c} U_i \ell_i\right]}$$

$$f' - f = \Delta\nu = \frac{\nu}{c} \frac{U_i \ell_i}{\left[1 - \frac{1}{c} U_i \ell_i\right]}$$

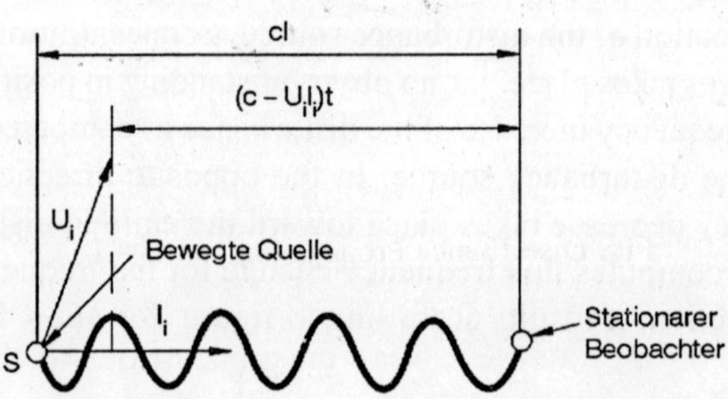

Fig. Frequency Change by Moving the Disturbance Source

$$f' = \frac{c}{\lambda'} = \frac{f}{1 - U/c} = \frac{f}{1 - Ma}$$

In negative x_1-direction it is computed:

$$f'' = \frac{f}{1 + U/c} = \frac{f}{1 + Ma}$$

With this the *Ma*-number proves to be an important quantity for characterizing wave propagations in fluids.

When the case occurs that the velocity of the moving disturbance source is equal to the propagation velocity of the disturbance, it results that the propagation of the disturbances in relation to the disturbance source is possible only in the half-plane from which the movement of the source takes place.

In the case that the velocity of the disturbance source exceeds the propagation velocity of the disturbances, a propagation image develo*ps* which is shown in figure. The propagation of the disturbances in relation to the moving disturbance source takes place within a cone, the socalled Mach cone. In front of the cone a disturbance-free area results which is strictly separated from the area with disturbances within the Mach cone. From considerations that are diagrammed in figure it results as half the aperture angle a of the cone:

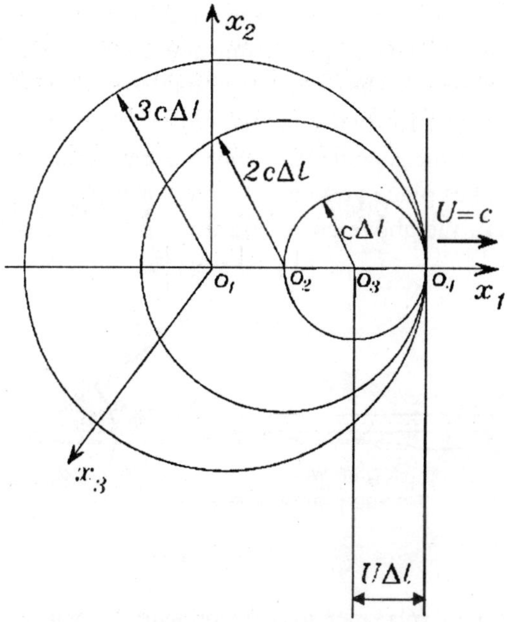

Fig. Disturbance Propagation at *Ma* = 1

$$\sin\alpha \frac{c\Delta t}{U\Delta t} = \frac{1}{Ma}$$ where α = half the angle of the Mach cone.

The above formula, employing Fig. is derived from the following quantities:

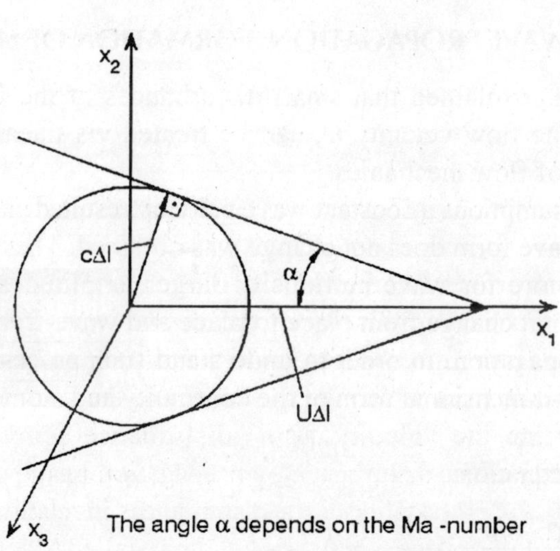

Fig. Aperture Angle of the Mach Cone

$c\Delta t$ = propagation of the disturbance in the time Δt

$U\Delta t$ = path of the disturbance source in the time Δt

When considerations are carried out in the two-dimensional sphere, the Mach cone represents two lines crossing one another which are defined as Mach lines or Mach waves The considerations stated above for spatial motions can easily be employed for two-dimensional problems also. They show that propagations of two-dimensional disturbances occur in the form of plane waves. The propagation takes place vertically to the wave planes.

With the aid of the above considerations observations can be explained that

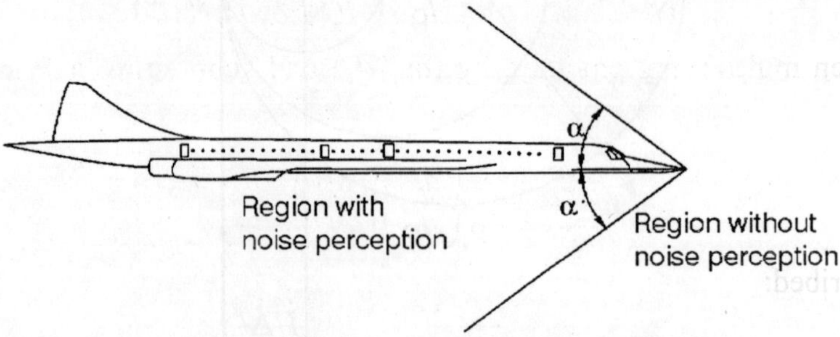

Fig: Explanation for Perception of Aeroplanes

can be made in relation to the flight of supersonic aeroplanes. Aeroplanes of this kind show a region in which the aeroplane cannot be heard, i.e. an observer can perceive an aeroplane flying towards him at supersonic speed much earlier with the eye than he can hear it. Only when the observer is within the Mach cone, he succeeds in seeing and hearing the aeroplane.

NON-LINEAR WAVE PROPAGATION, FORMATION OF SHOCK WAVES

There it was explained that small disturbances of the fluid properties ρ', P', T' or of the flow velocity u', can be treated via linearizations of the basic equations of flow mechanics.

On these assumptions a constant wave velocity resulted and a propagation where a given wave form does not change was obtained. These properties are not given any more for wave motions of larger amplitudes, so that wave velocities form that change from place to place and wave-fronts develop that deform with propagation. In order to understand such processes it is best to consider the one-dimensional form of the continuity and momentum equation with $U = U_1$, $x = x_1$:

Continuity equation:

$$\frac{\partial \rho}{\partial t} + \frac{\partial \rho}{\partial x} + \rho \frac{\partial U}{\partial x} = 0$$

Momentum equation:

$$\frac{\partial U}{\partial t}+U\frac{\partial U}{\partial x} = -\frac{1}{\rho}\frac{\partial P}{\partial x}$$

The following relation results for $\rho = \rho(U)$:

$$\frac{d\rho}{dU}\frac{\partial U}{\partial t}+U\frac{d\rho}{dU}\frac{\partial U}{\partial x}+\rho\frac{\partial U}{\partial x} = 0$$

Analogously, equation can be written:

$$\frac{\partial U}{\partial t}+U\frac{\partial U}{\partial t}+\frac{1}{\rho}\left(\frac{dP}{d\rho}\right)\left(\frac{d\rho}{dU}\right)\frac{\partial U}{\partial x} = 0$$

When multiplying equation by $(d\rho/dU)$ and subtracting it from one obtains:

$$\rho\frac{\partial U}{\partial x} = \frac{1}{\rho}\left(\frac{dP}{d\rho}\right)\left(\frac{d\rho}{dU}\right)^2\frac{\partial U}{\partial x}$$

or transcribed:

$$\frac{dU}{d\rho} = \pm\frac{1}{\rho}\sqrt{\left(\frac{dP}{d\rho}\right)} = \pm\frac{1}{\rho}\sqrt{\left(\frac{\partial P}{\partial \rho}\right)_{\text{ad}}}$$

This equation can now be integrated:

$$\int_0^{U_1} dU = \pm\int_{\rho_\infty}^{\rho}\sqrt{\left(\frac{dP}{d\rho}\right)}\frac{d\rho}{\rho}$$

In consideration of $P/\rho^k = \text{const}$ can be integrated::

$$U = \pm\int_{\rho_\infty}^{\rho}\sqrt{k\cdot\text{const}}\cdot\rho^{\frac{k-1}{2}}\frac{d\rho}{\rho} = \pm\frac{2}{k-1}\left[\sqrt{k\rho^{k-1}\text{const}}\right]_{\rho_\infty}^{\rho}$$

$$U = \pm\frac{2}{(k-1)}(a-c)$$

Thus for the propagation velocity of a wave of large amplitude:

$$a = c\pm\frac{(k-1)}{2}U$$

results a propagation velocity "“*a*”", which depends on the local flow velocity. Here c is the computed sound velocity for the undisturbed fluid.

When inserting equation one obtains the following relation:

$$\frac{\partial U}{\partial t} + U \cdot \frac{\partial U}{\partial x} \pm \sqrt{\left(\frac{dP}{d\rho}\right)} \frac{\partial U}{\partial x} = 0$$

or transcribed:

$$\frac{\partial U}{\partial t} + (U \pm a)\frac{\partial U}{\partial x} = 0$$

From the continuity equation one obtains:

$$\frac{\partial \rho}{\partial t} + (U \pm a)\frac{\partial \rho}{\partial x} = 0$$

so that for ρ the following general solution of the differential equation can be stated:

$$\rho = F_\rho\,(x_1 - (U_1 \pm a)) = F_\rho\left(x_1 - \left(c \pm \frac{k+1}{2}U_1\right)t\right)$$

where F_ρ () is a random function. Analogously it holds for the velocity:

$$U_1 = Fu\,(x_1 - (U_1 \pm a)) = F_u\left(x_1 - \left(c \pm \frac{k+1}{2}U_1\right)t\right)$$

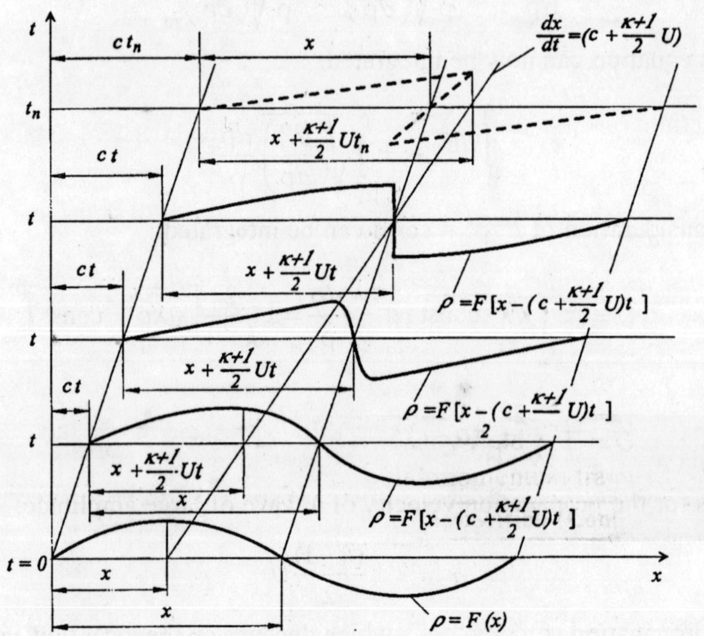

Fig: Wave Deformations and Formation of Compression Shocks

The relations allow to explain the propagation of a disturbance with a propagation velocity of $c \pm \frac{k+1}{2} U_1$ Because of this propagation velocity which depends on the local flow velocity, wave deformations develop as they are indicated. When considering the propagating part with the (+) –sign, then characteristic position changes in times t can be stated as follows:

$$x'_A = c \cdot t_\omega ;\; x'_B = x_A + ct_a + \frac{k+1}{2} Ut_a ;\; x'_C = x_B + ct_b .$$

The developing and progressive deformation of the wave is apparent. Thus the formation of compression forms comes about.

The local ambiguity of the density stated for t_n can of course not occur. When the wave front has built up in a way that all thermo-dynamic quantities of the fluid and also the velocity experience sudden changes, the maximum deformation possible of the propagating flow is reached. A compression shock has built up.

ALTERNATIVE FORMS OF THE BERNOULLI EQUATION

An introduction of the stream tube theory into the treatment of fluid flows one-dimensional isentropic flows were dealt with, employing the Bernoulli equation for incompressible flows:

$$\frac{1}{2} U_1^2 + \frac{k}{(k-1)} \frac{P}{\rho} = \frac{k}{(k-1)} \frac{P_H}{\rho H}$$

As thermo-dynamically possible maximum velocity it was determined, for $(P/\rho) \rightarrow 0$:

$$(U_{\max})^2 = \frac{2k}{(k-1)} \frac{P_H}{\rho H} = \frac{2k}{(k-1)} RT_H$$

so that it holds:

$$\frac{1}{2} U_1^2 = \frac{1}{2} U_{\max}^2 - \frac{k}{(k-1)} \frac{P}{\rho}$$

As the *Ma*-number represents a fundamental quantity in the treatment of gas-dynamic flow problems, it can be written:

$$1 = \left(\frac{U_{\max}}{U_1} \right)^2 - \frac{2k}{(k-1)} \frac{RT}{U_1^2} = \left(\frac{U_{\max}}{U_1} \right)^2 - \frac{2}{(k-1)} \frac{1}{Ma^2}$$

or transcribed:

$$\frac{1}{Ma} = \frac{k-1}{2} \left[\left(\frac{U_{\max}}{U_1} \right)^2 - 1 \right]$$

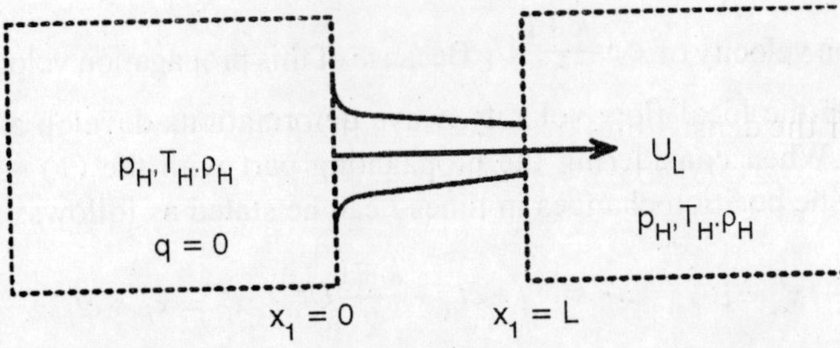

Fig: Compensation Flow between two Pressure Tanks

At the base of the above considerations was an expansion flow as it is indicated. For this flow results the so-called critical state, when $U_1 = c = U_c$ is reached, i.e. when the following relation holds:

$$\frac{1}{2}U_c^2 + \frac{U_c^2}{(k-1)} = \frac{c_H^2}{(k-1)} \qquad U_c^2 = \frac{2k}{(k+1)}RT_H$$

For the critical pressure it can be computed according to equation, in consideration of equation:

$$U_c^2 = \frac{2k}{(k+1)}RT_H\left[1-\left(\frac{P_c}{P}\right)^{\frac{k-1}{k}}\right] = \frac{2k}{(k+1)}RT_H$$

$$\frac{P_c}{P_H} = \frac{P*}{P_H} = \left[\frac{2}{(k+1)}\right]^{\frac{2k}{(k_1-1)}}$$

Employing the relations for isentropic density and temperature changes, one obtains:

$$\left(\frac{\rho_c}{\rho_H}\right) = \frac{\rho*}{\rho_H} = \left(\frac{P*}{P_H}\right)^{1/k} = \left[\frac{2}{(k+1)}\right]^{\frac{1}{(k-1)}}$$

$$\left(\frac{T_c}{T_H}\right) = \frac{T*}{T_H} = \left(\frac{P*}{P_H}\right)^{\frac{k-1}{k}} = \frac{2}{(k+1)}$$

The Mach number can now be employed to express the pressure, temperature and density changes, that are possible as a result of the Bernoulli equation for compressible media, as a function of the Mach number:

$$\frac{1}{2}U_1^2 + \frac{c^2}{(k-1)} = \frac{c_H^2}{(k-1)} \rightsquigarrow \frac{k-1}{2}Ma^2 + 1 = \frac{T_H}{T}$$

or transcribed:

$$\left(\frac{T}{T_H}\right) = \left[1 + \frac{(k-1)}{2} Ma^2\right]^{-1}$$

For the density and pressure variations the following relations can be derived:

$$\left(\frac{\rho}{\rho_H}\right) = \left(\frac{T}{T_H}\right)^{\frac{1}{k-1}} = \left[1 + \frac{(k-1)}{2} Ma^2\right]^{\frac{-1}{(k-1)}}$$

$$\left(\frac{P}{P_H}\right) = \left(\frac{T}{T_H}\right)^{\frac{k}{(k-1)}} = \left[1 + \frac{(k-1)}{2} Ma^2\right]^{\frac{-k}{(k-1)}}$$

For the sound velocity relation c/c_H it results:

$$\frac{c}{c_H} = \left(\frac{T}{T_H}\right)^{\frac{1}{2}} = \left[1 + \frac{k-1}{2} Ma^2\right]^{-2}$$

The above-cited relations are plotted in the diagram presented laterally. Each of the shown curves represents the Bernoulli equation, which expresses how

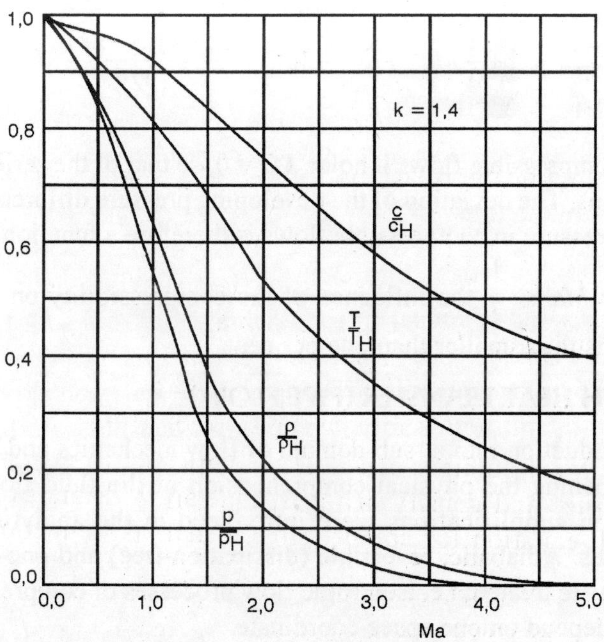

Fig. Diagram for Representing the Parameter Variations in the Bernoulli Equation

he sound velocity, the temperature, the pressure and the density, each standardized with the corresponding values, change in the high-pressure tank when the Mach-number changes are known. The temperature, density and pressure variations (U_1/U_{max}) were employed as a parameter of the representation.

It is a characteristic of compressible flows that the local stagnation pressure of a flow

$$\frac{1}{2}\rho U_1^2 = \frac{1}{2}\rho c^2 Ma^2 \frac{1}{2}\rho\left(\frac{kP}{\rho}\right)Ma^2 = \frac{1}{2}kPMa^2$$

depends on the locally existing pressure and the local Mach number. For the standardized pressure difference it holds:

$$\frac{P_H - P}{\frac{1}{2}\rho U_1^2} = \frac{2}{kMa^2}\frac{P_H - P}{P} = \frac{2}{kMa^2}\left[\frac{P_H}{P} - 1\right]$$

with (P_H/P) from equation we obtain:

$$\frac{P_H - P}{\frac{1}{2}\rho U_1^2} = \frac{2}{kMa^2}\left[\left(1+\frac{k-1}{2}Ma^2\right)^{\frac{k}{(k-1)}} - 1\right]$$

Via a series expansion for $Ma^2 < \frac{2}{(k-1)}$ it results:

$$\frac{P_H - P}{\frac{1}{2}\rho U_1^2} = 1 + \frac{1}{4}Ma^2 + \frac{2-k}{24}Ma^4 + \frac{(2-k)(3-2k)}{192}Ma^6 + \cdots$$

For incompressible flows it holds $Ma = 0$, so that of the series expansion only 1 remains. The deviation of the developing pressure differences from the stagnation pressure in compressible flows is therefore a function of the Mach number. For $Ma < \frac{1}{3}$ the influence of the compressibility on the pressure distribution is thus smaller than one per cent.

FLOW WITH HEAT TRANSFER (PIPE FLOW)

An introduction into a sub-domain of flow mechanics and in particular aims at deepening the physical comprehension of the fluid flows. For this purpose often simplifications were introduced in the analytical problem considerations. Adiabatic, reversible (dissipation-free) and one-dimensional fluid flows were treated, i.e. isentropic flow processes of compressible media which only depend on one space coordinate.

These considerations need some supplementary explanation in order to be able to understand special phenomena in the case of flowswith heat transfer. For dealing with such flows which can be considered as stationary and one-dimensional, i.e. experience changes only in the flow direction $x_1 = x$ the following basic equations are at disposal which are stated by $U_1 = U$

- Mass conservation:

$$\rho F U = \dot{m} = \text{const}$$

- Momentum equation:

$$\rho U \frac{dU}{dx} = -\frac{dP}{dx}$$

- Energy equation:

$$(dq) = c_v dT + P d\left(\frac{1}{\rho}\right) = c_p dT - \frac{1}{\rho} dP$$

- State equation for ideal gases:

$$\frac{P}{\rho} = RT$$

From the mass-conservation equation one obtains:

$$\frac{d\rho}{\rho} + \frac{dU}{U} + \frac{dF}{F} = 0$$

or for pipe flows with $\frac{dF}{F} = 0$:

$$\frac{dU}{U} = -\frac{d\rho}{\rho}$$

From the ideal gas equation it can be derived:

$$\frac{dP}{P} = \frac{d\rho}{\rho} + \frac{dT}{T}$$

and from the momentum equation one obtains $-\frac{d\rho}{\rho} = U\,dU$ or

$$-\frac{dP}{P} = \frac{\rho}{P} U_1 dU_1 = \frac{1}{RT} U^2 \frac{dU}{U}$$

With $kRT = c^2$ and from the momentum equation one obtains:

$$-\frac{dP}{P} = \frac{\kappa}{c^2} U^2 \frac{dU}{U} = \kappa Ma^2 \frac{dU}{U}$$

When finally including the energy equation into the considerations, it can be stated:

$$(dq) = c_p dT - \frac{dP}{\rho} = c_p dT + U dU$$

or transcribed:

$$\frac{dU}{U}=\frac{(dq)}{U^2}-\frac{c_p dT}{U^2}=\frac{1}{Ma^2}\left(\frac{c_p}{\kappa RT}\right)\frac{(dq)}{c_p}=\frac{1}{Ma^2}\frac{c_p}{\kappa RT}dT$$

i.e. it holds for the relative velocity change in a pipe flow as a result of heat supply:

$$\frac{dU}{U}=\frac{1}{(\kappa-1)Ma^2}\left(\frac{(dq)}{h}-\frac{dT}{T}\right)$$

where $h=c_p T$ was set. From equation it follows:

$$\frac{dT}{T}=\frac{dP}{P}-\frac{d\rho}{\rho}=-\kappa Ma^2\frac{dU}{U}+\frac{dU}{U}$$

or transcribed:

$$\frac{dT}{T}=(1-\kappa Ma^2)\frac{dU}{U}$$

This relation inserted in equation yields:

$$\frac{dU}{U}=\frac{1}{(\kappa-1)Ma^2}\left(\frac{(dq)}{h}-(1-\kappa Ma^2)\frac{dU}{U}\right)$$

Solved in terms of $\frac{dU}{U}$ one obtains:

$$\frac{dU}{U}=\frac{1}{(1-Ma^2)}\frac{(dq)}{h}$$

This relation inserted in) yields for the relative density change:

$$\frac{d\rho}{\rho}=\frac{-1}{(1-Ma^2)}\frac{(dq)}{h}$$

or for the relative changes in pressure and temperature it holds:

$$\frac{dP}{P}=\frac{-\kappa Ma^2}{(1-Ma^2)}\frac{(dq)}{h} \quad \text{and} \quad \frac{dT}{T}=\frac{(1-\kappa Ma^2)}{(1-Ma^2)}\frac{(dq)}{h}$$

For the local change of the *Ma* number it can also be derived:

$$\frac{d(Ma^2)}{Ma^2}=\frac{d(U^2/c^2)}{(U^2/c^2)}=\frac{T}{U^2}d\left(\frac{U^2}{T}\right)=2\frac{dU}{U}-\frac{dT}{T}$$

Thus for the change of the *Ma*-number with heat supply it holds:

$$\frac{dMa^2}{Ma^2}=\frac{(1+\kappa Ma^2)}{(1-Ma^2)}\frac{(dq)}{h}$$

As $(dq)=T\cdot ds$ and $h=c_p\cdot T$ it holds furthermore:

$$\frac{dMa^2}{Ma^2}=\frac{(1+\kappa Ma^2)}{(1-Ma^2)}\frac{ds}{c_p}$$

The above relations can now be employed for understanding how P, T, ρ, U and Ma change locally when one conveys heat to a pipe flow; i.e. $dq/h > 0$:

Subsonic Flow:

$\frac{dU}{U} > 0$; the flow velocity increases with heat supply

$\frac{d\rho}{\rho} > 0$ and $\frac{dP}{P} < 0$; density and pressure decrease with heat supply.

$\frac{dT}{T} > 0$; the temperature increases with heat supply for $Ma < \sqrt{\frac{1}{\kappa}}$.

$\frac{dT}{T} < 0$; the temperature decreases in spite of heat supply for $Ma > \sqrt{\frac{1}{\kappa}}$.

$\frac{dMa^2}{Ma^2} > 0$; the local Ma-number increases with heat supply

The above relations indicate that in spite of heat supply there is a decrease in temperature for $\sqrt{1/\kappa} < 1$.

Supersonic Flow:

$\frac{dU_1}{U_1} < 0$; the flow velocity decreases with heat supply.

$\frac{d\rho}{\rho} > 0$ and $\frac{dP}{P} > 0$; density and pressure increase with heat supply.

$\frac{dT}{T} > 0$; the temperature increases with heat supply.

$\frac{dMa^2}{ma^2} < 0$; the local Ma-number decreases with heat transfer.

The change of fluid-mechanical and thermo-dynamical state quantities in a pipe flow in principle takes place in a different way in the supersonic range than in the subsonic region.

When for deepening the physical comprehension one considers the occurring processes in the T-s-diagram for an ideal gas, one obtains:

$$(dq)_v = c_v \cdot dT_v = T \cdot ds_v \rightsquigarrow \left(\frac{\partial T}{\partial s}\right)_v = \frac{T}{c_v}$$

$$(dq)_P = c_P dT_P = T ds_P \rightsquigarrow \left(\frac{\partial T}{\partial s}\right)_P = \frac{T}{c_P}$$

From equation one obtains for the temperature change in a pipe flow with heat supply:

$$\frac{dT}{T}=\frac{(1-\kappa Ma^2)}{1-Ma^2)}\frac{dq}{h}=\frac{(1-\kappa Ma^2)}{(1-Ma^2)}\frac{Tds_R}{c_pT}$$

From this it is computed:

$$\left(\frac{\partial T}{\partial s}\right)_{Pipe}=\frac{T}{c_p}\frac{(1-\kappa Ma^2)}{(1-Ma^2)}=\left(\frac{\partial T}{\partial s}\right)_R$$

When introducing now an effective heat capacity $c_{Rohr}=c_R$ it holds, so gilt:

$$(dq)_R=c_R dT_R=Tds_R \rightsquigarrow \left(\frac{\partial T}{\partial s}\right)_R=\frac{T}{c_R}$$

and c_R is computed as:

$$c_R=c_p\frac{(1-Ma^2)}{(1-\kappa Ma^2)}=T\frac{1}{\left(\frac{\partial T}{\partial s}\right)_R}$$

With $k=\frac{c_p}{c_v}$ it can also be written:

$$c_R=\frac{c_p\cdot(1-Ma^2)}{\left(\frac{1}{\kappa}-Ma^2\right)}=T\frac{1}{\left(\frac{\partial T}{\partial s}\right)_R}$$

Thus it holds:

$$\frac{\left(\frac{\partial T}{\partial s}\right)_P-\left(\frac{\partial T}{\partial s}\right)_R}{\left(\frac{\partial T}{\partial s}\right)_V-\left(\frac{\partial T}{\partial s}\right)_R}=\frac{\frac{T}{c_p}-\frac{T}{c_R}}{\frac{T}{c_v}-\frac{T}{c_R}}=\frac{c_R-c_p}{c_R-c_v}$$

and further transcribed:

$$\frac{\left[\left(\frac{c_R}{c_p}\right)-1\right]}{\left[\frac{c_R}{c_p}-\frac{1}{\kappa}\right]}=\frac{1-Ma^2-1+\kappa Ma^2}{\kappa-\kappa Ma^2-1+\kappa Ma^2}=\frac{(\kappa-1)Ma^2}{(k-1)}=Ma^2$$

$$Ma^2=\frac{\left(\frac{\partial T}{\partial s}\right)_p-\left(\frac{\partial T}{\partial s}\right)_R}{\left(\frac{\partial T}{\partial s}\right)_v-\left(\frac{\partial T}{\partial s}\right)_R}=\frac{A}{B}$$

The relations expressed by equation are shown graphically. Here $(\partial T/\partial s)_p$

signifies the gradient of the isobars in the T-s-state di-agram and $(\partial T/\partial s)_v$ the gradient of the isochors and $(\partial T/\partial s)_R$ the change- of -state curve of the pipe flow with heat supply.It can now be shown that equation holds generally, not only for the flows of ideal gases generally treated in gas dynamics, but also for the flows of real gases.

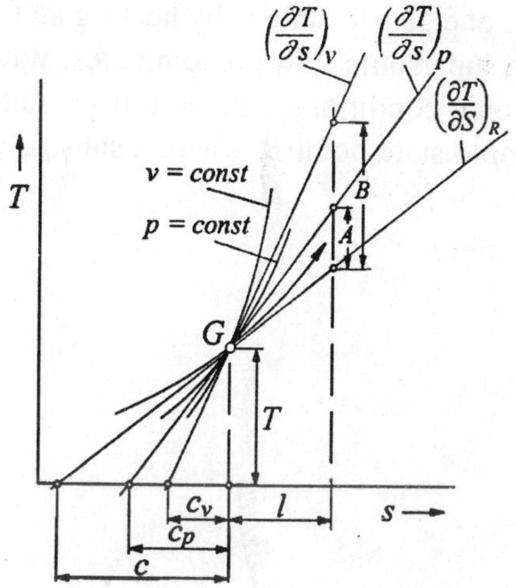

Fig. Change of state in the T-s-diagram for pipe flows with heat supply

In conclusion it shall be remarked that the relations for $\frac{dU}{U}$, $\frac{d\rho}{\rho}$, $\frac{dP}{P}$, $\frac{dT}{T}$ and $\frac{dMa^2}{Ma^2}$ for $Ma = 1$ lose their validity, if $(dq) = 0$ When one wants to get a subsonic flow via heat supply to sound velocity and then to supersonic flow, at the place $Ma = 1$ there has to be the heat supply $(dq) = 0$ After that it is necessary to cool the flow in order to obtain a further velocity increase.

Extended considerations show that the heat supply in the subsonic region leads to accelerating the flow, and in the supersonic region to delaying the flow.

For pipe flows with a radius R = const, a subsonic flow cannot be transferred/converted to a supersonic flow with steady heat supply.

When considering the course of the effective heat capacity of the pipe flow:

$$\frac{c_R}{c_v} = \frac{(Ma^2 - 1)}{(Ma^2 - 1/k)}$$

$0 \le Ma < \sqrt{1/\kappa}$ and $1 \le Ma < \infty$. In the *Ma*-number range $\sqrt{1/\kappa} < Ma <$ 1 The effective specific heat is thus positive for $Ma = \sqrt{1/\kappa}$ the effective heat capacity is negative. At $c_R \to \infty$. tends to go to infinity. At $Ma = \sqrt{1/\kappa}$ the local flow velocity has the value of the isothermal sound velocity. Because of the relation for the effective heat capacity c_R/c_v for the developments of subsonic and supersonic flows in heated and cooled pipe flows the thermodynamic state behaviours, at the same thermo-dynamical initial state. Starting from state A, one gets to state C by heating and after that by cooling up to state B, where a supersonic flow predominates. When on the other hand heating up at supersonic conditions, one gets from state point A to C, and after that by cooling into state point B, where a subsonic flow predominates.

Chapter 6

Hydrostatics and Aerostatics

HYDROSTATICS

Hydrostatics deals with the laws to which fluids are subjected that do not show motions in hydrostatic coordinate system in which the considerations are carried out, i.e. fluids which are at rest in the coordinate system employed for the considerations. General laws of fluid motions they are also applicable to the case of fluids at rest, i.e. non-flowing fluids. Thus from the continuity equation

$$\frac{\partial \rho}{\partial t}+\frac{\partial}{\partial x_i}(\rho U_i)=0$$

It can be shown that for $\rho = \text{const}$ and $U_i \neq f(x_i)$ the continuity equation is given by:

$$\underbrace{\frac{\partial \rho}{\partial t}+U_i\frac{\partial \rho}{\partial x_i}}_{D\rho/Dt=0}+\rho\underbrace{\frac{\partial U_i}{\partial x_i}}_{=0}=0.$$

This means that for $U_i = 0$ the simple partial differential equation holds:

$$\frac{\partial \rho}{\partial t}=0$$

whose general solution can be stated as follows:

$\rho = F\,(x_i)$.

The density ρ in a fluid at rest thus is only a function of the spatial coordinates x_i. When time variations of the density of the fluid occur, these lead inevitably to motions within the fluid because of the relation between the flow and density fields attributable to the continuity equation.

The general equations of momentum can be expressed as:

$$\rho\left[\frac{\partial U_j}{\partial t}+U_i\frac{\partial U_j}{\partial x_i}\right]=-\frac{\partial U_j}{\partial x_i}-\frac{\partial \tau_{ij}}{\partial x_i}+\rho g_j$$

and its special form is deduced for a fluid at rest ($U_j = 0$ and moleculardependent momentum transport $\frac{\partial \tau_{ij}}{\partial x_i}$ = 0) to the following system of partial 1606 Hydrostatics and Aerostatics differential equations. This represents the set of basic equations of hydrostatics and aerostatics:

$$\frac{\partial P}{\partial x_j} = \rho g_j \; (j = 1, 2, 3)$$

or written out for all three directions:

$$\frac{\partial P}{\partial x_1} = \rho g_1, \quad \frac{\partial P}{\partial x_2} = \rho g_2, \quad \frac{\partial P}{\partial x_3} = \rho g_3.$$

In this section the pressure distribution in a fluid, mainly defined by the field of gravity, will be considered more closely. Restrictions are made concerning the possible fluid properties; the fluid is assumed to be incompressible for hydrostatics, i.e ρ = const. This condition is in general quite well fulfilled by liquid, so that the following derivations can be considered as valid for liquids.

For the derivation of the pressure distribution in liquid at rest a rectangular Cartesian coordinate system is introduced, whose position is chosen such that the mass acceleration $\{g_i\}$ given by the field of gravity only shows one component in the negative x_2 direction, i.e. the following vector holds

$$\{g_i\} = \{0, -g, 0\}.$$

Then the differential equations given above generally for the pressure can be written as follows:

$$\frac{\partial P}{\partial x_1} = 0, \quad \frac{\partial P}{\partial x_2} = -\rho g, \quad \frac{\partial P}{\partial x_3} = 0.$$

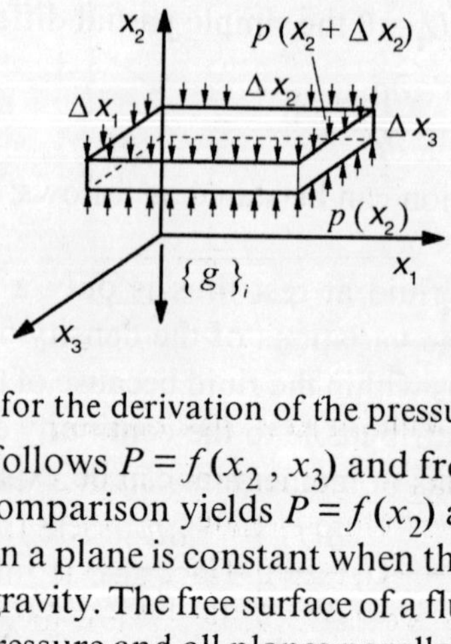

Fig. Coordinate system for the derivation of the pressure distribution in fluids

From $\partial P/\partial x_1 = 0$ follows $P = f(x_2, x_3)$ and from $\partial P/\partial x_3 = 0$ follows $P = f(x_1, x_2)$. Thus a comparison yields $P = f(x_2)$ and this shows that the pressure of a fluid within a plane is constant when the same is vertical to the direction of the field of gravity. The free surface of a fluid stored in a container is a plane of constant pressure and all planes parallel to it are also planes of constant pressure. The pressure increases in the direction that was defined by g_j i.e. in the g_j direction of the gravitational acceleration.

For the physical comprehension of hydrostatics it is also important to recognize that equation expresses that the increase in pressure in the negative x_2 direction is caused by the weight of the fluid element plotted in figure i.e.

$$-\rho g \overbrace{\Delta x_1 \Delta x_2 \Delta x_3}^{\Delta V} + p \overbrace{\Delta x_1 \Delta x_3}^{\Delta A} - (p + \frac{\partial p}{\partial x_2} \Delta x_2) \overbrace{\Delta x_1 \Delta x_3}^{\Delta A} = 0.$$

Employing the above physical insights and the resultant equations, the following statements can be made for a liquid of constant density located in a container. In the case that the field of gravity acts in the negative x_2 direction, i.e.

$$g_1 = 0, \quad g_2 = -g, g_3 = 0$$

the differential equations stated in with the solution $P = f(x_2)$ hold for this case.

Thus the partial differential $\partial P / \partial x_2$ can be written as total differential and one obtains for constant density fluids (ρ =const):

$$\frac{dP}{dx_2} = -\rho g \longrightarrow P - \rho g x_2 + C.$$

or rewritten

$$\frac{P}{\rho} + g x_2 = C$$

This relationship expresses that the sum of the "pressure energy" P/ρ and

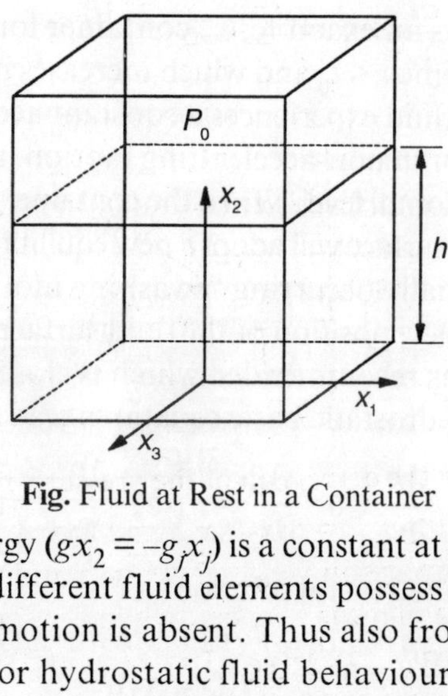

Fig. Fluid at Rest in a Container

the potential energy ($gx_2 = -g_j x_j$) is a constant at each point of a fluid at rest. As all points of different fluid elements possess the same total energy, the driving force for motion is absent. Thus also from the energy point of view the conditions for hydrostatic fluid behaviour exist.

When the fluid in the height h has a free surface on which an equally large pressure P_0 acts at all points, it represents, because of the relation $P = f(x_2)$, a plane x_2 = const, i.e. a horizontal plane.

For the pressure distribution one obtains with the boundary condition $P = P_0$ for $x_2 = h \rightsquigarrow C = P_0 + \rho g h$

$$P = P_0 + \rho g(h - x_2)\ 0 \leq x_2 \leq h.$$

This relationship expresses the known hydrostatic law, according to which the pressure in a fluid increases in a linear way with the depth below the free surface.

When one rewrites equation one obtains:

$$\frac{P_0}{\rho} + gh = \frac{P}{\rho} + gx_2 = \text{const}$$

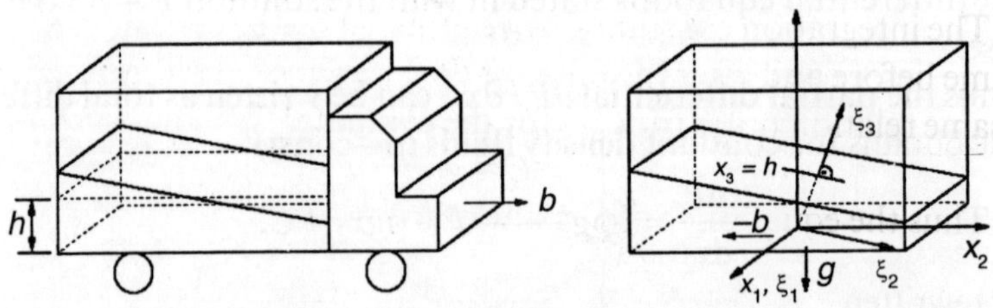

Fig. Position of the fluid level at constant acceleration

The laws of hydrostatics are often applicable also to fluids in moving containers when one treats these as "'accelerated reference systems"'. The externally imposed accelerative forces are then to be introduced as inertia forces. Figure shows as an example, a "container lorry" filled with a fluid which is at rest at the time $t < t_0$ and which increases its speed linearly at for all times $t \geq t_0$, i.e. the fluid experiences a constant acceleration.

At a state of rest or in non-accelerating motion, the fluid surface in the container forms a horizontal level. When the container experiences a constant acceleration b, the fluid surface will adopt a new equilibrium position, provided one disregards the initially occurring "swashing motions". When one now wants to compute the new position of the fluid surface, the introduction of a coordinate system x_i, is recommended which is closely connected with the container, where the hydrostatic basic equations read as follows:

$$\frac{\partial P}{\partial x_1} = 0\,; \quad \frac{\partial P}{\partial x_2} = -\rho b\,; \quad \frac{\partial P}{\partial x_3} = -\rho g\,.$$

From this results the general solution:

$$\frac{\partial P}{\partial x_1} = 0 \qquad p = f_1(x_2, x_3),$$

$$\frac{\partial P}{\partial x_2} = -\rho b \quad P = -\rho b x_2 + f_2(x_1, x_3),$$

$$\frac{\partial P}{\partial x_3} = -\rho g \quad P = -\rho g x_3 + f_3(x_1, x_2).$$

By comparing the solutions one obtains that f_1, f_2, f_3 can only be the sum of the terms obtained by partial integration plus a constant:

$$P = C - \rho(bx_2 + gx_3).$$

Along the free surface exists the pressure $P = P_0$ and thus the equation of the plane in which the free surface lies reads:

$$x_3 = -\frac{b}{g}x_2 + \frac{1}{g\rho}(C - P_0) \text{ for } -\infty < x_1 < +\infty.$$

The integration constant C is determined by the condition that the fluid volume before and after the onset of the acceleration is the same. Therefore the same relation holds for C as for the container with the fluid at rest:

$$C = g\rho h + P_0.$$

Thus the equation for the plane of the free surface reads:

$$x_3 = h - \frac{b}{g}x_2 \quad \text{for } -\infty < x_1 < +\infty.$$

As the solution of the problem has to be independent of the chosen coordinate system, a coordinate system ξ_i can be introduced which is rotated against the system x_i in such a way that the following equations for the coordinate transformations hold:

$\xi_1 = x_1$ (axis of rotation),

$$\xi_2 = \frac{1}{\sqrt{b^2 + g^2}}(gx_2 + bx_3)$$

and

$$\xi_3 = \frac{1}{\sqrt{b^2 + g^2}}(bx_2 + gx_3)$$

This is equivalent to the introduction of a resulting acceleration of the quantity $\sqrt{b^2 + g^2}$ in the direction ξ_3. Thus the basic hydrostatic equations read:

Thus $P = F(x_3)$ holds and $P = \rho\sqrt{b^2 + g^2}\xi_3$

The integration constant C results from the boundary condition:

$$P = P_0 + g\rho h \text{ for } \xi_3 = 0$$

$$P = P_0 + \rho g\left(h - \sqrt{1+\left(\frac{b}{g}\right)^2}\,\xi_3\right)$$

All further statements concerning the problem of the accelerated fluid container can also be made in the coordinate system ξ_i. Along the free surface $P = P_0$ holds and

$$\xi_3 = \frac{h}{\sqrt{1+\left(\frac{b}{g}\right)^2}}$$

Thus is the equation of the plane in which the free surface lies.

By the above treatment it becomes clear that it is possible to employ the hydrostatic basic laws also in accelerated reference systems, provided the inertia forces are taken into consideration that are attributable to the external motions. The occurring accelerations (inertial and gravitational) are to be added to a total acceleration in vectorial manner in order to obtain the direction and the quantity of the total acceleration. The free surface appears vertical to the vector of the total acceleration.

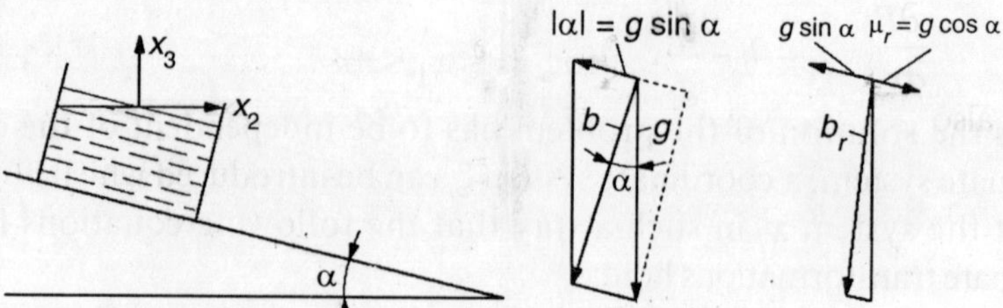

Fig. Water Container Sliding Down on an Inclined Plane.
Motion With and Without Friction

The example can also be categorized into the group of examples that can be treated by means of the basic laws of hydrostatics. This figure shows a water container which is sliding down an inclined plane with an angle of inclination α with respect to the horizontal plane. The container at rest posseses a water surface which is horizontal, as only the gravitational acceleration appears as inertia force per kg of fluid. When the fluid container is released and when the acceleration directed downwards is $|b| = g \sin\alpha$, the body starts moving and experiences in this way an acceleration which is parallel to the inclined plane. The resulting acceleration component acting on the fluid is composed of the component directed upwards with $|b| = g$ sina and the component directed downwards with $\mu_r\, g \cos\alpha$ Here μ_r is the friction coefficient which characterizes the interaction between the container bottom and the surface of the inclined plane.

When one treats at first the accelerated motion occurring downwards on the inclined plane without friction, one obtains in the coordinate system indicated the following set of hydrostatic basic equations:

$$\frac{\partial P}{\partial x_1} = 0$$

$$\frac{\partial P}{\partial x_2} = -\rho g \sin \alpha \cos \alpha$$

$$\frac{\partial P}{\partial x_3} = -\rho g(1 - \sin^2\alpha).$$

The position for the pressure distribution in the container sliding downwards and thus also the solution for the position/location of the fluid surface can be obtained by the solution of the equations.

From $\frac{\partial P}{\partial x_1} = 0$ follows on the one hand $P = f(x^2, x^3)$ and thus the following holds:

$$\frac{\partial P}{\partial x_2} = -\frac{1}{2}\rho g \sin(2\alpha) \longrightarrow P = f_1(x_3) - \frac{1}{2}\rho g \sin(2\alpha) x_2$$

and also

$$\frac{\partial P}{\partial x_3} = -\rho g \cos^2 \alpha \longrightarrow P = f_2(x_2) - \rho g(\cos^2 \alpha) x_3$$

By comparing the solutions one obtains:

$$P = C - \frac{1}{2}\rho g(\sin(2\alpha) x_2 + 2(\cos^2 \alpha) x_3)$$

Along the free surface $P = P_0$ holds and thus one obtains as solution for the location of the free surface:

$$x_3 = (\tan \alpha) x_2 + \frac{1}{\rho g \cos^2 \alpha}(C - P_0) \text{ for } -\infty < x_1 < +\infty.$$

As the origin of the coordinates also lies on the free surface $C = P_0$ follows and thus for the plane in which the free surface lies holds:

$$x_3 = -(\tan \alpha) x_2 \qquad \text{for } -\infty < x_1 < +\infty.$$

This equation shows that for friction-free sliding along the inclined plane the free surface lies parallel to the plane along which the container slides. This can also be derived from considerations of the left acceleration diagram in which it can be seen that the resulting acceleration "'b"' is located vertically with respect to the inclined plane.

When one adds for the downwards motion the occurring frictional force, one obtains the following set of hydrostatic basic equations:

$$\frac{\partial P}{\partial x_1} = 0$$

$$\frac{\partial P}{\partial x_2} = -\rho g(\sin \alpha - \mu_r \cos \alpha) \cos \alpha$$

$$\frac{\partial P}{\partial x_3} = -\rho g[1 - (\sin \alpha - \mu_r \cos \alpha) \sin \alpha].$$

Thus the solution corresponding to the equation reads:
$P = C - \rho g[(\sin \alpha - \mu_r \cos \alpha) \cos \alpha]x_2 - \rho g[1-(\sin \alpha - \mu_r \cos \alpha) \sin \alpha]x_3$.

If one puts on the one hand $P = P_0$, for the free surface, one obtains the equation for the plane in which the free surface lies. When one takes further into consideration, that the origin of the coordinates lies again on the free surface, i.e. $C = P_0$, one obtains as final equation:

$$x_3 = -\left[\frac{(\sin\alpha - \mu_r \cos\alpha)\cos\alpha}{1-(\sin\alpha - \mu_r \cos\alpha)\sin\alpha}\right]x_2.$$

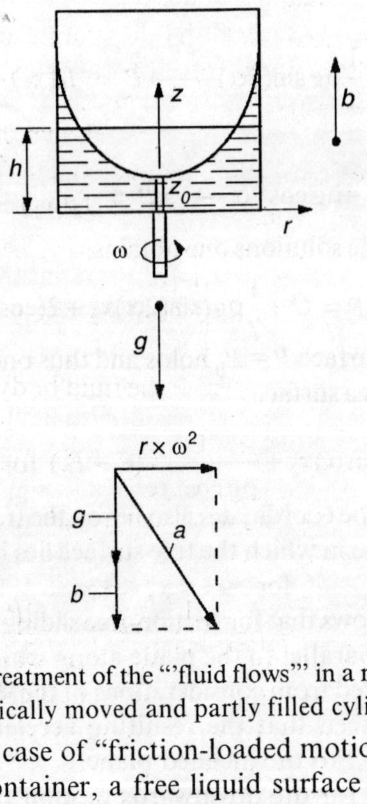

Fig. Treatment of the "'fluid flows'" in a rotating vertically moved and partly filled cylinder

For this general case of "friction-loaded motion" along the inclined plane of the fluid container, a free liquid surface appears which is less inclined when compared to the horizontal plane than the inclined plane.

Attention has to be paid, however, to the fact that the derivations only hold when $\mu_r \leq \tan\alpha$.

For $\mu_r \geq \tan\alpha$ one obtains the limiting case of a container at rest i.e. the frictional force is higher then the forward accelerating force.

As a last example to show the employment of hydrostatic laws in accelerated reference systems. It shows a rotating cylinder closed on the top and at the bottom, which is partly filled with a liquid.

When the cylinder is at rest, the free surface of this liquid assumes a horizontal position, as the different liquid particles only experience the gravitational force as mass force. When the cylinder is put into rotation, one observes a deformation of the liquid surface which progresses until as a final form paraboloid. When now on this rotating motion an additional accelerated vertical motion is superimposed, one detects that the hyperboloid can assume different

shapes, depending on the magnitude of the vertical acceleration and on the direction in which it takes place. In the following it shall be shown that the issue of the shape of the hyperboloid can be answered on the basis of the basic equations of hydrostatics. For this purpose a coordinate system is chosen, which is firmly coupled to the walls of the rotating and vertically accelerated cylinder and which thus experiences the rotating motion as well as the accelerated vertical motion.

The above mentioned examples have shown that the hydrostatic basic equations are applicable, provided that no fluid motion occurs in the chosen coordinate system and that the external acceleration forces are taken into consideration as inertia forces.

It is shown that for the following derivations the horizontally occurring centrifugal acceleration $\omega^2 r$, as well as the "'vertical acceleration"' b, have been taken into account.

If one considers the processes in the fluid body in a coordinate system (r, φ, z), rotating with the cylinder, one finds that all fluid particles are at rest after having reached the stable final state of motion. With reference to the chosen coordinate system the prerequisite for the employment of the hydrostatic basic equation is fulfilled, which in cylindrical coordinates adopt the form indicated below:

$$\frac{\partial P}{\partial r} = \rho g_r\,; \quad \frac{1}{r}\frac{\partial P}{\partial \varphi} = \rho g_\varphi\,; \quad \frac{\partial P}{\partial z} = \rho g_z\,.$$

For $g_r = r\omega^2$, $g_\varphi = 0$ and $g_z = -(g + b)$ one obtains for the problem to be treated the following set of basic equations and their general solution.

$$\frac{\partial P}{\partial r} = \rho r\omega^2 \quad \longrightarrow \quad P = \frac{1}{2}\rho\omega^2 r^2 + f_1(\varphi, z)\,,$$

$$\frac{1}{r}\frac{\partial P}{\partial \varphi} = 0 \quad \longrightarrow \quad P = f_2(r, z),$$

$$\frac{\partial P}{\partial z} = -\rho(g + b) \quad \longrightarrow \quad P = -\rho(g + b)z + f_3(r, \varphi).$$

By comparing the solutions it results in:

$$P = C + \frac{\rho}{2}\omega^2 r^2 - \rho(g+b)z \text{ for } 0 \le \varphi \le 2\pi.$$

When one introduces on the axis $r = 0$, for the position of the parabolic apex $z = z_0$, $P = P_0$ holds at the location $r = 0$ and $z = z_0$. This yields for the integration constant:

$$C = P_0 + \rho(g + b)z_0.$$

Therefore the equation for the pressure distribution in the liquid body:

$$P = P_0 + \frac{\rho}{2}\omega^2 r^2 - \rho(g+b)(z - z_0) \text{ for } 0 < \varphi < 2\pi.$$

Along the free surface of the liquid the following holds for the pressure $P = P_0$, so that the free surface employing, can be represented as follows:

$$z = z_0 + \frac{\omega^2}{2(g+b)}r^2 \quad \text{for } 0 \le \varphi \le 2\pi.$$

The introduced apex position z_0 can be determined from the condition that the liquid volume before the rotations starts, i.e $\pi R^2 h$, has to be equal to the liquid volume which exists, in rotation between the free surface of the liquid and the cylinder walls. Thus the following holds:

$$pR^2 h = 2\pi\int_0^R rz\,\mathrm{d}r = 2\pi\int_0^R r\left[z_0 + \frac{\omega^2}{2(g+b)}r^2\right]\mathrm{d}r$$

and carrying out the integration yields:

$$\frac{1}{2}R^2 h = \left[\frac{1}{2}z_0 r^2 + \frac{\omega^2}{8(g+b)}r^4\right]_0^R = \frac{1}{2}R^2\left[z_0 + \frac{\omega^2}{4(g+b)}R^2\right]$$

$$z_0 = h - \frac{\omega^2}{4(g+b)}R^2$$

$$z = h - \frac{\omega^2}{4(g+b)}(R^2 - 2r^2)$$

On the basis of the above indicated relationship the different forms of the free liquid surface can now be looked at. Some typical cases. These will be discussed in the following on the basis of the above derivations and the derived final relationship. It is hoped that it becomes thus clear for the reader how physical information can be obtained by derivations on basic equations of fluid mechanics e.g. the form of the free surface of a liquids in containers can be calculated.

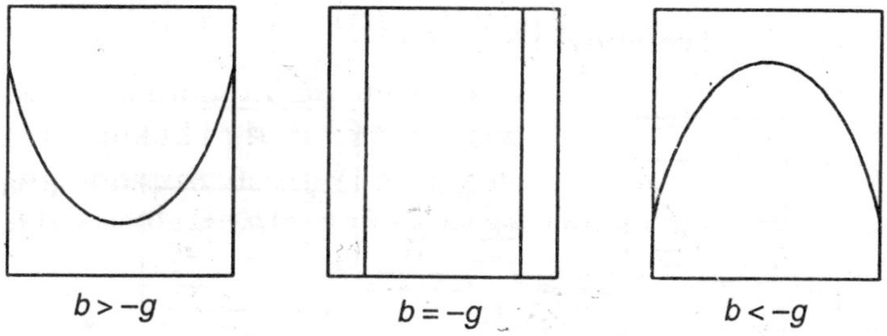

Fig. Examples of Possible Forms of the Fluid Surface in a Rotating Vertically Accelerated Cylinder

The positions of the liquid surface indicated in Figure can be stated by the indicated relationship in the relative magnitudes of b and g:

$b > -g$: When the vertical acceleration of the container takes place upwards and the resultant b points downwards, respectively, with $0 > b > -g$, the "'opening of the parabola"' is positive according to equation. The liquid touches the bottom and side areas of the container.

$b = -g$: When the vertical acceleration of the container takes place downwards with $b = -g$, the entire fluid rests at the side wall of the container.

$b < -g$: When the vertical acceleration of the container takes place all downwards with $b < -g$, the "opening of the parabola" is negative according to equation. The fluid touches the ceiling and side areas of the container this can be taken from equation.

COMMUNICATING CONTAINERS AND PRESSURE-MEASURING INSTRUMENTS

Communicating Containers

In many fields of engineering one has to deal with fluid systems that are connected to one another by transverse pipelines. Special systems are those in which the fluid is at rest, i.e. in which the fluid does not flow. Figure represents schematically such a system which consists of two containers with "fluids at rest" that are connected with one another by a pipeline with a valve.

When the valve is opened, both these systems can interact with one another in such a way that a flow takes place from the container with higher pressure at the entrance of the communication line to the container with lower pressure. When this balancing flow fails to materialize, the same fluid pressure exists on both sides of the tap, i.e. it holds:

$$p_{01} + \rho_{1g}\,(H_1 - h_1) = p_{02} + \rho_2 g\,(H_2 - h_2).$$

When there is the same fluid in both containers with $\rho_1 = \rho_2 = \rho$ and thus:

$$p_{02} - p_{01} = \rho g\,[(H_1 - H_2) - (h_2 - h_1)].$$

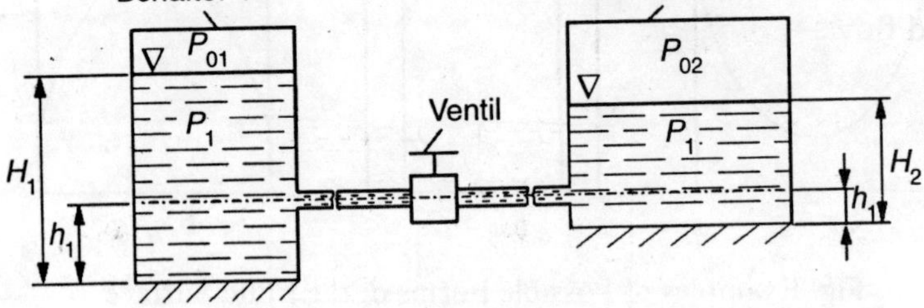

Fig. Sketch for the explanation of the pressure conditions with communicating containers

For the containers and open on top surfaces:

$$p_{02} = p_{01} = p_0$$

Here we assume that the pressure over both the free surfaces is equal and thus:

$$(H_1 - h_1) = (H_2 - h_2)$$

i.e. in open communicating containers filled with the same fluid the fluid levels take the same height with respect to a horizontal plane.

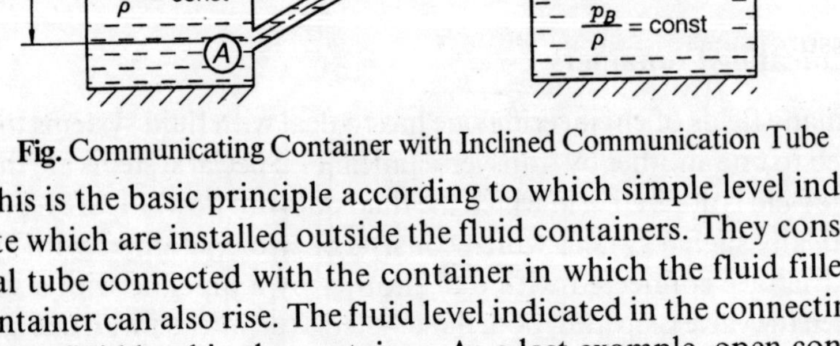

Fig. Communicating Container with Inclined Communication Tube

This is the basic principle according to which simple level indicators operate which are installed outside the fluid containers. They consist of a vertical tube connected with the container in which the fluid filled in in the container can also rise. The fluid level indicated in the connecting tube shows the fluid level in the container. As a last example, open containers are considered that are connected to one another by means of an inclined tube that is directed upwards. For these containers one finds that the fluid surfaces in both containers adopt the same level. When this final state is reached (equilibrium state) no equalizing flow takes place between the containers, although the pressure at the deeper lying end of the pipe shows a higher hydrostatic pressure at the connecting point. The reasons for the fact that equalizing flow does not come up in spite of a higher hydrostatic pressure at the deeper lying end of the pipe. The energy considerations carried out there show that the total energies of the fluid particles are the same at both ends of the pipe and thus the basic prerequisite for the start of fluid flows is missing.

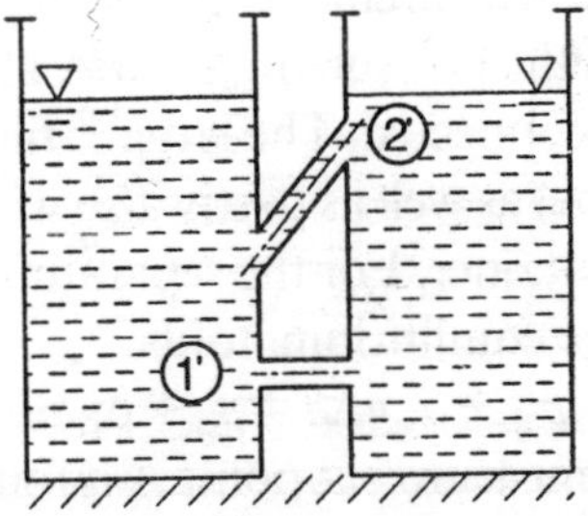

Fig. Sketch for the Consideration of the Influence/action on Fluids at Rest

The behaviour of communicating containers that are filled with fluids at rest can often be understood easily by making it clear to oneself that the pressure influence of a fluid on walls is identical at each point with the pressure influence on fluid elements which one installs instead of walls. For example the pressure distributions in the fluid container are identical with those of the same container when components are installed to obtain two partial containers connected with one another, in the case that the fluid surfaces are kept at the same level as the original level.

Owing to the installed walls the pressure conditions do not change in the right container as compared to the left container. The container areas installed at the left replace the pressure influence of the fluid particles omitted by the walls.

Pressure-measuring Instruments

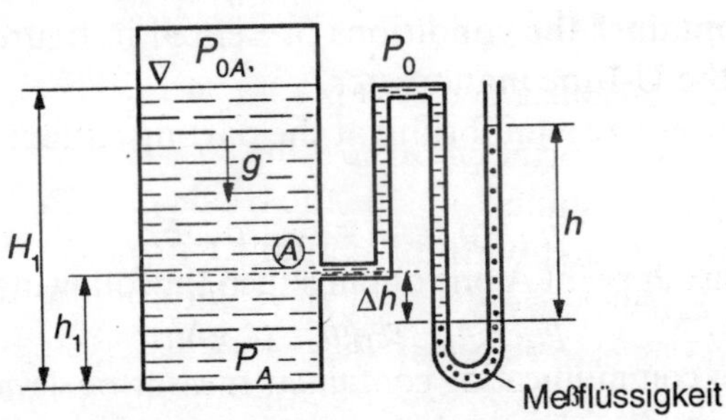

Fig. Diagram for Explaining the Basic Principle of Pressure Measurements by Communicating Systems

The insights into pressure distribution in containers gained are based on pressure relationships that were described for communicating systems. From the statements that were made about the pressures in the containers, relationships between the fluid levels could be derived.

In return it is now possible, in the case that the established fluid levels are known, to employ the general pressure relationships, in order to obtain information on the pressures occurring in containers. The basic principle according to which pressure measurements are carried out by communicating systems. To be measured is the pressure in point A of the container to which a "'U tube manometer"' is connected. The latter is filled with a measurement fluid (dark part of the U tube) as well as partly also with the fluid which enters into the U tube from the container. For the separating plane between the two fluids the following pressure equilibrium holds:

$$p_A + \rho_A g\Delta h = p_0 + \rho_F gh.$$

For the pressure to be measured at point A it follows:

$$p_A = p_0 + \rho_F gh - \rho_A g\Delta h.$$

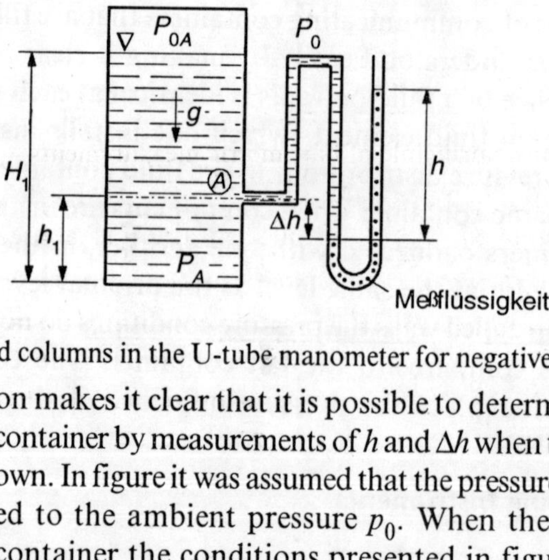

Fig. Fluid columns in the U-tube manometer for negative pressure

This equation makes it clear that it is possible to determine the pressure at point A in the container by measurements of h and Δh when the fluid densities ρ_F and ρ_A are known. In figure it was assumed that the pressure in the container is high compared to the ambient pressure p_0. When there is a negative pressure in the container the conditions presented in figure will exist for the fluid level in the U-tube manometer.

Thus for the pressure equilibrium at the parting surface of both fluids holds:

$$p_A - \rho_A g\Delta h = p_0 - \rho_F gh.$$

For the pressure at point A one obtains then the following relation:

$$p_A = p_0 - \rho_F gh + \rho_A g\Delta h.$$

On the basis of communicating containers measuring devices can also be created and employed to measure the atmospheric pressure, i.e. to carry out barometric measurements.

A system can in principle be produced as follows:

- A glass tube of a length of more than 1 m, at the lowest end of which a spherical extension of the tube section has been made, is filled with mercury to the top.
- The glass tube filled with mercury, is turned upside down into a container also filled with mercury.

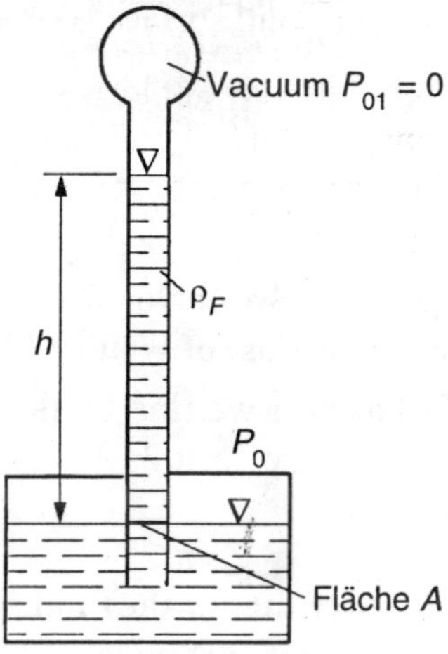

Fig. Basic principle of barometric measurements

- The level of the mercury column in the glass tube over the surface of the mercury in the external container is a measure of the barometric pressure.

$$p_0 = \rho_F g h.$$

A barometer, can be employed to verify experimentally the pressure distributions in the atmosphere.

FREE FLUID SURFACES

Surface Tension

A special characteristic of fluids is that in contrast to solids, they have no form of their own, but always adopt the form of the container in which they are put. While doing this, a free surface forms that the same shows a position which is perpendicular to the vector of the gravitational acceleration.

In this way the fluid properties under gravitational influence were formulated which are known from phenomena of every day life. It was always assumed that the fluid, at disposal, possesses a total volume having the same order of magnitude as the larger container at disposal. The fluid properties hold only when these conditions are met. This is known from observations of small quantities of liquids which form drops when put on surfaces. It is seen that different shapes of drops can form, depending on which surface and which fluid for forming drops is used. More detailed considerations show moreover that the gas surrounding the fluid and the solid surface all have an influence on the forming shape of a drop. The latter is often neglected and one differentiates considerations of fluid-solid combinations with reference to their wetting possibility, depending on whether the establishing angle of contact between fluid surface and solid surface is smaller than $\pi/2$ or larger.

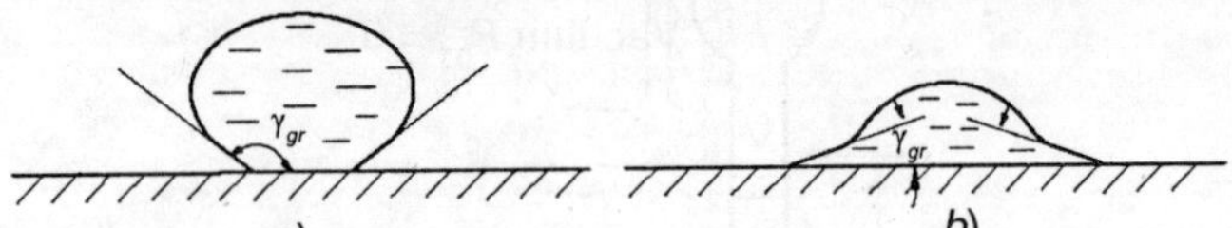

Fig. (a) Shape of Drop in the Case of Non-Wetting Fluid Surfaces; (b) Shape of Drop in the Case of Wetting Fluid Surfaces

The surface is classified as non-wetting by the fluid when

$$\gamma_{gr} > \pi/2$$

It holds furthermore that for

$$\gamma_{gr} > \pi/2$$

the surface is classified as wetting for the fluid.

Surfaces covered by a layer of fat are known as examples of surfaces that cannot be wetted by water. Cleaned glass surfaces are to be classified as wetting for many fluids. The above phenomena can be explained by the fact that different "'actions of forces"' can act on fluid elements. Equivalent physical considerations can be made also owing to the surface energy that can be attributed to free fluid surfaces. When a fluid element is located in a layer that is far away from a free fluid surface, it is surrounded from all sides by homogenous fluid molecules and one can assume that the cohesion forces occurring between the molecules annul each other.

This is, however, no longer the case when one considers fluid elements in the proximity of free surfaces. As the forces exerted by gas molecules on the water particles are negligible in comparison to the cohesion forces of the liquid, a particle lying at the free surface experiences an action of forces in direction of the fluid. "Lateral forces" also act on the fluid element which thus finds itself in an interphase boundary surface in a state of tension that attributes special characteristics to the free surface.

It is thus for example possible to deposit carefully applied flat metal components on free surfaces without fluid penetrating into them. The carrying of razor blades on water surfaces is an experiment that is often presented in basic courses of physics. In nature "pond skaters" make use of this particular property of the water surface to cross pools and ponds skillfully and quickly.

When a drop of fluid gets into contact with a firm support adhesion forces also occur in addition to the internal cohesion forces. When these adhesion forces are stronger than the cohesion forces that are typical for the fluid, we have the case of a wetting surface and water drops form. If, however, the cohesion forces are stronger, we have the case of a non-wetting surface and the shapes of the drops.

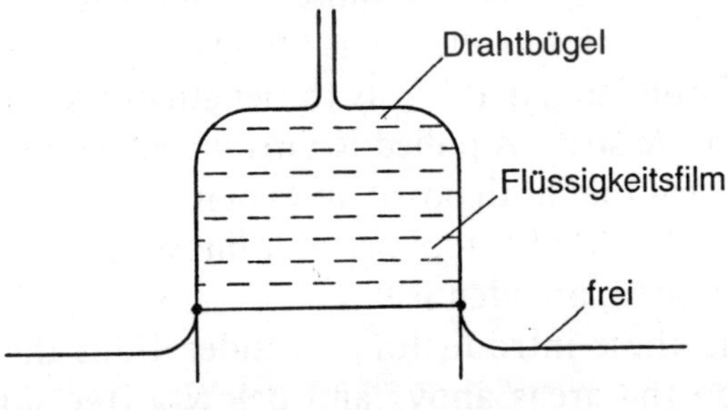

Fig. Strap experiment to prove the action of forces as a consequence of surface tension

More detailed considerations of the processes in the proximity of the free surface of a fluid show that we have to do there with a complicated transit domain (with finite extension vertical to the fluid surface) from a fluid area to a gas area. It suffices, however, for many considerations to be made in fluid mechanics to introduce the surface as a layer with a thickness of $\delta \to 0$. To the same are attributed the properties that comprise the complex transit layers between fluid and gas.

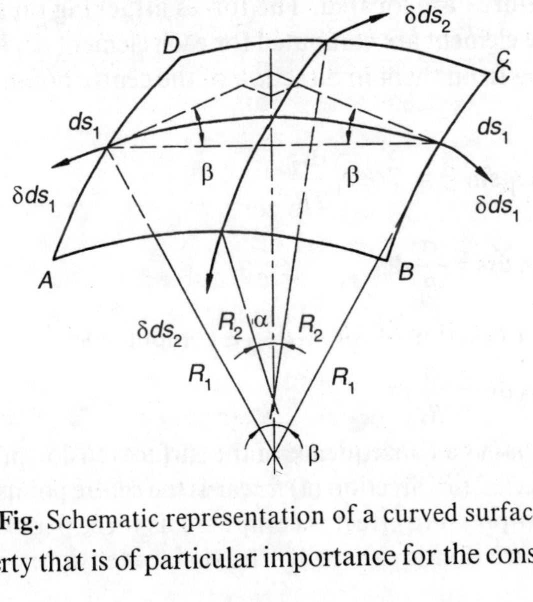

Fig. Schematic representation of a curved surface

The property that is of particular importance for the considerations to be carried out here is the surface tension. This surface tension can be proven by immersing a strap, in a fluid. When pulling the strap through the free surface upwards, one observes that this requires an action of forces which is proportional to the distance between the strap arms. The proportionality constant describing this fact is defined as surface constant. The surface tension represents thus an action of forces of the free surface per unit linear measure. It can also be introduced as the energy that is required to build up the tension in the liquid film in figure. Both introductions are identical as in both types of energy equation formulated in this way the length of the liquid film in the direction in which the strap is pulled is introduced from the energy set-up. This makes it clear that both possibilities of introduction of the surface tension, one as the action of forces per unit linear measure and other as the energy per unit area, are identical.

In concluding these introductory considerations the effect of the surface tension on the areas above and below a free surface shall be investigated. From observations of free surfaces in the middle of large containers one can infer that the surface tension there has no influence on the fluid and the gas area lying above it, as the free surface forms vertically to the field of gravity of the earth, as stated in Figure. From this, it follows that considerations of fluids with free surfaces can be carried out far away from fluid boundaries (container walls) without consideration of the wall effects. When one considers a curved surface element, one understands that as a consequence of the occurring surface tensions actions of forces are directed to the side of the surface on which the centre points of the "circles of curvature" are located. The forces attacking on sides AD and BC of the surface element are computed for each element ds_1 and the action of forces resulting from them in direction of the centre points of the circles of curvature is:

$$dK_1 = 2\sigma\, ds_1 \sin\beta = 2\sigma ds_1 \frac{ds_2}{2R_1}$$

$$dK_1 = \frac{\sigma}{R_2} ds_1\, ds_2 = \frac{\sigma}{R_1} dO$$

Accordingly the action of forces dK_2 is computed as

$$dK_2 = \frac{\sigma}{R_2} ds_2\, ds_1 = \frac{\sigma}{R_2} dO\,.$$

This shows that as a consequence of the surface tension pressure effects occur that are directed (in direction of) towards the centre points of the circles of curvature. This pressure effect is computed as force per unit area, i.e. a differential pressure that is caused by the surface tension:

$$\Delta p_\sigma = \frac{dK_2}{dO} = \sigma\left(\frac{1}{R_1} + \frac{1}{R_2}\right).$$

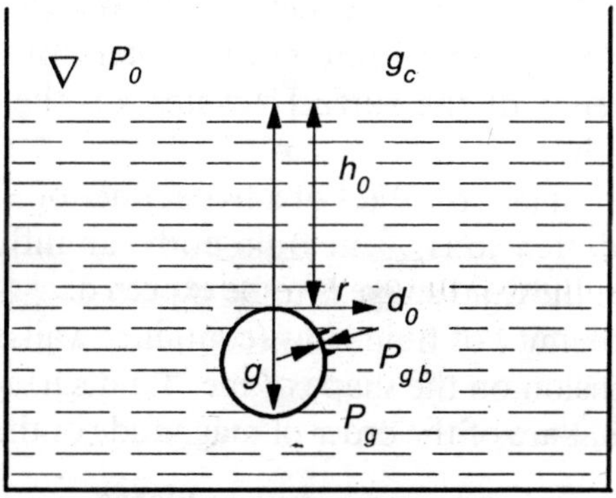

Fig. Diagram for the Consideration of Pressure in Bubbles

When there is a spherical surface it holds

$$R_1 = R_2 = \longrightarrow \Delta p_\sigma = \frac{2\sigma}{R}.$$

This relation means that the gas pressure in a spherical bubble is larger than the fluid pressure imposed from outside:

$$p_F + \frac{2\sigma}{R} = P_g.$$

For very small bubbles this pressure difference can be very large.

When one considers the equilibrium state of a surface element of a bubble, the following relation can be written for the pressure in the upper apex:

$$p_0 + \rho_F g h_0 + \sigma\left(\frac{2}{R_0}\right) = p_{g,0}$$

For a surface element of any height the following pressure equilibrium holds:

$$p_0 + \rho_F g(h_0 + y) + \sigma\left(\frac{1}{R_1} + \frac{1}{R_2}\right) = p_{g,0} + r_g g y.$$

When one now forms the difference of these pressure relations one obtains:

$$\left(\frac{1}{R_1} + \frac{1}{R_2}\right) - \frac{2}{R_2} + \frac{1}{\sigma}(\rho_F + \rho_g) g y = 0\,.$$

Thus the characteristic quantity for the standardization of equation is to be introduced

$$\alpha = \sqrt{\frac{2\sigma}{g(\rho_F - \rho_g)}}.$$

which is known as Laplace constant or capillary constant. It has the dimension of a length and indicates in orders of magnitude when a perceptible influence of the surface tension on the surface shape of a medium exists. It holds:

- When the Laplace constant of a free surface of a liquid is comparable with the dimensions of the fluid body, an influence of the surface tension on the fluid shape is to be expected.
- In the proximity of fluid rims (container walls) an influence of the surface tension on the shape of the "fluid surface" is to be expected in areas that are of the order of magnitude of the Laplace constant.

Heights of Throw in Tubes and between Plates

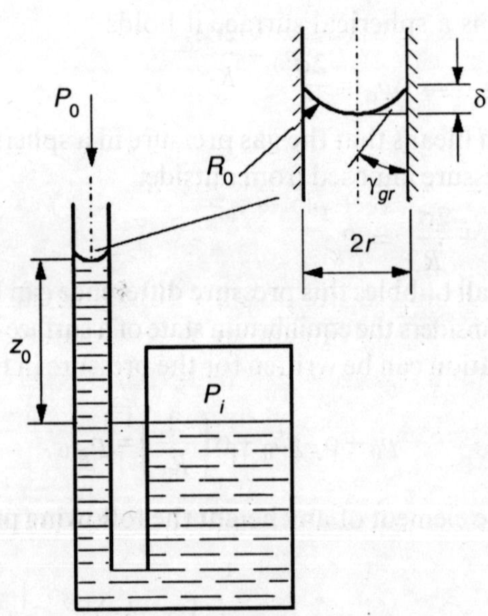

Fig. Diagram for Considerations of Heights of Throw in Tubes and between Plates

From the final statements consequences result for considerations of heights of throw of fluids. Such considerations were carried out, but influences of the boundary surfaces between fluid, solid and gaseous media remained unconsidered there, i.e. the influence of the boundary surface tension or surface tension was not taken into consideration. One sees that the considerations stated for communicating systems only hold when the dimensions of the systems are larger than to the Laplace constant of the fluid boundary surfaces. Moreover, the considerations only hold far away from fluid rims. In the immediate proximity of the rim there exists an influence of the surface tension which remained unheeded.

The processes taking place in fluid containers of small dimensions can be treated easily when carrying out a division of the container walls in as "wetting" ones and "non-wetting" ones. When making the considerations at first for wetting walls, experiments show that for such surfaces, in small tubes and between plates with small distances/gaps, the fluid in the tube or between the plates assumes a height which is above the height of the surface of a larger container.

From equilibrium considerations it follows:

Pressure between plates $$p_0 - \frac{\sigma}{R_0} = p_F = p_i - \rho_F g z_0 ,$$

Pressure in tubes $$p_0 - \frac{2\sigma}{R_0} = p_F = p_i - \rho_F g z_0$$

or in other form:

Height of throw between plates $$z_0 = \frac{1}{\rho_F g}(p_i - p_0) + \frac{\sigma}{\rho_F g R_0},$$

Height of throw in tube $$z_0 = \frac{1}{\rho_F g}(p_i - p_0) + \frac{2\sigma}{\rho_F g R_0}.$$

Here the radius of curvature R_0 is to be considered as an unknown for the determination of which two possibilities exist. To simplify the derivations one can assume with a precision that is sufficient in practice that the surface in the rising pipe adopts the form of a partial sphere for the tube and that of a partial cylinder for the gap of plate. The angle of contact between fluid surface and tube wall or plate wall has to be known from statements on the possibility of wetting. When one defines this angle as γ_{gr}, one obtains the following relation:

$$r = R_0 \cos \gamma_{gr}.$$

For the final relation of the height of throw z_0 for the plates and the tube thus holds:

Plates $$z_0 = \frac{1}{\rho_F g}(p_i - p_0) + \frac{\sigma}{\rho_F g r} \cos \gamma_{gr},$$

Tube $$z_0 = \frac{1}{\rho_F g}(p_i - p_0) + \frac{2\sigma}{\rho_F g r} \cos \gamma_{gr}.$$

This final relation now shows that even in the case of pressure equality, i.e. $p_i = p_0$, the height of throw assumes finite values if $\gamma_{gr} < \frac{\pi}{2}$. This fact has to be considered when employing communicating systems for measurements of the height of throw and when measuring pressures. The second possibility to compute the height of pressure is given by the fact that it is experimentally possible, although with a bigger inaccuracy, to determine the quantity δ by means of the following considerations.

$$r^2 + (R_0 - \delta)^2 = R^2{}_0 \; R_0 = \frac{r^2 + \delta^2}{2\delta}.$$

The height of throw z_0 is computed from this as follows:

$$z_0 = \frac{1}{\rho g}(p_i - p_0) + \frac{4\sigma\delta}{\rho_F g(r^2 + \delta^2)}.$$

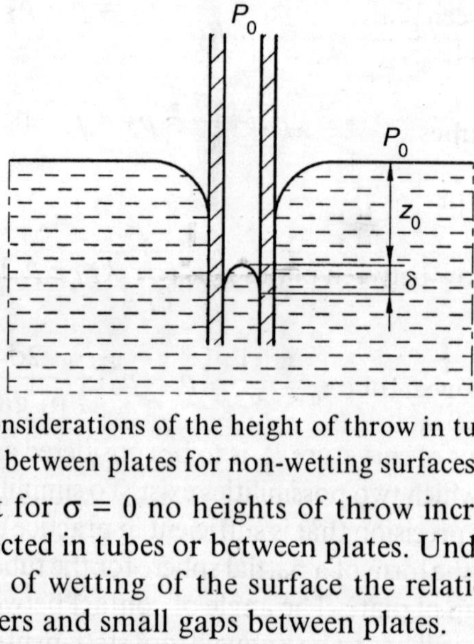

Fig. Considerations of the height of throw in tubes and between plates for non-wetting surfaces

It proves that for σ = 0 no heights of throw increased by surface e ects are to be expected in tubes or between plates. Under such conditions for the possibility of wetting of the surface the relations hold also for small tube diameters and small gaps between plates.

In the case of non-wetting surfaces it is observed that the fluid in the interior of a rising tube or the gap of a plate does not reach the height which the fluid outside the tube or the gap of the plate assumes. Analogous to the preceding considerations for wetting fluids it can be stated:

$$z_0 = \frac{2\sigma}{R_0 \rho g}$$

where R_0 can be introduced again.

The relation thus obtained indicates that the final relations derived for the wetting surfaces can often be applied also to non-wetting media, if one considers the sign of γ_{gr} and d Thus δ is for example to be introduced positively for wetting fluids in the above relations, whereas for non-wetting surfaces d has to be inserted negatively.

Bubble Formation at Nozzles

The injection of gases into fluids for chemical reactions or for an exchangeof-materials represents a process which is employed in many fields of process engineering. Thus bubble formation on nozzles as an introducing process is of interest for these applications. Moreover, the simulation of boiling processes, where the steam bubbles are replaced by gas bubbles, represents another field, where precise knowledge of bubble formation is required.

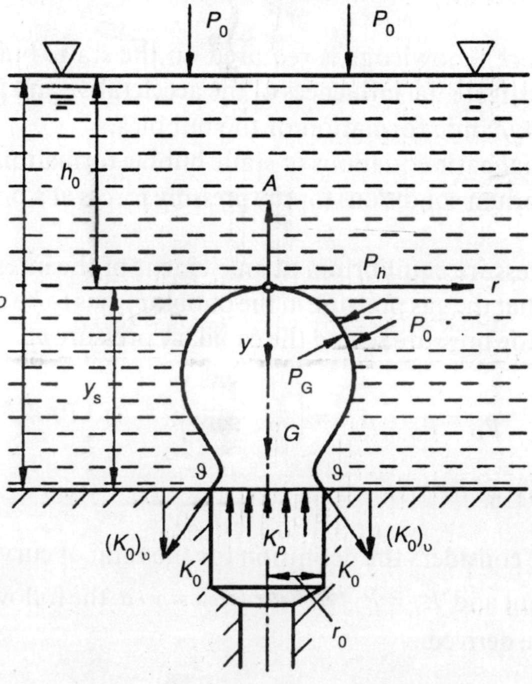

Fig. Equilibrium of forces at a bubble (A buoyancy force, G gravity, h_D distance of the nozzle from the fluid surface, h_0 distance of the bubble vertex from the fluid surface, K_O surface forces, K_p pressure forces, P_h hydrostatic pressure, p_0 atmospheric pressure on the fluid surface)

While gas bubbles form at nozzles during the gassing of liquids, the pressure in the interior of bubbles is changes. For the theoretically conceivable static bubble formation, this is attributed to different curvatures of the bubble boundary surface which are traversed during the formation of bubbles and thus to changes of the capillary pressure. Superimposed upon these are changes in pressure which have their origin in the upward movement of the bubble vertex taking place during the formation. With the dynamic formation of bubbles additional changing pressure e ects are to be expected which are essentially based on accelerative and frictional forces.

By static bubble formation one understands the formation of bubbles under pressure conditions, which allow to neglect the pressure effects on an element of the interface boundary surface due to accelerative and frictional forces. Although in practice this is the case only to a very limited extent, the static bubble formation has a certain importance.

As it is theoretically conceivable, some important basic knowledge can be gained from it which contributes to the general understanding of bubble formation.

Furthermore, knowledge is required on the static bubble formation in order to investigate the influences of the accelerative and frictional forces in the case of dynamic formation of the bubbles.

The essential basic equations of static bubble formation can be derived from the equilibrium conditions for the pressure forces at a boundary-surface element.

For the pressure equilibrium at an element of the interface boundary surface holds, that the gas pressure in the bubble p_G has to be equal to the sum of the hydrostatic pressure p_h and the capillary pressure p_σ

$$p_G = p_\sigma + p_h = \left(\frac{1}{R_1} + \frac{1}{R_2}\right)\sigma + p_0 + \rho_F g (h_0 + y).$$

Here the gas pressure is

$$p_G = p_{G,0} + \rho_G g y$$

When one considers the definition for the radii of curvature, with a as Laplace constant and $\overline{R}_j = R_j / a, \overline{r} = r / a\ \overline{y} = y / a$ the following differential equation can be derived:

$$\frac{\overline{y}''}{(1+\overline{y}'^2)^{3/2}} + \frac{\overline{y}'}{\overline{r}(1+\overline{y}'^2)^{1/2}} = 2\left(\frac{1}{\overline{R}_0} - \overline{y}\right).$$

By the substitution of

$$\overline{z} = \frac{\overline{y}'}{\sqrt{1+\overline{y}'^2}} = \sin\vartheta$$

the differential equation of second order can be replaced by a system of two differential equations of first order

$$\frac{d}{d\overline{r}}(\overline{r}\,\overline{z}) = 2\overline{r}\left(\frac{1}{\overline{R}_0} - \overline{y}\right),$$

$$\frac{d\bar{y}}{d\bar{r}} = \frac{\bar{z}}{\sqrt{1-\bar{z}^2}} = \tan\vartheta,$$

which are used for integration. The desired bubble volume $\bar{V}$ is obtained in dimensionless form by the following partial integration

$$\bar{V} = \pi \int_0^{\bar{y}} \bar{r}^2 d\bar{y} = \pi\bar{r}^2\bar{y} - 2\pi \int_0^{\bar{y}} \bar{r}\,\bar{y} d\bar{r}$$

and with the use of equation

$$\bar{V} = \pi\bar{r}\left[\bar{z} + \bar{r}\left(\bar{y} - \frac{1}{\bar{R}l0}\right)\right].$$

If one introduces again dimension-possessing quantities, the equation can be written as follows:

$$\frac{V}{a^3} = \pi\left(\frac{r}{a}\right)\left[\bar{z} + \frac{r}{a}\left(\frac{y}{a} - \frac{a}{R_0}\right)\right]$$

$$V = a^3\pi r\left[\bar{z} + \frac{r}{a^2}\left(y - \frac{a^2}{R_0}\right)\right].$$

With a and equation the bubble volume V can be written as:

$$V = \frac{2\sigma}{g(\rho_F - \rho_G)}\pi r\left\{\sin\vartheta + \frac{r}{2\sigma} g(\rho_F - \rho_G)\left[y - \frac{2\sigma}{g(\rho_F - \rho_G)R_0}\right]\right\}.$$

Equation represents an integral form of the differential equation system which allows considerations on the equilibrium of forces on bubbles. For the forces acting on a bubble, the equilibrium condition can be written in the form

$$Vg\rho_F - Vg\rho_G + \pi r^2\left[\frac{2\sigma}{R_0} - g(\rho_F - \rho_G)y\right] = 2\pi r\sigma\sin\vartheta$$

where the first two terms represent the buoyancy force and the weight of the bubble and the third term on the left side is the pressure force on the bubble crosssection πr^2 and the height y The surface forces are indicated on the right side. Equation should be employed in such cases where the bubble volume is to be computed from the conditions of the equilibrium of forces. For the computation of the pressure changes the pressure in the bubble vertex, owing to a transformation of the equations, can be expressed as stated below:

$$p_{G,0} = \frac{2\sigma}{R_0} + p_0 + \varrho_F g(h_D - y_s);$$

For the pressure at the nozzle mouth varies according to equations

$$p_{G,D} = \frac{2\sigma}{R_0} + p_0 + \varrho_F g h_D - g(\varrho_F - \varrho_g) y_s;$$

Equation can be written in dimensionless form:

$$\Delta\bar{p}_D = \frac{1}{\sqrt{2g\sigma(\varrho_F - \varrho_G)}}[p_{G,D} - p_0 - \varrho_F g h_D] = \frac{1}{R_0} - \bar{y}_s.$$

Although the differential-equation system permits the com-putation of all bubble forms of the static bubble formation and by means of equations, the corresponding bubble volumes and pres-sure differences can be obtained as important quantities of the bubbles, the problem with regard to the single steps of the bubble formation is indefinite/uncertain. The solution of the equations only allows the computation of a one-parameter set of curves, where the vertex radius R_0 is introduced into the derivations as a parameter. It does not permit to predict in which order the different values of the parameter are traversed. This has to be introduced into the considerations as an additional information in order to obtain a set of bubble forms that are traversed in the course of the bubble formation. Theoretically it is now possible to choose any finite, ordered quantity of $R_{0,i}$ values and to compute for these the corresponding bubble forms. Of practical importance, however, is only one $R_{0,i}$ variation, which is given by most of the experimental conditions and for which conditions have been formulated as follows:

- All bubbles form above a nozzle with the radius $\bar{r}_D$.
- $\bar{R}_{0,i} = \infty$. As starting point of the static bubble formation the horizontal position of the interface boundary surface above the nozzle is chosen.

All further vertex radii are selected according to the condition

$$\bar{V}_D[\bar{R}_{0,i+1}] \geq \bar{V}_D[R_{0,i}]$$

This means that the theoretical investigations are restricted to the bubble formation which comes about through a slow and continuous gas feeding through nozzles having a radius of $\bar{r}_D$. Gas refluxes through the nozzles, and thus a decrease of the bubble volume with mounting vertex radius, as equations would make possible, are excluded by relation c) from the considerations. The consequent application of this relation leads to the formation of a maximum bubble volume. Same has to be considered as volume of the bubble at the start of the separation process, i.e.

$$\bar{V}_A = (\bar{V}_D)_{\max}.$$

In the computations the differential equation system was solved numerically for different vertex radii, considering the indicated conditions, and thus the bubble form was ascertained. They can be consulted for the comprehension of the static bubble formation on nozzles in fluids.

Figure shows bubble forms that represent different stages of bubble formation with slow gas feeding through nozzles. The results are reproduced for $\overline{r}_D$ = 0, 4 and this corresponds to a nozzle radius of $r_D \approx$ 1, 6 in the case of air bubbles in water.

The change of the bubble volume during the formation of gas bubbles on nozzles of different radii $\overline{r}_D$, where the vertex radius $\overline{R}_0$ was

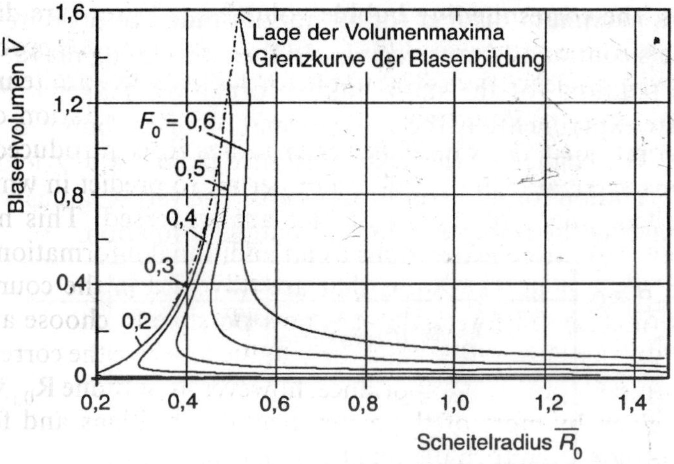

Fig. Bubble forms of the static bubble formation $\overline{r}_D$ = 0,4 ascertained through integration of the equation systems

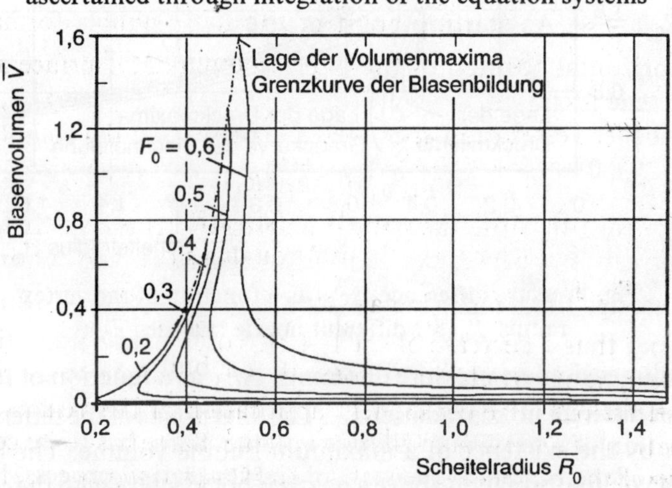

Fig. Bubble volume $\overline{V}$ as a function of the vertex radius $\overline{R}_0$ for the different nozzle radiuses $\overline{r}_D$

chosen for designating the respective formation stage. From this diagram it can be inferred that a large part of the bubble forms at an almost constant vertex radius and it is an important property for larger nozzle radii. For smaller nozzle radii, stronger/larger changes of the vertex radius are to be expected during the formation of the gas bubbles.

The pressure difference $\Delta\overline{P}_D$ as a function of the vertex radius $\overline{R}_0$ is represented for different nozzle radii $\overline{r}_D$ From this representation it can be gathered that for the static bubble formation on nozzles initially a continuous pressure increase at the nozzle mouth is necessary. After having reached a maximum distinct for all nozzles radii the pressure decreases again. This continuously increasing and then decreasing pressure change, which is required for the static bubble formation, makes it difficult to investigate experimentally the static formation of gas bubbles on nozzles in fluids. The change of the vertex distance from the nozzle during the bubble formation for different nozzle radii. The vertex radius was chosen for designating the respective stage of bubble formation.

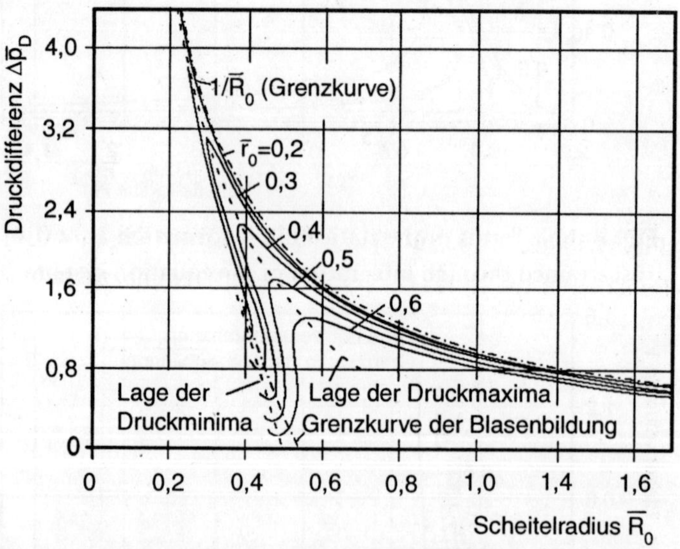

Fig. Pressure difference $\Delta\overline{P}_D$ as a function of the vertex radius $\overline{R}_0$ for different nozzle radiuses $\overline{r}_D$

The dimensionless pressure difference $\Delta\overline{P}_D$ as a function of the bubble volume for different nozzle radiuses $\overline{r}_D$. The final points of the different curves are given by the existence of a maximum bubble volume. The bubble of maximum volume formation above a nozzle is not identical with the one having minimum pressure. Reference is made to the fact, however, that the latter is excluded from the possible bubble forms owing to the definition of the static bubble formation, as it could only be verified by the gas reflux to the nozzles. For general comprehension of the problem of static bubble formation, two further facts can be referred to:

- For the static bubble formation a basic radius $\overline{r}_{D,gr} = 0.648$, exists which splits up the static bubble formation into two different domains. Theoretical investigations for nozzle radiuses $\overline{r}_D \geq \overline{r}_{D,gr}$ were not carried out.

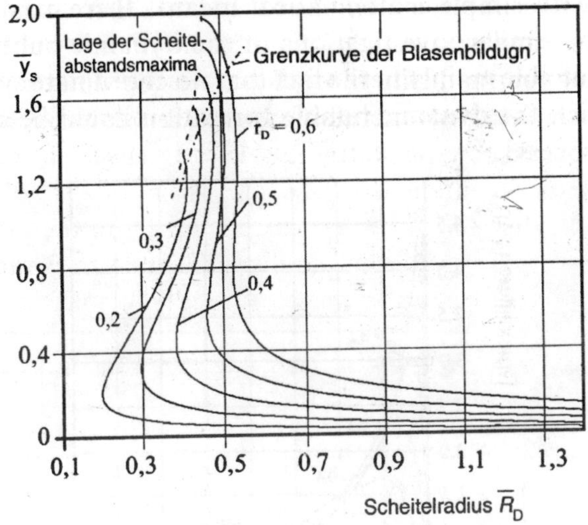

Fig. Distance of the bubble vertex $\overline{y}_s$ ys from the nozzle top as a Function of the vertex radius $\overline{R}_0$ for different nozzle radiuses $\overline{r}_D$

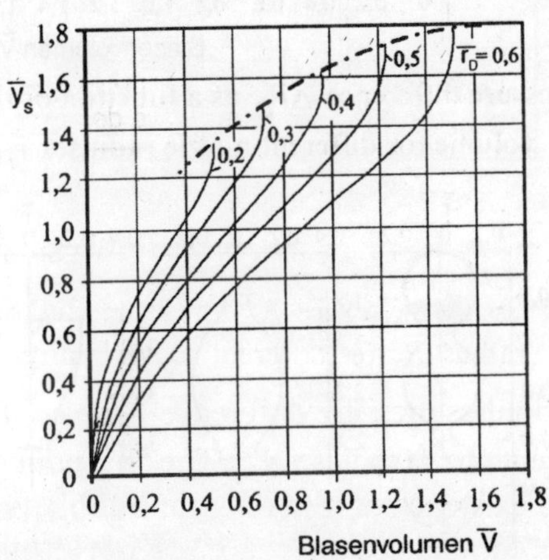

Fig. Distance of the bubble vertex $\overline{y}_s$ from the nozzle top as a function of the nozzle volume $\overline{V}$ for different nozzle radiuses $\overline{r}_D$

here, as they are not of importance for the introduction of the results intended here for better comprehension of the bubble formation.

- The differential equation system allows the computation of bubble chains. These bubble chains were not investigated further, as they are not in accordance with the above-stated definition of the air feed/supply for the examined static bubble formation.

While the static nozzle formation can theoretically be understood essen-tially with simple mathematical means, there exist considerable difficulties for similar investigations of the dynamic bubble formation. This is attribut able mainly to the fact that no coordinate system could be found in which the dynamic bubble formation could be described as a stationary process.

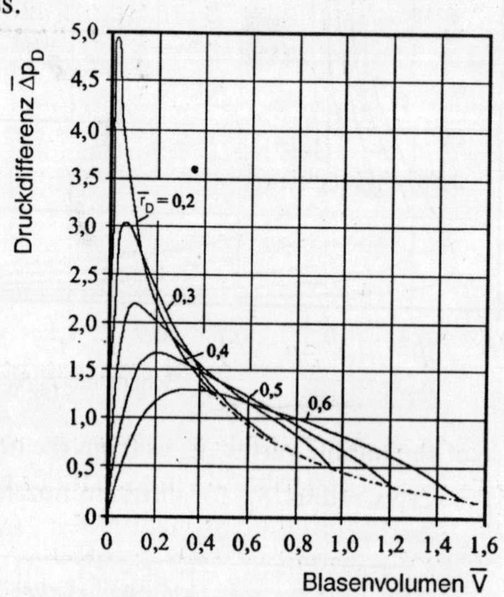

Fig. Pressure difference $\Delta\overline{P}_D$ as a function of the nozzle volume for different nozzle radiuses $\overline{r}_D$

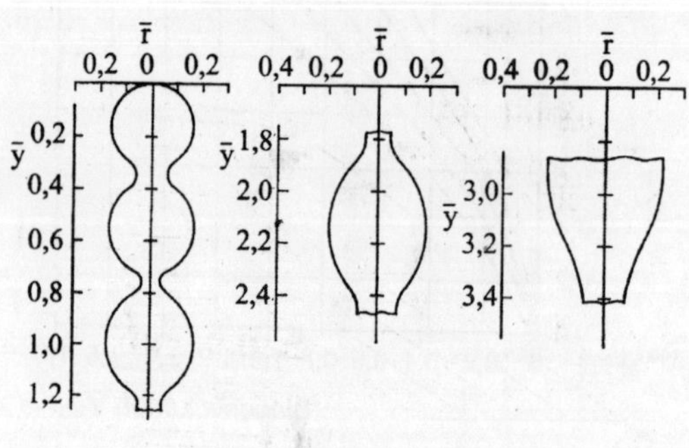

Fig. Bubble chains $\overline{R}_0$ = 1, 60 ascertained from the Differential equation system

Moreover, for the dynamic bubble formation the pressure on an element of the interface boundary surface is dependent on the fluid motions during bubble formation and thus is computable only by solving the non-stationary Navier-Stokes equation. This, however, is solvable only with difficulty and big computing e orts.

AEROSTATICS

Pressure in the Atmosphere

Aerostatics differs from hydrostatics in that for designating the fluid, the partial differential equations for fluids at rest derived from the Navier-Stokes equations are applied along with the equation of state for an ideal gas rather than for an ideal fluid ρ = const.

$$\frac{P}{\rho} = RT$$

Thus the valid partial differential equations in aerostatics read:

$$\frac{\partial P}{\partial x_j} = \rho \hat{g}_j = \frac{P}{RT} \hat{g}_j$$

or written out for j = 1, 2, 3:

$$\frac{\partial P}{\partial x_1} = \frac{P}{RT} \hat{g}_1, = \frac{\partial P}{Rx_2} = \frac{P}{RT} \hat{g}_2, \frac{\partial P}{\partial x_3} = \frac{P}{RT} \hat{g}_3.$$

This theorem of partial differential equations can now be employed for the computation of the pressure distribution in such fluids whose state designation is possible with a precision that is sufficient for the considerations to be made due to the ideal gas equation.

When one considers the atmosphere of the earth as consisting of a compressible fluid at rest, whose state can be described thanks to the ideal gas equation with a precision sufficient for the derivations at issue, an approximated relation between the height above the surface of the earth and the

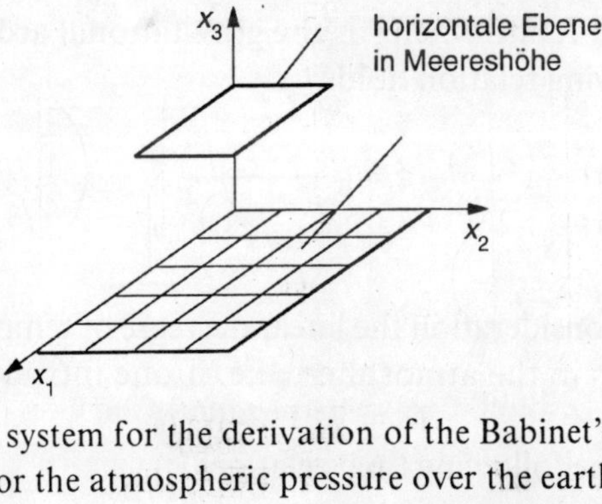

Fig. Coordinate system for the derivation of the Babinet's approximation formula for the atmospheric pressure over the earth's surface

pressure of the atmosphere in a considered height H can be stated, which in general is defined as barometric height formula. In particular the relation known as Babinet approximate formula shall be derived here.

When one uses the coordinate system indicated, in which the plane x_1, x_2 forms a horizontal area at the level of the sea surface, the above stated partial differential equations can be written as follows.

Differential equations:

$$\frac{\partial P}{\partial x_1} = 0 \rightsquigarrow P = f(x_2, x_3)$$

$$\frac{\partial P}{\partial x_2} = 0 \rightsquigarrow P = f(x_1, x_3)$$

i.e. $P = f(x_3)$ and thus the relation holds $\frac{\partial P}{\partial x_3} \to \frac{dP}{dx_3}$:

$$\frac{dP}{dx_3} = -\frac{P}{RT}\hat{g}$$

The differential equation which is to be solved, can be written as follows:

$$\frac{dP}{dx_3} = -\frac{P}{RT}\hat{g} \qquad \frac{dP}{P} = -\frac{1}{RT}\hat{g}dx_3$$

with the general solution:

$$\ln\frac{P_2}{P_1} = -\frac{1}{R}\int_{(x_3)_1}^{(x_3)_2}\left(\frac{\hat{g}}{T}\right)dx_3.$$

The integral appearing in the above equation can be solved only when it is known how the gravitational acceleration g changes with height and when further the temperature variation as a function of x_3 can be stated. For the gravitational acceleration g, it is known that it changes with the height x_3 inversely in proportion to the square of the distance from the centre of the earth. This follows directly from Newton's law of gravitation, when the influence of the rotation of the earth is neglected. When one designates the radius of the earth R and when g is the gravitational acceleration at the sea surface, the following relation holds:

$$\hat{g} = g\frac{1}{\left(1+\frac{x_3}{R}\right)^2}.$$

Taking into consideration the linear decrease of temperature with height which often exists in the atmosphere, i.e. if one introduces:

$$T = T_0\,(1-\alpha x_3)$$

one obtains the following final relation:

$$\ln \frac{P_H}{P_0} = -\frac{g}{RT_0} \int_0^H \frac{dx_3}{(1-\alpha x_3)\left(1+\frac{x_3}{R}\right)^2}.$$

When one now imposes restrictions concerning the height above which the above-indicated integration is to take place and when one chooses them such that the following relations hold:

$$\alpha x_3 \ll 1 \frac{x_3}{R} \ll 1$$

one can obtain by exponential-series development :

$$\ln \frac{P_H}{P_0} = -\frac{g}{RT_0} \int_0^H (1+\alpha x_3 + ...)\left(1-\frac{2x_3}{R}+...\right)dx_3$$

or the approximate formula by neglecting the terms of higher order:

$$\ln \frac{P_H}{P_0} = -\frac{g}{RT_0} \int_0^H \left[1+\left(\alpha - \frac{2}{R}\right)x_3\right]dx_3.$$

By solving this equation the following final relation results for the pressure distribution in the lower atmosphere:

$$P_H = P_0 \exp\left\{-\frac{gH}{RT_0}\left[1+\frac{1}{2}\left(\alpha-\frac{2}{R}\right)H\right]\right\}.$$

This pressure relation describes with good precision the standard pressure distribution existing in the atmosphere.

The approximate formulae stated by Babinet for the pressure distribution in the atmosphere can be derived from the general differential equation

$$\frac{dP}{P} = -\frac{\hat{g}}{RT}dx_3$$

by introducing the following hypotheses:

$$P = \frac{P_H + P_0}{2} = \text{const} \quad \hat{g} = g = \text{const}$$

$$T = \frac{T_0 + T_H}{2} = \text{const.}$$

When one introduces these simplifications/reductions, the following solution for the differential equation results:

$$\frac{2}{P_H + P_0}(P_H - P_0) = -\int_0^H \frac{g}{RT}dx_3$$

or resolved:

$$\frac{P_H - P_0}{P_H + P_0} = -\frac{gH}{R(T_0 + T_H)}$$

$$H = -\frac{R}{g}T_0\left(1+\frac{T_H}{T_0}\right)\left(\frac{P_H - P_0}{P_H + P_0}\right).$$

When one rearranges this equation according to P_0 it results:

$$P_H = P_0 \frac{\frac{R}{g}T_0\left(1+\frac{T_H}{T_0}\right) - H}{\frac{R}{g}T_0\left(1+\frac{T_H}{T_0}\right) + H}.$$

The two relations in comparison to the standard pressure distribution in the atmosphere.

Rotating Containers

As a further example for the employment of the aerostatic basic equations. By a single pressure measurement in the upper centre point of the cylinder ceiling surface and by employment of the partial differential equation of aerostatics the entire pressure load on the circumferential surface of a rotating cylinder filled with a compressible medium shall be defined.

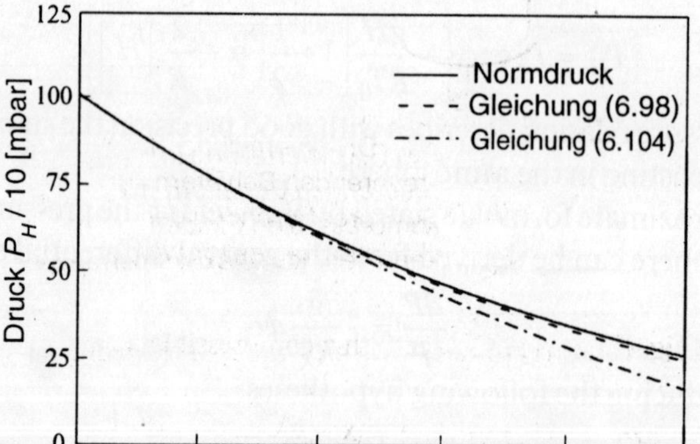

Fig. Standard pressure distribution in the atmosphere and distributions computed on the basis of approximate formulas

Contrary to the example treated here the cylinder shall experience a pure rotational motion only, so that the partial differential equations of aerostatics can be written for T = const as follows:

$$\frac{\partial P}{\partial r} = \frac{P}{RT} r\omega^2 \rightarrow \ln P = \frac{\omega^2}{2RT} r^2 + F(\varphi, z),$$

$$\frac{1}{r}\frac{\partial P}{\partial \varphi} = 0 \rightarrow P = F(r, z),$$

$$\frac{\partial P}{\partial z} = -\frac{P}{RT} g \rightarrow \ln P = -\frac{g}{RT} z + F(r, \varphi).$$

By comparing the solutions one obtains:

$$\ln P = C + \frac{\omega^2}{2RT} r^2 - \frac{g}{RT} z.$$

The pressure P_0, measured at the coordinate origin permits the definition of the integration constant C as:

$$C = \ln P_0.$$

From this P is computed:

$$P = P_0 \exp\left\{\frac{\omega^2}{2RT} r^2 - \frac{g}{RT} z\right\}$$

Along the floor space $z = -L$ the pressure distribution is computed as:

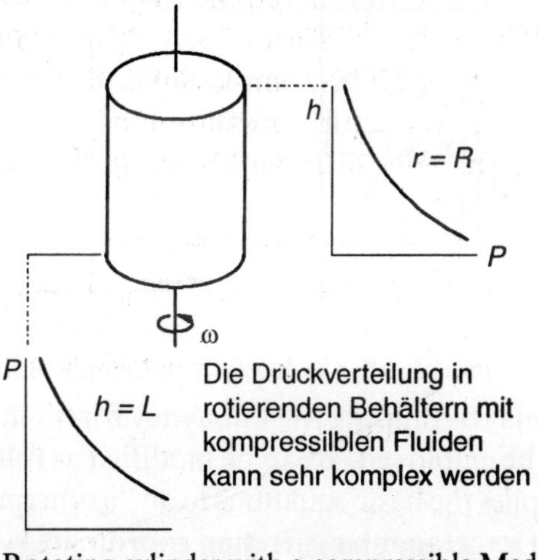

Fig. Rotating cylinder with a compressible Medium

$$P(r, z = -L) = P_0 \exp\left\{\frac{\omega^2}{2RT} r^2 + \frac{gL}{RT}\right\}.$$

Along the vertical circumferential surface $r = R$ holds:

$$P(r = R, z) = P_0 \exp\left\{\frac{\omega^2}{2RT} r^2 - \frac{g}{RT} z\right\}.$$

Aerostatic Buoyancy

When employing aerostatic laws, mistakes are often made by using relations whose validity strictly speaking is only in hydrostatics. As an example is cited at this place the buoyancy of bodies which is computed according to figure as the difference of the pressure forces on the lower and upper sides of a immersed body. The basic equations for these computations.

They lead to the generally known relation which expresses that the buoyancy force experienced by a body immersed in a fluid is equal to the weight of the fluid displaced by the body. The derivations on the basis make it clear that the relation holds only for fluids with ρ = const. For the buoyancy force on a body element holds:

$$\Delta A_i = \Delta P_i \Delta F_i$$
$$\Delta P_i = \rho_F g h_i (x_1).$$

Thus for the buoyancy force holds:

$$A = \sum_{i_1}^{N} \rho_F g h_i (x_1) \Delta F_i = \rho_F g V$$

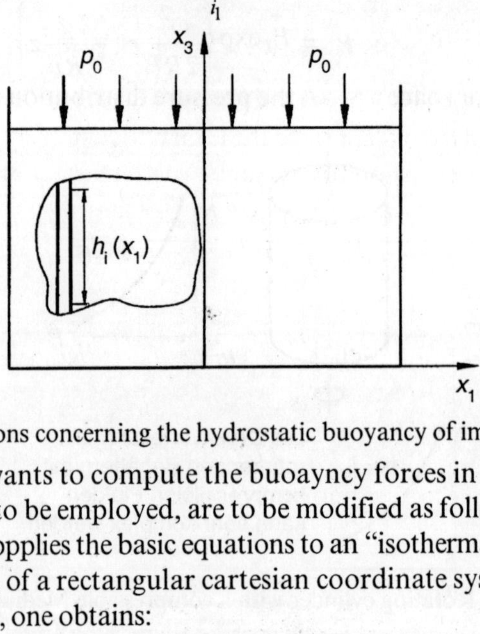

Fig. Explications concerning the hydrostatic buoyancy of immersed bodies

When one wants to compute the buoayncy forces in gases, the laws of aerostatics have to be employed, are to be modified as follows.

When one appplies the basic equations to an "isothermal atmosphere", in which the axis x_3 of a rectangular cartesian coordinate system points in the upward direction, one obtains:

$$\frac{dP}{dx_3} = -\frac{P}{RT} g.$$

When one sets g = const and $T = T_0$ = const, one obtains

$$\frac{dP}{P} = -\frac{g}{RT_0} dx_3$$

or integrated:

$$\ln \frac{P}{P_0} = -\frac{g}{RT_0} x_3.$$

In this relation the pressure P_0 prevails at the level $x_3 = 0$. For the pressure path in an isothermal atmosphere can thus be written:

$$P = P_0 \exp\left\{-\frac{g}{RT_0} x_3\right\}$$

Employing now again the relation for the buoyancy occurring on a body element:

$$\Delta A_i = \Delta P_i \Delta F_i$$

and writing with employment of (6.116)

$$\Delta P_i = P_0 \exp\left\{-\frac{gx_{3i}}{RT_0}\right\} - P_0 \exp\left\{-\frac{g}{RT_0}(x_{3i} + h_i)\right\}$$

i.e.

$$\Delta P_i = P_0 \exp\left\{-\frac{gx_{3i}}{RT_0}\right\}\left[1 - P_0 \exp\left\{-\frac{g}{RT_0} h_i\right\}\right]$$

it becomes quickly evident that for the total buoyancy in the present case, simple relation stated for ρ = const for the total buoyancy force does not result. The Taylor's series expansion for P_i can be written as:

$$\Delta P_i = \frac{\partial P}{\partial x_3} h_i + \frac{1}{2}\frac{\partial^2 P}{\partial x_3^2} h_i^2 + \frac{1}{6}\frac{\partial^3 P}{\partial x_3^3} h_i^3 + \cdots$$

$$= -\frac{P_0 g}{RT_0} \exp\left\{-\frac{gx_3}{RT_0}\right\} h_i + \frac{P_0 g^2}{2R^2T_0^2} \exp\left\{-\frac{gx_3}{RT_0}\right\} h_i^2 -$$

$$\Delta P_i \approx \rho(x_3) g h_i \left[1 - \frac{gh_i}{2RT_0} + \frac{1}{6}\left(\frac{gh_i}{RT_0}\right)^2 - + \cdots\right].$$

Thus one obtains for the buoyancy force:

$$A \simeq \rho(x_3) g \left[V - \sum_{i=1}^{N} \frac{gh_i^2 \Delta F_i}{2RT_0} + - \cdots\right].$$

The first term of this expression (6.123) corresponds to the buoyancy force of hydrostatics.

Conditions for Aerostatics: Stable Laminations

A fluid can find itself in mechanical equilibrium (i.e. show no macroscopic motion) without being in thermal equilibrium. Equation, the condition for the mechanical equilibrium, can also be given when the temperature in the fluid is not constant. Here the question arises, however, whether such an equilibrium is stable. It appeared that the equilibrium shows the demanded stability only under a certain condition. When this condition is not met,the static state of the fluid is unstable and in the fluid occur flows of random order which endeavor to mix the fluid such that no constant temperature is achieved (in it). This motion is defined as free convection. The stability condition for the mechanical equilibrium is, in other words, the condition for the absence of the convection. It can be derived as follows.

We consider a fluid at the level z and with the specific volume $V(p', s)$; here p' and s are the equilibrium pressure and the equilibrium entropy of the fluid at this level. We assume that this fluid element is displaced upwards by the small stretch x in an adiabatical way. Its specific volume thus becomes $V(p', s)$, where p' is the pressure at the level $z + \xi$. For the stability of the chosen equilibrium state it is necessary (although in general not sufficient) that the force occurring here drives back the element to the starting position. The considered volume element therefore has to be heavier than the fluid "displaced" by it in the new position. The specific volume of the fluid is $V(p', s)$; here s is the equilibrium entropy of the fluid at the level $z + \xi$. Thus we have as a stability condition

$$V(p',s') - V(p',s) > 0.$$

This difference is expanded in powers of $s' - s = \dfrac{ds}{dz}\xi$ and we obtain

$$\left(\frac{\partial V}{\partial s}\right)_p \frac{\partial s}{\partial z} > 0.$$

In accordance with thermodynamic relations because of $T\,ds = dh - v\,dP$ holds

$$\left(\frac{\partial V}{\partial s}\right)_p = \frac{T}{c_p}\left(\frac{\partial V}{\partial T}\right)_p;$$

c_p being the specific heat at constant pressure. The specific heat c_p is always positive like the temperature T therefore we can transform (6.124)into

$$\left(\frac{\partial V}{\partial T}\right)_p \frac{ds}{dz} > 0.$$

Most materials expand with warming, i.e. it holds $\left(\dfrac{\partial V}{\partial T}\right)_p > 0$. The condition for the absence of the convection is reduced then to the inequality

$$\frac{ds}{dz} > 0.$$

i.e. the entropy has to increase with the level.

From this one can easily find a condition for the temperature gradient $\dfrac{\partial T}{\partial z}$. From the derivation of $\dfrac{ds}{dz}$ we can write:

$$\frac{ds}{dz} = \left(\frac{\partial s}{\partial T}\right)_p \frac{dT}{dz} + \left(\frac{\partial s}{\partial p}\right)_T \frac{dP}{dz} = \frac{c_p}{T}\frac{dT}{dz} - \left(\frac{\partial V}{\partial T}\right)_p \frac{dp}{dz} > 0.$$

Finally we insert $\frac{dp}{dz} = -\rho g$ according to and obtain

$$\frac{dT}{dz} > -\frac{gT\rho}{c_p}\left(\frac{\partial V}{\partial T}\right)_p .$$

Convection will occur when the temperature in the direction top to bottom decreases and the gradient is then larger in value than $\frac{gT}{c_p}\rho\left(\frac{\partial v}{\partial T}\right)_p$.

When one investigates the equilibrium of a gas column and can assume the gas to be ideal, $\frac{T}{V}\left(\frac{\partial V}{\partial T}\right)_p = 1$, holds and the stability condition for the equilibrium simply reads

$$\frac{dT}{dz} > -\frac{g}{c_p}.$$

When this stability requirement (I assume that the stability condition is meant and not the condition) in the atmosphere is not met, the present temperature lamination is unstable and it will give way to a convective temperature-compensation flow as soon as the smallest disturbances occur.

Chapter 7

Integral Forms of the Basic Equations

The basic equations of flow mechanics were derived in a form valid for all flow problems. In order to obtain a generally valid form of the equations that is valid for all flows, these were formulated as differential equations for field quantities. They represent local formulations of mass, momentum and energy conservation. Employing these equations to special flow problems, it is advantageous and often absolutely necessary to derive and employ the integral forms of the equations. The momentum equation in direction j - and the mechanical and caloric forms of the energy equation. The applications of the derived integral forms of the fluid-mechanical basic equations. It shall thus be shown how it is possible to solve fluid-mechanical problems. As in this book only an introduction to the solution of problems is envisaged, simplified assumptions are made in the course of the solutions, to which the attention is drawn in order to ensure that the reader is aware of the limits of the validity of the derived results. On the basis of the exemplary applications independent solutions of more extensive and complicated problems should be possible. Depending on the problem the integral form of the momentum equation or the mechanical energy equation can be employed.

INTEGRAL FORM OF THE CONTINUITY EQUATION[H]

The continuity equation, i.e. the mass-conservation equation in local formulation expressed in field variables, was stated as follows

$$\frac{\partial \rho}{\partial t} + \frac{\partial (\rho U_i)}{\partial x_i} = 0$$

Applying to the equation the integral operator $\int_{V_K} () \, dV$ i.e. integrating this equation via a given control volume $V = V_K$, one obtains:

$$\int_{V_K} \left(\frac{\partial \rho}{\partial t} \right) dV + \int_{V_K} \left(\frac{\partial (\rho U_i)}{\partial x_i} \right) dV = 0$$

Here V_K is an arbitrary control volume which is to be selected for the solution of flow problems in such a way that simple solution paths can be found.

Considering that the integration applied to (∂r/∂t) and the partial differentiation carried out for r can be done in any sequence, one obtains:

$$\frac{\partial}{\partial t}\left(\int_{V_K} \rho dV\right)+\int_{V_K}\left(\frac{\partial(\rho U_i)}{\partial x_i}\right)dV = 0$$

Applying now to the second term of the above equation Gauss' integral theorem, it results:

$$\frac{\partial}{\partial t}\left(\int_{V_K} \rho dV\right)+\int_{O_K} \rho U_i \, dF_i = 0.$$

Here the second integral is to be carried out over the entire surface of the control volume, where the direction of dF_i is considered positive from the inside of the volume to its outside.

In equation the following consideration was carried out:

$$\int_{V_K} \frac{\partial(\rho U_i)}{\partial x_i} dV \Leftrightarrow \int_{O_K} (\rho\vec{U}) d\vec{A} \quad \underset{\text{Theorem}}{\overset{\text{Gauss'}}{\Leftrightarrow}} \quad \int_{O_K} \rho U_i dF_i$$

- In this relation the surface vector dF_i represents a directed quantity, i.e. it contains the normal vector n of the surface element with its absolute value $| dF_i |$.
- Because of the double index we have a scalar product of the velocity vector U_i with the surface vector dF_i.

In equation the resulting integrals have the following meaning/ significance:

$$M = \int_{V_K} \rho dV = \text{Total mass in the control volume,}$$

$$\dot{m}_{\text{aus}} - \dot{m}_{\text{ein}} = \int_{O_K} \rho U_i \text{d}F_i = \begin{array}{l}\text{Difference of the mass outflows and inflows}\\ \text{over the surface of the control volume.}\end{array}$$

Thus the integral form of the continuity equation yields:

$$\frac{\partial M}{\partial t} = \dot{m}_{\text{ein}} - \dot{m}_{\text{aus}}.$$

For the volume flow through a surface with the velocity component U_i normal to this surface results because of the above sign convention for the surface vector dF_2 (outer normal to the surface) for the inflow in the control volume or the outflow from it:

$$\dot{V}_{\text{in}} = -\int_F |U_i||\text{d}F_i| \qquad \dot{V}_{\text{out}} = +\int_F |U_i||\text{d}F_i|.$$

The surface-averaged flow velocity can thus be computed for any point in time at

$$\tilde{U} = \frac{\dot{V}_{\text{ein}}}{|F|} = \frac{1}{|F|}\int_F U_i \, \mathrm{d}F_i \, .$$

When one considers, however, that for the area-averaged density $\tilde{\rho}$ holds:

$$\tilde{\rho} = \frac{1}{|F|}\int_F \rho \left| \mathrm{d}F_i \right|,$$

the mass flow through a surface can also be written

$$|\dot{m}| = \tilde{\rho}\tilde{U}F \quad \text{mit} \quad F = |F|.$$

For moderate velocities where ρ is changing little, for internal flows thus a surface decrease is connected to an increase in velocity.

When carrying out considerations in flow mechanics, the above derivations, taking into account the physical conditions of mass conservation, lead as required to the following mathematical representations.

Differential form:

$$\boxed{\frac{\partial \rho}{\partial t} + \frac{\partial (\rho U_i)}{\partial x_i} = 0}$$

Integral form:

$$\boxed{\frac{\partial M}{\partial t} + \dot{m}_{\text{ein}} - \dot{m}_{\text{aus}}}$$

Inner flows:

$$\frac{\partial M}{\partial t} = 0 \rightsquigarrow \boxed{\dot{m} = \tilde{\rho}\tilde{U}F = \text{const}}$$

can furthermore be derived by differentiating:

$$\frac{d}{dx}(\dot{m}) = \frac{d}{dx}\left(\tilde{\rho}\tilde{U}F\right) = 0 \rightsquigarrow \boxed{\frac{d\tilde{\rho}}{\tilde{\rho}} + \frac{d\tilde{U}}{\tilde{U}} + \frac{dF}{F} = 0}.$$

All above indicated framed equations represent different forms of mass conservation that can be employed for the solution of flow problems.

THE INTEGRAL FORM OF THE MOMENTUM EQUATION

The momentum equation for local considerations was formulated and derived for each component j – of the momentum as follows:

$$\rho\left(\frac{\partial U_j}{\partial t} + U_i \frac{\partial U_j}{\partial x_i}\right) = -\frac{\partial P}{\partial x_j} - \frac{\partial \tau_{ij}}{\partial x_i} + \rho g_j \, .$$

When one adds to this equation the continuity equation multiplied by U_j in the form i.e. adding the following terms:

$$U_j \frac{\partial \rho}{\partial t} + U_j \frac{\partial (\rho U_i)}{\partial x_i} = 0,$$

one obtains

$$\frac{\partial (\rho U_j)}{\partial t} + \frac{\partial (\rho U_i U_j)}{\partial x_i} = -\frac{\partial P}{\partial x_j} - \frac{\partial \tau_{ij}}{\partial x_i} + \rho g_j .$$

When integrating via a given control volume, i.e. when applying the operator $\int_{V_k} (\mathrm{d}V)$. to all terms of the equation, one obtains the integral form of the component j – of the momentum equation of flow mechanics:

$$\int_{V_K} \frac{\partial (\rho U_j)}{\partial t} dV + \int_{V_K} \frac{\partial (\rho U_i U_j)}{\partial x_i} dV$$

$$= -\int_{V_K} \frac{\partial P}{\partial x_j} dV - \int_{V_K} \frac{\partial \tau_{ij}}{\partial x_i} dV + \int_{V_K} \rho g_j dV \sum K_j$$

The term ΣK_j is added as "integration constant", i.e. all those forces in the direction j - have to be included that act as external forces on the boundaries of the chosen control volume. Considering that the integration and differentiation represent in their sequence exchangeable mathematical operators and employing Gauss' integration theorem, the following form of the integral momentum theorem can be derived:

$$\underbrace{\frac{\partial}{\partial t} \int_{V_K} \rho U_j \mathrm{d}V}_{\text{I}} + \underbrace{\int_{O_K} \rho U_i U_j \mathrm{d}F_i}_{\text{II}} =$$

$$\underbrace{-\int_{O_K} P \,\mathrm{d}F_j}_{\text{III}} \underbrace{- \int_{O_K} \tau_{ij} \,\mathrm{d}F_i}_{\text{IV}} + \underbrace{\int_{V_K} \rho g_j \,\mathrm{d}V}_{\text{IV}} + \underbrace{\sum K_j}_{\text{IV}} .$$

This equation comprises 6 terms whose physical significance is stated below:

- I: Temporal change of the momentum j -Impulses in the interior of a control volume.
- II: Sum of inflows and outflows of flow momentum per time unit in direction j – summed up over the entire surface surrounding the considered
  control volume.
- III: Resulting pressure force in the direction j – ., pressure distribution summed up over the entire surface surrounding the considered control volume.

IV: Sum of the momentum inflows and outflows j – occurring per time unit by molecular momentum transport over the entire surface of the control volume.

V: j – Component of the inertia force acting on the control volume.

VI: Sum of all external (not flow-mechanically induced) forces acting in the j – direction on the boundaries of the control volume.

The integral form of the momentum equation can be employed for a large number of problems of flow mechanics, in order to determine actions of forces caused by fluid motions on walls, flow aggregates etc. On the basis of selected representations it shall be made clear how the above derived integral form of the momentum equation is to be used in the case of the flow problems that serve as examples. Here it is important to perceive the universal validity of the integral form of the momentum equation to safeguard its general use in solving flow problems, beyond the considered examples.

INTEGRAL FORM OF THE MECHANICAL ENERGY EQUATION

It was shown that the momentum equation j :

$$\rho\left[\frac{\partial U_j}{\partial t}+U_i\frac{\partial U_j}{\partial x_i}\right]=-\frac{\partial P}{\partial x_j}-\frac{\partial \tau_{ij}}{\partial x_i}+\rho g_j,$$

can be transferred to the mechanical energy equation by multiplication by

U_j :

$$\rho\left[\frac{\partial\left(\frac{1}{2}U_j^2\right)}{\partial t}+U_i\frac{\partial\left(\frac{1}{2}U_j^2\right)}{\partial x_i}\right]=$$

$$-\frac{\partial\left(PU_j\right)}{\partial x_j}+P\frac{\partial U_j}{\partial x_j}-\frac{\partial\left(\tau_{ij}U_j\right)}{\partial x_i}+\tau_{ij}\frac{\partial U_j}{\partial x_i}+\rho g_j U_j.$$

Multiplying the continuity equation by $\left(\frac{1}{2}U_j^2\right)$, it results:

$$\left(\frac{1}{2}U_j^2\right)\frac{\partial\rho}{\partial t}+\left(\frac{1}{2}U_j^2\right)\frac{\partial\left(\rho U_i\right)}{\partial x_i}=0,$$

which can be added to so that one obtains:

$$\frac{\partial}{\partial t}\left(\frac{1}{2}\rho U_j^2\right)+\frac{\partial}{\partial x_i}\left(\rho U_i\frac{1}{2}U_j^2\right)$$

$$=-\frac{\partial\left(PU_j\right)}{\partial x_j}+P\frac{\partial U_j}{\partial x_j}-\frac{\partial\left(\tau_{ij}U_j\right)}{\partial x_i}+\tau_{ij}\frac{\partial U_j}{\partial x_i}+\rho g_j U_j.$$

When one integrates this equation via a given control volume, one obtains by employing Gauss' integral theorem and considering the mathematically possible inversion of the integration and differentiation sequence:

$$\underbrace{\frac{\partial}{\partial t}\int\limits_{V_K}\frac{1}{2}\rho U_j^2 \mathrm{d}V}_{\text{I}} + \underbrace{\int\limits_{O_K}\rho U_i \frac{1}{2}U_j^2 \mathrm{d}F_i}_{\text{II}} = -\underbrace{\int\limits_{O_K} PU_j \,\mathrm{d}F_j}_{\text{III}}$$

$$+\underbrace{\int\limits_{V_K} P\frac{\partial U}{\partial x_j}\mathrm{d}V}_{\text{IV}} - \underbrace{\int\limits_{O_K}\tau_{ij}U_j \,\mathrm{d}F_i}_{\text{V}} + \underbrace{\int\limits_{V_K}\tau_{ij}\frac{\partial U_j}{\partial x_j}dV}_{\text{VI}}$$

$$+\underbrace{\int\limits_{V_K}\rho g_j U_j dV}_{\text{VII}} + \underbrace{\sum \dot{E}}_{\text{VIII}}$$

This equation comprises 8 terms having the below-stated physical significations:

- I: Temporal change of the entire kinetic energy within the limits determining the control volume.
- II: Outflow minus inflow of the kinetic energy of the fluid per time unit over the entire surface of the considered control volume.
- III: Inflow minus outflow of "pressure energy" per time unit over the entire surface of the considered control volume.
- IV: Work done during expansion per time unit which is done by the entire control volume.
- V: Molecule-dependent input of kinetic energy of the considered fluid per time unit over the entire surface of the control volume.
- VI: The kinetic energy per time unit dissipated over the entire control volume which is transferred into heat.
- VII: Potential energy per time unit of the total mass in the entire control volume.
- VIII: Energy input per time unit over the surface of the control volume or power supplied to the fluid by flow-mechanical machines.

The differential form of the momentum equation j –. and the differential form of the mechanical energy equation do not represent independent equations, as the latter emanated from the first by multiplication by U_j, followed by various mathematical derivations and rearrangements of the different terms. This statement holds only in a restricted/limited way for the integral form of the basic equations. By addition of the term ΣK_j in equation and the term $\Sigma \dot{E}$ in equation it is possible that independent forms of the momentum equation and the mechanical energy equation come about. This is known from the treatment of impacts of spheres from mechanics for which the known momentum and energy equations from the equations can be derived as follows:

- The left side of equation yields for ρ = const for an integration over the entire sphere volume

$$\int_{V_K}\left[\rho\frac{\partial U_j}{\partial t}+\rho U_i\frac{\partial U_j}{\partial x_i}\right]\mathrm{d}V=\int_{V_K}\frac{D}{Dt}(\rho U_j)\,\mathrm{d}V=\frac{D}{Dt}\int_{V_K}\rho U_j\,dV\,,$$

and thus

$$\frac{D}{Dt}\int_{V_K}\rho U_j\,dV\;\frac{d}{\mathrm{d}\,t}(m_K U_j).$$

- the spheres 1 and 2 can be written:

$$\frac{d}{\mathrm{d}t}(m_K U_j)_1=(K_j)_1 \text{ and } \frac{d}{\mathrm{d}t}(m_K U_j)_2=(K_j)_2,$$

or transcribed, because of $(K_j)_1 = -(K_j)_2$:

$$\frac{d}{\mathrm{d}t}\left[(m_K U_j)_1+(m_K U_j)_2\right]=0\rightsquigarrow \boxed{(m_K U_j)_1+(m_K U_j)_2=\text{const}}.$$

- The left side of yields for ρ = const

$$\int_{V_K}\left[\rho\frac{\partial}{\partial t}\left(\frac{1}{2}U_j^2\right)+\rho U_i\frac{\partial}{\partial x_i}\left(\frac{1}{2}U_j\right)^2\right]dV$$

$$=\int_{V_K}\frac{D}{Dt}\left(\rho\frac{1}{2}U_j^2\right)=\frac{D}{Dt}\int_{V_K}\rho\frac{1}{2}U_j^2\mathrm{d}V\,,$$

and thus

$$\frac{\mathrm{d}}{\mathrm{d}t}\left(m_K\cdot\frac{1}{2}U_j^2\right)=\frac{D}{Dt}\int_{V_K}\rho\frac{1}{2}U_j^2\,\mathrm{d}V\,.$$

- equation yields for the spheres:

$$\frac{d}{\mathrm{d}t}\left(m_K\frac{1}{2}U_j^2\right)_1=(\dot{E})_1 \text{ and } \frac{d}{\mathrm{d}t}\left(m_K\cdot\frac{1}{2}U_j^2\right)_2=(\dot{E})_2,$$

or transcribed because of $(\dot{E})_1(\dot{E})_2=0$:

$$\frac{d}{\mathrm{d}t}\left[\left(m_K\frac{1}{2}U_j^2\right)_1+\left(m_K\frac{1}{2}U_j^2\right)_2\right]=(\dot{E})_1(\dot{E})_2=0,$$

$$\boxed{\left(m_K\frac{1}{2}U_j\right)_1+\left(m_K\frac{1}{2}U_j\right)_2=\text{const}}.$$

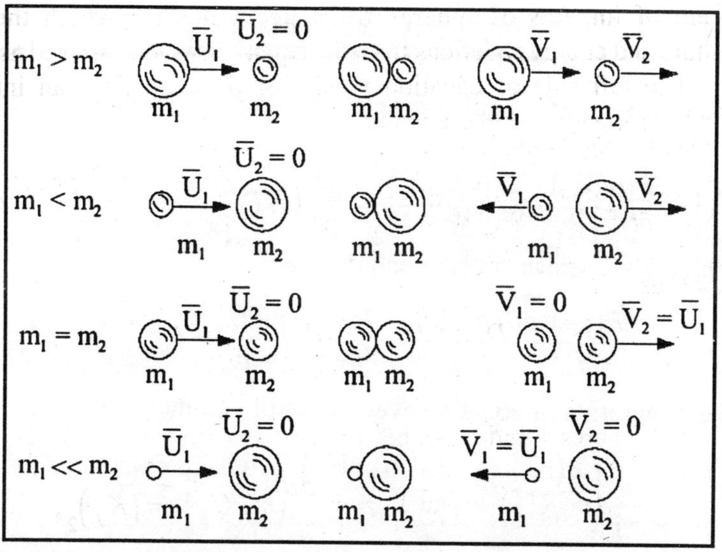

Fig. Possible Motions of Spheres following an Elastic Impact

The insights gained on the elastic impact by employing the formulas. The representations are stated for different mass ratios of the spheres. From the integral forms of the basic equations of flow mechanics result thus the impact laws for spheres which are known from lectures of mechanics in physics. This makes clear the general applicability of the integral form of the mechanical energy equation stated in equation.

INTEGRAL FORM OF THE THERMAL ENERGY EQUATION

The thermal energy equation was derived and stated for an ideal gas in equation as follows:

$$\rho c_u \left[\frac{DT}{Dt}\right] = \lambda \frac{\partial^2 T}{\partial x_i^2} - P\frac{\partial U_i}{\partial x_i} - \tau_{ij}\frac{\partial U_j}{\partial x_i}.$$

For an ideal fluid it was stated with equation:

$$\rho c_u \left[\frac{DT}{Dt}\right] = \lambda \frac{\partial^2 T}{\partial x_i^2} - \tau_{ij}\frac{\partial U_j}{\partial x_i}.$$

When one chooses equation for the further considerations, this equation can also be written:

$$\rho c_u \left[\frac{\partial T}{\partial t} + U_i \frac{\partial T}{\partial x_i}\right] = -\frac{\partial \dot{q}_i}{\partial x_i} - P\frac{\partial U_i}{\partial x_i} - \tau_{ij}\frac{\partial U_j}{\partial x_i}$$

Adding to equation the continuity equation multiplied by $c_v T$:

$$c_u dotT \frac{\partial \rho}{\partial t} + c_u T \frac{\partial (\rho U_i)}{\partial x_i} = 0$$

one obtains the initial equation for the derivation of the integral form of the thermal energy equation:

$$\frac{\partial(\rho c_u T)}{\partial t} + \frac{\partial(\rho c_u T U_i)}{\partial x_i} = -\frac{\partial \dot{q}_i}{\partial x_i} - P\frac{\partial U_i}{\partial x_i} - \tau_{ij}\frac{\partial U_j}{\partial x_i}.$$

With $c_v T = e$ (inner energy) one obtains:

$$\frac{\partial(\rho e)}{\partial t} + \frac{\partial(\rho e U_i)}{\partial x_i} = -\frac{\partial \dot{q}_i}{\partial x_i} - P\frac{\partial U_i}{\partial x_i} - \tau_{ij}\frac{\partial U_j}{\partial x_i}$$

The integration of equation over a control volume yields:

$$\int_{V_K} \frac{\partial(\rho e)}{\partial t} dV + \int_{V_K} \frac{\partial(\rho e U_i)}{\partial x_i} dV = -\int_{V_K} \frac{\partial \dot{q}_1}{\partial x_i} dV$$

$$-\int_{V_K} P\frac{\partial U_i}{\partial x_i} dV - \int_{V_K} \tau_{ij}\frac{\partial U_j}{\partial x_i} dV + \sum(\dot{Q} + \dot{E}).$$

Transcribed, in consideration of Gauss' integral theorem and the reversibility of the sequence of integration and differentiation, one obtains:

$$\underbrace{\frac{\partial}{\partial t}\left(\int_{V_K} \rho e \, dV\right)}_{\text{I}} + \underbrace{\int_{O_K} \rho e U_i \, dF_i}_{\text{II}} = -\underbrace{\int_{O_K} \dot{q}_i dF_i}_{\text{III}}$$

$$-\underbrace{\int_{V_K} P \cdot \frac{\partial U_i}{\partial x_i} dV}_{\text{IV}} - \underbrace{\int_{O_K} \tau_{ij}\frac{\rho U_j}{\partial x_i} dV}_{\text{V}} + \underbrace{\sum(\dot{Q} + \dot{E})}_{\text{VI}}.$$

The terms of the resulting equation can be interpreted as follows:

I: Temporal change of the inner energy within the control volume V_K.

II: Convective outflow and inflow of inner energy per time unit over the surface O_K of the control volume.

III: Molecular heat flow per time unit, i.e. the sum of the outflow and inflow, over the surface O_K of the control volume.

IV: The work done during expansion by the total volume per time unit

V: The mechanical energy dissipated per time unit in the total volume

VI: External heat and energy flow per time unit which is supplied to the total volume

The above equation holds likewise for an ideal fluid, but term IV is equal to zero, as no work can be done during expansion because of ρ = const.

APPLICATIONS OF THE INTEGRAL FORM OF THE BASIC EQUATIONS

The importance of the integral forms of the basic equations of flow mechanics becomes clear from applications that are listed below. Many manuals on the basics of flow mechanics treat flow problems of this kind. Typical examples are treated that make clear that the derived integral form of the basic equations represent the basis for a variety of problem solutions, where attention has to be paid to that solutions often can be derived only by employing simplifications. Reference is made to these simplifications and their implications for the obtained solutions in the framework of the derivations.

In order to introduce the reader into the methodically correct handling of the integral form of the equations, each of the problems treated below is solved by starting from the employed basic equation in each case. Then those terms in the integral form of the used basic equation are deleted which are equal to zero for the treated problem. In addition, by introducing simplifications terms are removed which have very little influence on the treated problem, so that easily comprehensible solutions are obtained.

Outflow from Containers

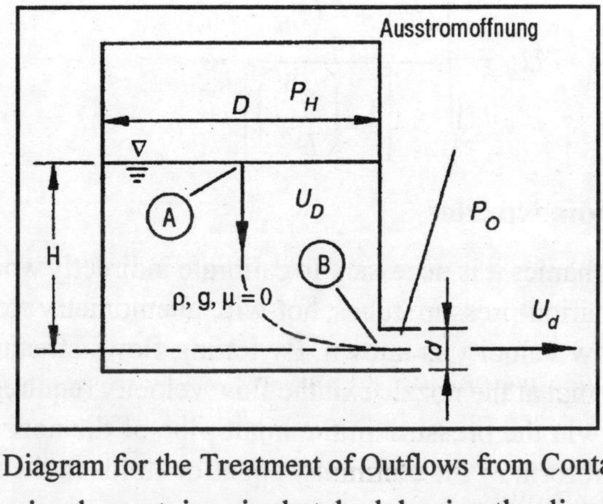

Fig. Diagram for the Treatment of Outflows from Containers

In Figure a simple container is sketched, having the diameter *D*, which is partly filled with a fluid and is assumed to be closed at the top. Between the fluid surface and the container lid there is a gas having the pressure P_H.

The fluid height is *H* and at the bottom of the container there is an opening with the diameter d. Sought is the outflow velocity from the container, i.e. the velocity U_d.

From the opposite diagram, i.e. from Figure it can be seen that the water surface is moving downwards with the velocity U_D. because of the fluid flowing out, which exits with U_d from the exhaust/escape opening. Via the integral form of the continuity equation holds:

$$\tilde{\rho}\tilde{U}F = \text{const} \rightsquigarrow \rho U_D \frac{\pi}{4} D^2 = \rho U_d \frac{\pi}{4} d^2 \rightsquigarrow \boxed{U_D = \frac{d^2}{D^2} U_d}.$$

By employment of the Bernoulli equation between the points (A) and (B) one obtains:

$$\frac{1}{2}U_D^2 + \frac{P_H}{\rho} + gH = \frac{1}{2}U_d^2 + \frac{P_0}{\rho}.$$

Thus it results:

$$\frac{1}{2}U_d^2 = \frac{1}{2}U_D^2 + gH + \frac{1}{\rho}(P_H - P_0),$$

or after insertion of:

$$\frac{1}{2}U_d^2 = \frac{1}{2}\left(\frac{d^4}{D^4}\right)U_d^2 + gH + \frac{1}{\rho}(P_H - P_0),$$

$$\frac{1}{2}U_d^2 = \frac{1}{2}\left(\frac{d^4}{D^4}\right)U_d^2 + gH + \frac{1}{\rho}(P_H - P_0),$$

$$U_d = \sqrt{\frac{2gH + \frac{2}{\rho}(P_H - P_0)}{\left[1 - \left(\frac{d^4}{D^4}\right)\right]}}.$$

Exit Velocity from Nozzles

In flow mechanics it is necessary to calibrate indirectly working measuring processes (stagnation-pressure tubes, hot-wire anemometry etc.) in flow fields in which the flow velocity is known. By letting flows stream into nozzles it can be achieved that at the nozzle exit the flow velocity required for calibration can be adjusted via the pressure in the input pipe of the nozzle.

The integral form of the continuity equation holds in the following form:

$$\dot{m} = \tilde{\rho}\tilde{U}F = \text{const} \rightsquigarrow \rho U_A \frac{\pi}{4} D^2 = \rho U_B \frac{\pi}{4} d^2,$$

i.e. one can write for U_A:

$$U_A = \frac{d^2}{D^2} U_B.$$

For the planes (A) and (B) it can be written as a result of the Bernoulli equation:

$$\frac{1}{2}U_A^2+\frac{P_A}{\rho}=\frac{1}{2}U_B^2+\frac{P_B}{\rho}=\frac{1}{2}\frac{d^4}{D^4}\cdot U_B^2+\frac{P_A}{\rho}.$$

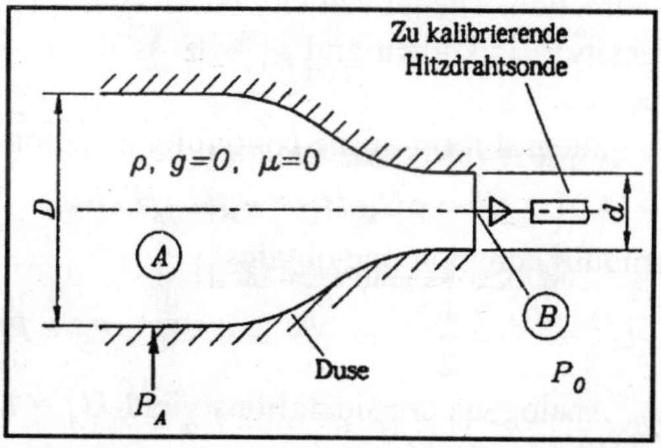

Fig. Diagram of a Nozzle-Calibrating Length for Velocity-Measuring Tubes

From this follows:

$$U_B=\sqrt{\frac{2(P_A-P_B)}{\rho\left(1-\frac{d^4}{D_4}\right)}}.$$

When one chooses $D \approx d$ one obtains for U_B, in good approximation:

$$U_B=\sqrt{\frac{2}{\rho}(P_A-P_B)}=\sqrt{\frac{2}{\rho}(P_A-P_0)}.$$

By adjusting different P_A – values, the entire velocity regime required for the calibration of measuring tubes can be set.

Momentum on a Plane Vertical Plate

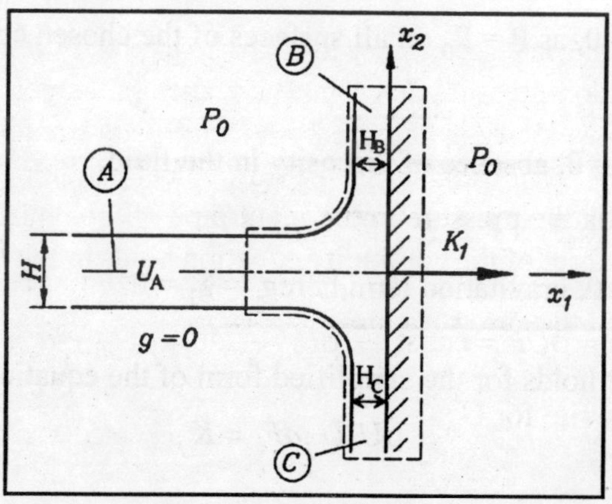

Fig. Diagram for the Consideration of the Momentum on a Vertical Plate

When flowing fluid jets are decelerated, forces appear which are used in many technology fields. The question remains as to which force is to be applied in the x_1 – direction to prevent the deflection of the plate due to the momentum of the flat fluid jet. The flat jet has the thickness H in the x_2 –direction and the width b in the x_3 –direction. The jet velocity far away from the plate is known and is U_A The density ρ is known and $g_1 = 0$, as the x_1 -direction axis is horizontally placed.

Employing the integral form of the continuity equation it results:

$$\rho U_A Hb = \rho U_C H_C b + \rho U_B H_B b.$$

From the Bernoulli equation one obtains:

$$\frac{1}{2}U_A^2 + \frac{P_A}{\rho} = \frac{1}{2}U_B^2 + \frac{P_B}{\rho} \text{ and } P_A = P_B = P_0$$

and thus $U_A = U_B$. Analogous considerations yield $U_A = U_C$. Owing to the symmetry of the problem one obtains:

$H = 2H_C = 2H_B$; d.h. $H_C = H_B$.

For the solution of the problem the integral form of the momentum equation can be employed, as it is stated in equation

$$\frac{\partial}{\partial t}\int_{V_K} \rho U_j dV + \int_{O_K} \rho U_i U_j dF_i = -\int_{O_K} P \cdot dF_j$$

$$-\int_{O_K} \tau_{ij} dF_i + \int_{V_K} \rho g_i dV + \sum K_j .$$

For the present problem the following simplifications of this universally valid equation hold:

$\frac{\partial}{\partial t}\int_{V_K} \rho U_j dV = 0$, stationary flow problem,

$\int_{O_K} P\, dF_j = 0$, as $P = P_0$ on all surfaces of the chosen control volume,

$\int_{O_K} \tau_{ij} \cdot dF_i = 0$, absence of viscosity in the fluid,

$\int_V \rho g_j dV = 0$, gravitation term here$g_j = g_1 = 0$.

Therefore it holds for the simplified form of the equation:

$$\int_{O_K} \rho U_i U_j dF_i = K_j$$

and thus one obtains by integration for $j = 1$:

$$K_1 = -\rho U_A^2 HB.$$

The result of the above derivations shows that K_1. must act in the negative $x1$ -direction, in order to prevent the deflection of the plane plate by the incoming flat fluid jet. This gives an example of which kind the force terms are ΣK_j that occur in equation. All forces are included that act on the considered control volume.

Momentum on an Inclined Plane Plate

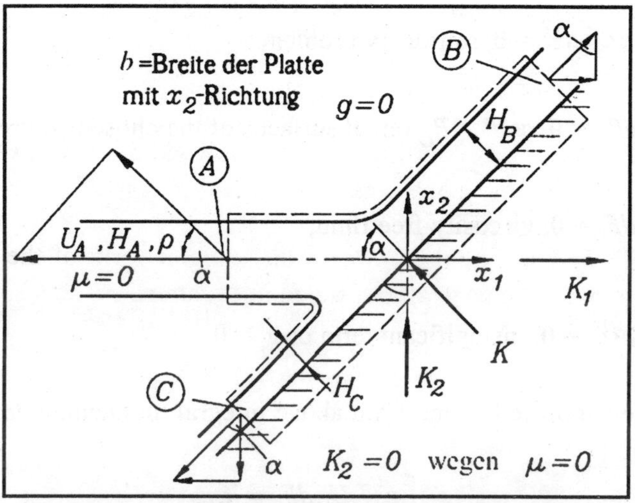

Fig. Diagram for Explaining the Independence of the Considerations from the Coordinate System

For a fluid jet hitting an inclined plane plate, because of the inclination of the plate, which encloses the angle a with the axis of the incoming flat fluid jet, the jet splits up in two jets of unequal thickness. The thicker jet goes upwards and has the height $h_B = \varepsilon H_A$. The thinner jet goes downwards and has the height $h_C = (1 - \varepsilon)H_A$.

This results from the continuity equation. $U_A = U_B = U_C$ results because $g = 0$ from the employment of Bernoulli equation.

When employing the continuity equation in integral form, i.e. when considering that for the solution of the problem the mass conservation can be used, it follows:

$\rho U_A H_A b = \rho U_B h_B b + \rho U_C h_C b$, or else because of $U_A = U_B = U_C$ from the Bernoulli equation

$U_A H_A = U_B h_B + U_C h_C \rightsquigarrow H_A = h_B + h_C$.

For the two split jets forming on the plate it can therefore be stated:

$h_B = \varepsilon H_A$ and $h_C = (1-\varepsilon)H_A$.

When one employs the integral momentum equation:

$$\frac{\partial}{\partial t}\int_{V_K} \rho U_j dV + \int_{O_K} \rho U_i U_j dF_i = -\int_{O_K} P\, dF_j$$

$$-\int_{O_K} \tau_{ij} dF_i + \int_{V_K} \rho g_i dV + \sum K_j .$$

the following simplifications can be introduced:

$\frac{\partial}{\partial t}\int_{V_K} \rho U_j dV = 0$, stationary problem,

$\int_{O_K} P\, dF_j = 0$, as $P = P_O$ on all surfaces of the chosen control volume,

$\int_{O_K} \tau_{ij} dF_i = 0$, viscosity-free fluid,

$\int_{V_K} \rho g_i dV = 0$, insignificant term or $g_j = 0$

For the simplified form of the above integral momentum equation thus holds:

$$\int_{O_K} \rho U_i U_j dF_i = K_j .$$

When one chooses a coordinate system oriented by the plate, then for K_p result by integration over the planes (*A*), (*B*) and (*C*) three contributions which are stated below:

$$K_P = -\rho U_A^2 H_A b \sin\alpha + \rho U_A^2 \varepsilon H_A b - \rho U_A^2 (1-\varepsilon) H_A b .$$

As in the present problem $\mu = 0$ was set, for the force $K_P = 0$. acting along

the plate attacked by flow, From this results from equation

$$-1-\sin\alpha + 2\varepsilon = 0,$$

or one obtains for ε:

$$\varepsilon = \frac{1}{2}(1+\sin\alpha).$$

For the force acting vertically to the plate it is computed K_S:

$$K_S = \rho U_A^2 H_A\, b \cos\alpha.$$

It is evident that the considerations carried out above have to be independent from the chosen coordinate system. When one chooses the coordinate system one obtains for K_1 the below-stated contributions by integration over the planes (*A*), (*B*) and (*C*):

$$K_1 = -\rho U_A^2 H_A b + \rho U_A^2 \varepsilon H_A b \ \sin\alpha - \rho U_A^2 (1-\varepsilon) H_A b \sin\alpha.$$

For the force K_2 yields:

$$K_2 = \rho U_A^2 \varepsilon H_A b \ \cos\alpha - \rho U_A^2 (1-\varepsilon) H_A b \sin\alpha.$$

As because of $\mu = 0$ the total force on the plate K resulting from K_1 and K_2 has to act on the plate vertically, it holds:

$$\tan\alpha = \frac{K \sin\alpha}{-K_1} = \frac{K_2}{K \cos\alpha} = \frac{2\in\cos\alpha - \cos\alpha}{1 - 2\varepsilon\sin\alpha + \sin\alpha}.$$

From this it is computed $2\varepsilon \cos^2\alpha - \cos^2\alpha = \sin\alpha - 2\varepsilon \sin^2\alpha + \sin^2\alpha$ or for

$$\varepsilon = \frac{1}{2}(1 + \sin\alpha).$$

Jet Deflection at an Edge

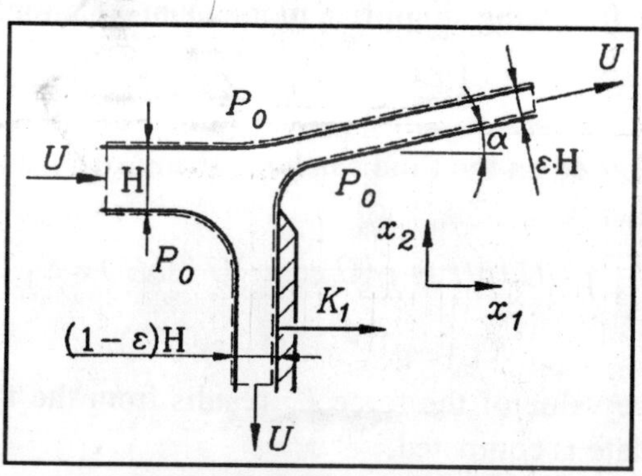

Fig. Jet Deflection at an Edge

When a fluid jet (height H, width b) hits with part of its cross-sectional area a plate standing vertically to the jet, the arriving fluid is partitioned in two partial jets.

One of the two partial jets runs vertically to the original jet direction downwards along the plate, the other partial jet is deflected upwards around the angle α opposite the original jet direction. Neglecting viscosity forces and gravitational forces and assuming a constant ambient pressure from the Bernoulli equation, it results that the two partial jets have the same velocity each, which is equal to the velocity of the fluid in the original jet. Because of the continuity equation the two partial jets have the jet heights εH and $(1 - \varepsilon)H$. In the momentum equation:

$$\frac{\partial}{\partial t}\int_{V_K}\rho U_j dV + \int_{O_K}\rho U_i U_j dF_i =$$
$$-\int_{O_K} P\, dF_j - \int_{O_K}\tau_{ij}\, dF_i + \int_{V_K}\rho g_j dV + \sum K_j$$

the following simplifications can be made:

$\frac{\partial}{\partial t}\int_{V_K}\rho U_j dV = 0$, stationary problem,

$\int_{O_K} P\, dF_j = 0$, constant pressure along the surface of the control volume,

$\int_{O_K}\tau_{ij}\, dF_i = 0$, viscosity forces are neglected,

$\int_{O_K}\rho g_j dV = 0$, gravitation is neglected..

It results the following simplified momentum equation:

$$\int_{O_K}\rho U_i U_j dF_i = \sum K_j .$$

The force exerted on the fluid can be determined by the equation for the x_1 – components.

$$-\rho UU\,(bH) + \rho(U\cos\alpha)U\,(\varepsilon bH) = K_1.$$

$$K_1 = -\rho U^2 bH\,(1-\varepsilon\cos\alpha)$$

The negative value of the force K_1 results from the fact that the force exerted on the plate is computed.

From the equation for K_2

$\rho U(1-\varepsilon)Hb(-U) + \rho U\varepsilon bH(U\sin\alpha) = 0$, results for $K_2 = 0$ the connection between the deflection angle α and the ratio ε:

$$-\rho U^2 bH\,[(1-\varepsilon) - \varepsilon\sin\alpha] = 0,$$

$$\varepsilon = \frac{1}{1+\sin\alpha}.$$

Thus the splitting up of the jet in can be determined from the deflection of the jet from the horizontal position, i.e. by measuring the angle α.

Mixing Process in a Pipe of Constant Cross-section

In a pipe two flows at a constant velocity (U_A, U_B). each are mixed with one another($\mu = 0$). The pressure at point 1 and the partial areas (U_A, U_B). and A_B. shall be given. Sought is the pressure P_1 at point 2 where a constant velocity $\vec{U}_C$. over the pipe cross-section has set-in.

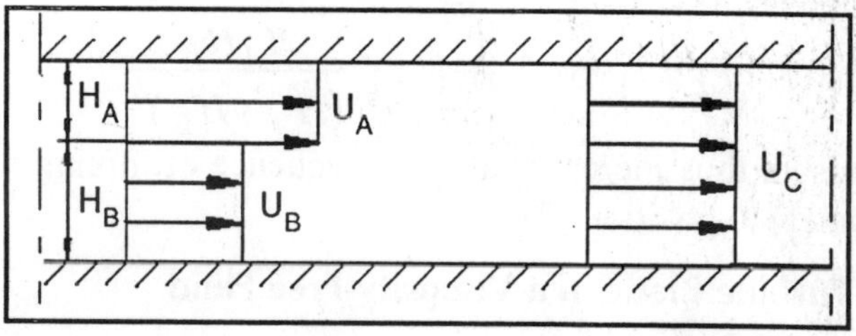

Fig. Diagram for Explaining the Mixing Process

From the integral form of the continuity equation one obtains:

$$b\left(-\rho H_A U_A - \rho H_B U_B + P\left(H_A + H_B\right)\vec{U}_C\right) = 0,$$

or rearranged for $\vec{U}_C$:

$$\vec{U}_C = \frac{U_A H_A + U_B H_B}{H_A + H_B}.$$

The momentum equation

$$\frac{\partial}{\partial t}\int_{V_K} \rho U_j dV + \int_{O_K} \rho U_i U_j dF_i = -\int_{O_K} P dF_j$$

$$-\int_{O_K} \tau_{ij} dF_i + \int_{V_K} \rho g_i dV + \sum K_j .$$

can be simplified as follows:

$\frac{\partial}{\partial t}\int_{V_K} \rho U_j dV = 0$, stationary flow problem,

$\int_{O_K} \tau_{ij} dF_i = 0, \mu = 0$, i.e. the assumption of absence of viscosity,

$\int_{V_K} \rho g_j dV = 0$, no component of gravitation in horizontal direction,

$\sum K_j = 0$, no external forces act on the control volume.

From this follows:

$$\int_{O_K} \rho U_i U_j dF_i = -\int_{O_K} P\, dF_j .$$

For the present problem this results in:

$$-\rho U_A^2 H_A - \rho U_A^2 H_B + \rho \vec{U}^2 (H_A + H_B) = (P_1 - P_2)(H_A + H_B).$$

When one now inserts the above expression for $\vec{U}$ and solves the equation for P_2 one obtains:

$$P_2 = P_1 + \rho (U_A - U_B)^2 \frac{H_A H_B}{(H_A + H_B)^2}.$$

The pressure thus increases as a consequence of mixing the two flows from position (1) to position (2).

Force on a Turbine Blade in a Viscosity-Free Fluid

In the flow machines blade wheels are used to exploit the momentum of fluid jets for propulsion purposes. A jet from a rectangular nozzle (flat jet having the width H, and the depth b)hits a stationary blade which deflects the jet symmetrically to two sides around the angle 180 ± β. The sizes of the inflowing and out flowing fluid flows are given. The pressure along the surface of the diagrammed control volume is equal all over to the ambient pressure, so that the integral form of the momentum equation

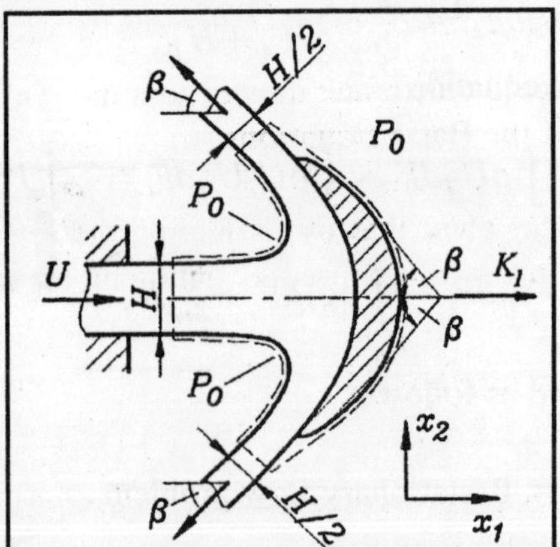

Fig. Diagram for Explaining the Force on a Turbine Blade

$$\frac{\partial}{\partial t}\int_{V_K} \rho U_j dV + \int_{O_K} \rho U_i U_j dF_i = -\int_{O_K} P\, dF_j$$

$$-\int_{O_K} \tau_{ij} dF_i + \int_{V_K} \rho g_i dV + \sum K_j$$

can be simplified in the following way, the gravitation being negligible.

$$\frac{\partial}{\partial t}\int_{V_K} \rho U_j dV = 0, \text{ Stationary flow problem,}$$

$$\int_{O_K} P\, dF_j = 0, \text{ as } P = P_0 \text{ on all surfaces of the control volume,}$$

$$\int_{O_K} \tau_{ij} dF_i = 0, \text{ as} \mu = 0 \text{ was set to zero,}$$

$$\int_{V_K} \rho g_i dV = 0, \text{ as gravitation is negligible.}$$

Thus it results:

$$\int_{O_K} \rho U_i U_j \, dF_i = \sum K_j .$$

As the problem is symmetrical, only the horizontal component $j = 1$ has to be considered, i.e. in $j = 2$ no resultant force appears so that the conservation of momentum can be written as follows:

$$-\rho H b U^2 + 2\frac{1}{2}\rho H b U\left(-U \cdot \cos \beta\right) = K_1 .$$

The resultant force on the blade thus results in

Integral Forms of the Basic Equations

$K_S = -K_1 = \rho H b U^2 (1 + \cos \beta)$.

The formula makes clear that by deflecting the jets in direction of the incoming flow, an increase of the force K_1 acting on the turbine blade can be obtained.

Force on a Periodical Blade Grid

Assumptions: $g = 0$, $\mu = 0$

Two-dimensional blade grids are used in flow machines in order to exploit forces caused by flows for propulsion purposes, e.g. in turbines for driving mechanical aggregates.

$$\vec{U}_A = \{(U_1)A, (U_2)A, 0\}$$

or U_3 – component of the flow field is zero. Because of the deflection of the flow by the blade grid the flow-o velocity differs from the attack velocity, so that it holds:

$$\vec{U}_B = \left\{(U_1)_B, (U_2)_B, 0\right\} \neq \vec{U}_A .$$

The statements for $\vec{U}_A$ and $\vec{U}_B$, for which always holds

$$(U_3)_A = (U_3)_B = 0$$

make clear that the flow field remains "two-dimensional" and also has two components. Thus for the attack-flow/inflow and off-flow the following

equation can be derived from the continuity equation, where B is the width of the blade grid in x_3 –direction:

$$\rho(U_1)_A tB = \rho(U_1)_B tB.$$

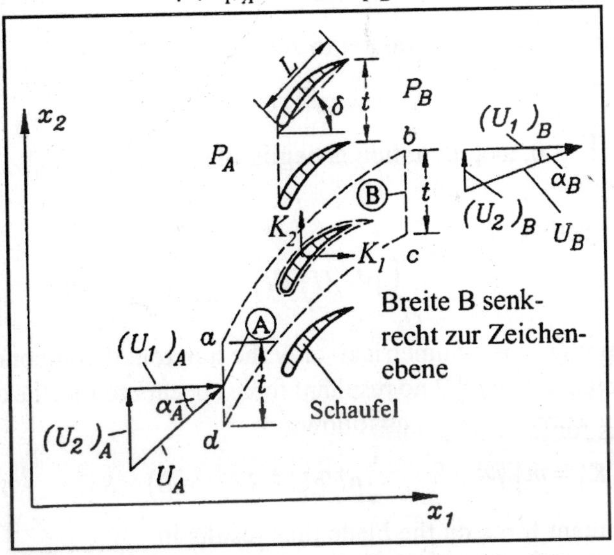

Fig. Diagram Explaining the Effect of Blade Grids

For the attack-flow/inflow and off-flow thus the same velocity components in x_1 -direction result. The purpose of blade grids is therefore to change the velocity component $(U_2)_A$ of the attack-flow.

For the computation of the force on a grid blade a control volume (a, b, c, d). is chosen. In x_2 – direction two flow lines are chosen as boundary of the control volume which are positioned along $(a-b)$ and $(d-c)$. with the distance of the blade grid t. Thus the inflow and outflow to the chosen control volume take place only over the areas $(d-a)B$. and $(c-d)B$. When one neglects the gravitational effects, the Bernoulli equation yields the following connection:

$$\frac{P_A}{\rho}+\underbrace{\frac{1}{2}\left[(U_1)_A^2+(U_2)_A^2\right]}_{\text{Entire kinetic enery in (A)}}=\frac{P_B}{\rho}+\underbrace{\frac{1}{2}\left[(U_1)_B^2+(U_2)_B^2\right]}_{\text{Entire kinetic enery in (B)}}.$$

For computation of the forces on the blade the integral form of the momentum equation is at disposal:

$$\frac{\partial}{\partial t}\left(\int_{V_K}\rho U_j dV\right)+\int_{O_K}\rho U_i U_j dF_i =$$

$$-\int_{O_K} P\, dF_j - \int_{O_K}\tau_{ij}\, dF_i + \int_{V_K}\rho g_j dV + \sum K_j$$

With the assumptions one obtains from :

$$\frac{\partial}{\partial t}\int_{V_K} \rho U_j dV_P = 0,$$ as there is a stationary flow,

$$\int_{O_K} \tau_{ij}\, dF_i = 0,$$ as the is a viscosity-free fluid,

$$\int_{V_K} \rho g_j dV = 0,$$ as the gravitational forces are negligible,

the momentum equation results in the simplified relation:

$$\int_{O_K} \rho U_i U_j dF_i = -\int_{O_K} P\, dF_j + \sum K_j \quad \text{Oder}$$

$$K_j = \int_{O_K} \rho U_i \cdot U_j\, dF_i + \int_{O_K} P\, dF_j$$

Thus for the forces in the directions $j = 1$ and $j = 2$ results:

$$K_1 = \dot{m}\underbrace{\left[(U_1)_B - (U_1)_A\right]}_{=0} + (P_B - P_A)B\cdot t = (P_B - P_A)B\cdot t,$$

$$K_2 = \dot{m}\left[(U_2)_B - (U_2)_A\right].$$

From the Bernoulli equation one obtains:

$$(P_B - P_A) = \frac{\rho}{2}\left[(U_2)_A^2 - (U_2)_B^2\right].$$

Thus for K_1 results:

$$K_1 = \frac{\rho}{2}Bt\left[(U_2)_A^2 - (U_2)_B^2\right],$$

or expressed with the attack-flow angles and the flow-off/off -flow) angles:

$$K_1 = \frac{\rho}{2}Bt\left[U_A^2 \sin^2\alpha_A - U_B^2 \sin^2\alpha_B\right],$$

and

$$K_2 = -\rho\cdot Bt\left[U_A \sin\alpha_A - U_B \sin\alpha_B\right].$$

Here K_1 and K_2 are the forces acting on the control volume. The forces acting on the blades are:

$$(K_1)_s = -K_1 = -\frac{\rho}{2}Bt\left[U_A^2 \sin^2\alpha_A - U_B^2 \sin^2\alpha_B\right],$$

$$(K_2)_s = -K_2 = -\rho Bt\left[U_A \sin\alpha_A - U_B \sin\alpha_B\right].$$

The blade thus experiences a force $(K_1)_s$ in the negative x_1 – direction and a force $(K_B)_2$ in the positive x_2 – direction.

Euler's Turbine Equation

The considerations carried out related to a blade grid arranged in a plane, which is located in a plane flow. When disposing the blade grids radially on a rotating running wheel as diagrammed, one obtains the basic arrangement of a radial turbine. The first part of the term radial turbine, i.e. the word "radial" designates the main flow direction, in which the flow through the turbine blades takes place, namely radially from the inside to the outside. When employing the continuity equation, one obtains:

$$\rho(U_r)_A\,(2\pi r_A)\,B = \rho(U_r)_B\,(2\pi r_B)B$$

or otherwise stated, the demand for mass conservation results:

$$(U_r)_B = \frac{r_A}{r_B}(U_r)_A \,.$$

From the Benoulli equation results:

$$\frac{P_A}{\rho} + \frac{1}{2}\left[(U_r)_A^2 + (U_t)_A^2\right] = \frac{P_B}{\rho} + \frac{1}{2}\left[(U_r)_B^2 + (U_t)_B^2\right].$$

From the integral form of the momentum equation can be determined for the force in radial direction:

$$K_r = \dot{m}\left[(U_r)_B - (U_r)_A\right] + 2\pi(r_3 P_B - r_A P_A)b.$$

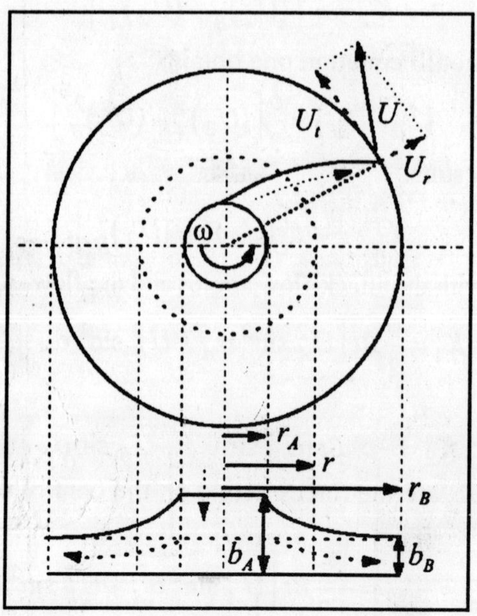

Fig. Diagram of the Flow through a Radial Turbine

Furthermore, for the force in tangential direction holds:

$$K_t = \dot{m}\left[(U_t)_B - (U_t)_A\right].$$

For the moment imposed on the control volume by the running wheel results:

$$M_t = \dot{m}\left[r_B\left(U_t\right)_B - r_A\left(U_t\right)_A\right].$$

The mechanical power output connected with the turbine amounts to:

$$P_{\text{Turb}} = -M_t\omega = -\dot{m}\omega\left[r_B\left(U_t\right)_B - r_A\left(U_t\right)_A\right].$$

Equation as well as equation are called Euler's turbine equation in the literature. Here for $r_A = r_e =$. the "inflow radius" of the blade rim is assumed and for $r_B = r_a =$. the "outflow radius".

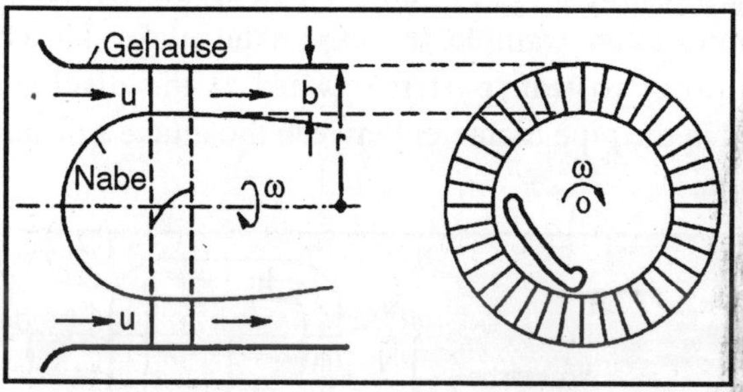

Fig. Diagram of the course of the Flow through an Axial Turbine

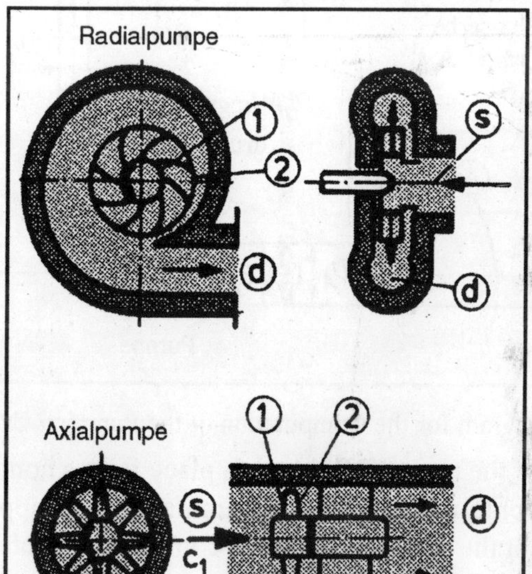

Fig. Schematic Representation of Radial and Axial Pumps

The resulting equation does not only hold for turbines, but generally for flow machines, like compressors, air blowers (ventilators), pumps etc. Here pumps and turbines differ in the considerations carried out only with regard to the sign of the energy exchange between running wheel and flowing fluid.

In a turbine energy is extracted from the fluid flow, so that one can collect a usable moment at the shaft of this power engine, or the corresponding energy is at disposal, respectively. In a pump on the other hand energy is supplied to the fluid flow via the "running wheel, i.e. a torque for driving the machine is exerted at the shaft. Thus the energy necessary for pumping is supplied.

Finally it is mentioned that Euler's turbine formula can also be applied to flow machines through which flows pass axially.

Power of Flow Machines

A typical application of a flow machine, in the chosen case the application of a pump, is shown as an example. It sucks on the intake side a certain quantity of water ṁ in order to transport it upward at the discharge side. Here, differences exist in the pipe diameter between the intake side and the discharge side of the pump.

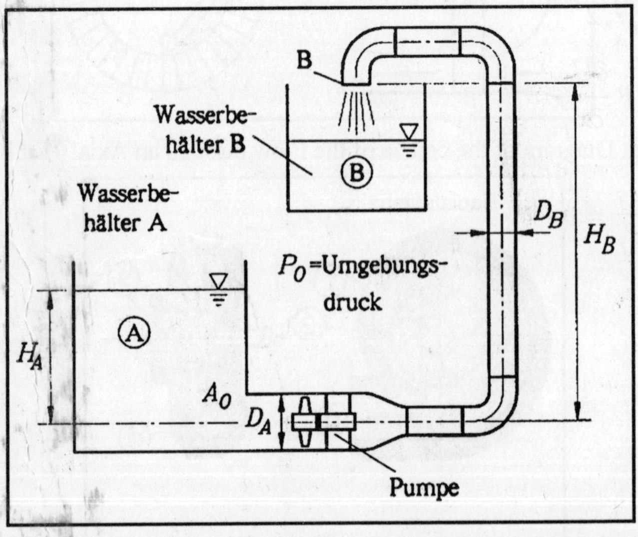

Fig. Diagram for the Computation of the Pumping Capacity

The suction of the pumped fluid takes place from a container (A) and the transport into a container (B), as diagrammed in the opposite figure. All quantities located at the suction side of the considered problem are designated by index A, the quantities on the discharge side of the pump by index B.

From the integral form of the continuity equation results for the considered problem:

$$\rho_A \frac{\pi}{4} D_A^2 \tilde{U}_A = \rho_B \frac{\pi}{4} D_B^2 \tilde{U}_B$$

For $\rho_A = \rho_B = \rho = \text{const}$ results:

$$D_A^2\tilde{U}_A = D_B^2\tilde{U}_B = \frac{\dot{m}}{\rho} = \dot{V}$$

For the computation of the pump capacity it is recommended to employ the integral form of the mechanical energy equation, as stated in equation:

$$\frac{\partial}{\partial t}\int_{V_K}\frac{1}{2}\rho U_j^2 dV + \int_{O_K}\rho U_i\left(\frac{1}{2}U_j^2\right)dF_i = -\int_{O_K}PU_j dF_j + \int_{V_K}P\frac{\partial U_j}{\partial x_j}dV$$

$$-\int_{O_K}\tau_{ij}U_j dF_j + \int_{V_K}\tau_{ij}\frac{\partial U_j}{\partial x_i}dV + \int_{V_K}\rho g_j\dot{V}\,dx_j + \sum\dot{E}$$

When reducing this equation by assumptions that apply to the present problem:

$\frac{\partial}{\partial t}\int_{V_K}\frac{1}{2}\rho U_j^2 dV = 0$ stationary pumping conditions,

$\int_{V_K}P\frac{\partial U_j}{\partial x_j}dV = 0$ no work done during expansion, as r = const and thus

$\frac{\partial U_j}{\partial x_j} = 0,$

and neglecting the t_{ij} -terms in the above integral energy equation, then it holds:

$$\int_{O_K}\rho U_i\left(\frac{1}{2}U_j^2\right)dF_i = -\int_{O_K}PU_j dF_j + \int_{V_K}\rho g_j U_j d\dot{V} + \sum\dot{E}$$

or

$$-\dot{m}\left[\frac{1}{2}\tilde{U}_A^2 + \frac{1}{2}\tilde{U}_B^2\right] = \dot{V}_A P_A - \dot{V}_B P_B + pg\dot{V}H_C + p_m,$$

where P_m represents the power introduced by the pump into the fluid. For the capacity of the pump thus holds:

$$P_m = \dot{V}\left\{\left(\frac{\rho}{2}\tilde{U}_B^2 + P_B + \rho g H_C\right) - \left(\frac{\rho}{2}\tilde{U}_A^2 + P_A\right)\right\}.$$

When one considers now P_B and P_A more in detail, it holds:

$P_B = P_0$ sowie $P_A = P_0 + \rho g H_A - \frac{\rho}{2}U_A^2$, and $\tilde{U}_A \approx 0$ and $U_A = \frac{\dot{V}}{F_A}$

sowie $U_B = \frac{\dot{V}}{F_B}$,

So that one obtains the following relation:

$$P_m = \dot{V}\left\{\frac{\rho}{2}\left(\frac{\dot{V}^2}{F_B^2}-\frac{\dot{V}^2}{F_A^2}\right)-\rho q H_A+\frac{\rho}{2}\left(\frac{\dot{V}}{F_A}\right)^2+\rho g H_C\right\},$$

or summarized:

$$P_m = \frac{\rho}{2}\left(\frac{\dot{V}^3}{F_B^2}\right)+\rho q\left(H_C - H_A\right)\dot{V}.$$

For the electrical capacity P_e of the pump results, with η as efficiency factor:

$$P_e = \frac{1}{\eta}P_m = \frac{1}{\eta}\left[\frac{\rho\dot{V}^3}{2F_B^2}+\rho q\dot{V}\left(H_C - H_A\right)\right]$$

For the pumping capacity results that the stated electrical power is required to supply the kinetic energy of the fluid leaving the pipe per time unit, plusthe power required per time unit for surmounting the hydrostatic pressure level.

Chapter 8

Stream Tube Theory

General Considerations

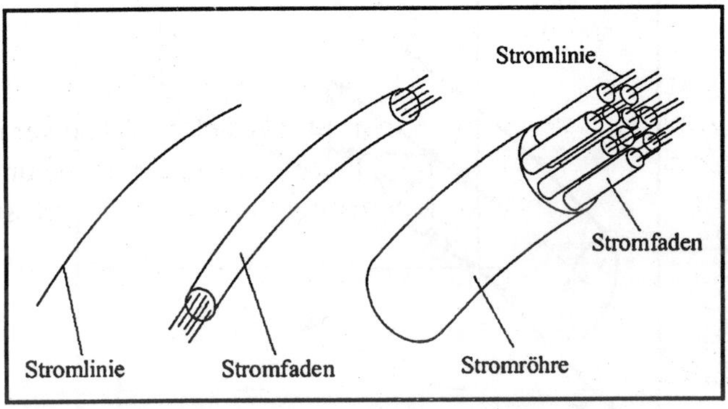

Fig: Stream Line, Stream Filament and Stream Pipe

The preceding considerations on the derivation of the integral form of the basic equations can also be used advantageously to derive simplified equations applicable to so-called stream filaments and to put them to use for the solution of flow problems. Here one starts the considerations from flow lines that are introduced as lines of a flow area and which at a certain point in time possess/control the direction of the flow at each point of the flow field. One can imagine a stream filament built up from a bundle of such flow lines and one can make a subdivision of the entire flow field into a multitude of stream filaments. Furthermore, it is possible to bundle stream filaments to obtain stream pipes.

Here, attention has to be given to that considerations on properties of flows applied to stream filaments can only be employed advantageously when the stream quantities assigned to each area of the stream filament can be considered to be constant over the cross-section of the stream filament. This makes it occasionally necessary to choose the cross-sectional area of a stream filament su ciently small, so that for the considered problem the assumption of uniform state and flow quantities over the cross-sectional area of the stream filament can be fulfilled su ciently precisely.

For the stationary stream tube theory it results that the fluid elements appertaining to a stream filament appertain to this stream filament permanently. Fluid particles that are located outside a stream filament at a certain point in time, can never become components of the considered filament. Each fluid particle of a stationary flow area belongs to a certain stream filament though, so that it is possible to describe the properties of the flow area by the properties of the considered stream filaments.

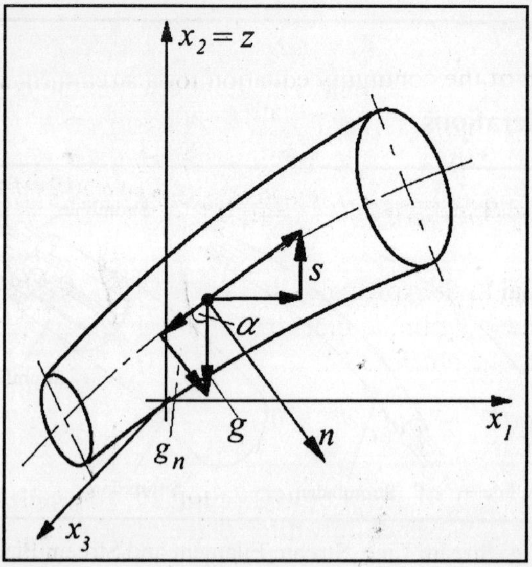

Fig: Stream Filament with Introduced $n - s$ – coordinate system

To simplify the considerations on stream filaments the below-stated assumptions on stream filaments are introduced:

- A stream filament is always completely filled with the fluid for which the flow considerations are carried out.
- The section changes permissible along a stream filament are small
- A stream filament is assumed to be weakly/slightly curved in flow direction

Although the assumptions introduced above for stream filaments represent considerable limitations, the derivations given in the following sections show that the introduction of stream filaments into flow-mechanical considerations leads to equations via which physically clear/illustrative solutions of flow problems can be stated.

The considerations carried out on the basis of stream filaments show that in some cases the properties of entire flow fields can be described by the properties of stream filaments. When the flow quantities change only little over the entire cross-sections of internal flows, the basic equations derived for stream filaments of small dimensions can also be employed to acquire/

record the most important properties of internal flows by a one-dimensional flow theory. To this purpose the internal flow is treated as a single stream pipe.

The justification for this is given only, however, when friction influences are small or can be neglected for first considerations of flows.

DERIVATIONS OF THE BASIC EQUATIONS

Continuity Equation

The derivation of the continuity equation for a stream filament builds up on the differential form of mass conservation and which, after integration over a control volume, having employed Gauss' integral theorem:

$$\frac{\partial \rho}{\partial t}+\frac{\partial(\rho U_i)}{\partial x_i}=0 \rightsquigarrow \int_{V_c}\left(\frac{\partial \rho}{\partial t}\right) dV + \int_{O_c} \rho U_i dF_i = 0$$

Here V_c is equal to the volume of the considered control volume and O_c is its surface. Exchanging in the first term of the above equation integration and differentiation, one obtains:

$$\frac{\partial}{\partial t}\int_{V_c} \rho dV + \int_{O_c} \rho U_i dF_i = 0 \rightsquigarrow \frac{\partial M_c}{\partial t} = -\int_{O_c} \rho U_i dF_i$$

Applying the form of mass conservation to a stream filament and considering that the flow passes only through the cross-sectional areas of the stream filament, for $\frac{\partial M_c}{\partial t}= 0$ (stationary flow conditions) it results that the mass inflows and outflows for a stream filament are the same.

$$\int_{F_A} \rho U_i df_i = \int_{F_B} \rho U_i df_i \rightsquigarrow F_A U_{s,A\rho A} = F_B U_{s,B\rho B}$$

where the plane of the area F stands vertically to the flow direction s Therefore it results that the mass flow $\dot{m} = \rho F\, U_s$ along a stream filament is constant.

In the considerations carried out above it was already said that because of small cross-sectional stream filaments F, ρ and U_s can be set equally constant over F When one wants to apply the considerations also to stream pipes, the considerations have to be carried out in a refined/improved way.They have to consider that the assumption of constant density and velocity in the presence of large cross-sectional areas is only given conditionally. The introduction of cross-sectionally averaged quantities into the considerations carried out is necessary as has been shown below.

When carrying out the following averaging, with employment of the meanvalue theorem:

Stream Tube Theory

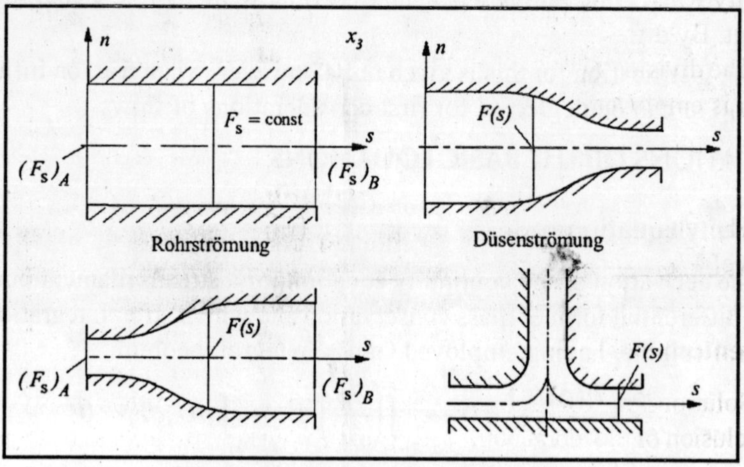

Fig. Flows which can be Computed One-Dimensionally by the Approximation Method

$$\widetilde{\rho U_s} = -\frac{1}{F_s}\int_{F_a} \rho U_s df_s$$

it can be stated for stationary flow considerations :

$$\overbrace{\frac{\partial M_c}{\partial t}}^{=0} = -\int_{V_c} \rho U_i dF_i \rightsquigarrow \left(\widetilde{\rho U_s}\right)_A F_A = \left(\widetilde{\rho U_s}\right)_B F_B$$

This relation for the mass flows is often simplified further in stream-tube theory by assuming $\widetilde{\rho U_s} = \tilde{\rho}\tilde{U}_s$ i.e. the relation is employed as follows:

$$\tilde{\rho}_A \tilde{U}_{s,A} F_A = \tilde{\rho}_B \tilde{U}_{s,B} F_B = \rho U_s F_s$$

where $\tilde{\rho}_A$ and $\tilde{\rho}_B$, as well as $\tilde{U}_{s,A}$ and $\tilde{U}_{s,B}$ are defined as follows:

$$\tilde{\rho}_A \frac{1}{F_s}\iint_{F_a} \rho df \ \ and \ \tilde{U}_s = \frac{1}{F_s}\iint_{F_a} U_s df$$

The above derivations make clear that the employment of the simplified integral form of the continuity equation $\tilde{\rho}\tilde{U}_s F_s$ = const is only justified for such flows that have no strong variation in density or velocity over the flow cross-section of a stream pipe. Equation finds employment in the following sections on this assumption. As now strong variations of the quantities ρ and U_1 over the cross section are excluded, it is also justified to introduce local quantities in the above relation mechanics a number of questions arise that aim at recording infinitesimal changes of a thermodynamic state or flow quantity when infinitesimal changes of other parameters exist/take place/are present. By differentiation of equation one obtains:

The division of equation leads to a further form of the continuity equation which is employed in the following sections:

$$\frac{d\tilde{\rho}}{\tilde{\rho}}+\frac{d\tilde{U}_s}{U_s}+\frac{dG}{F}=0$$

The equation expresses how e.g. the velocity of a fluid will change relatively, when common relative changes in density and cross-section area occur.

Momentum Equation

Solutions of flow problems on the basis of the stream-tube theory requires the inclusion of the momentum equations, however, these have to be transcribed to the stream-filament coordinates. Starting from the general momentum equations

$$\rho\left[\frac{\partial U_j}{\partial t}+U_i\frac{\partial U_j}{\partial x_i}\right]=-\frac{\partial P}{\partial x_j}-\frac{\partial \tau_{ij}}{\partial x_i}+\rho g_j$$

and neglecting the molecular momentum-transport terms for the s-direction of the stream filament, the following form of the momentum equations results:

$$\rho\left[\frac{\partial U_s}{\partial t}+U_s\frac{\partial U_s}{\partial x_s}\right]=-\frac{\partial P}{\partial s}+\rho g_s \text{ mit } g_s=-g\cos\alpha$$

with $\cos\alpha=\dfrac{\partial z}{\partial s}$, so that it holds:

$$\rho\left[\frac{\partial U_s}{\partial t}+U_s\frac{\partial U_s}{\partial s}\right]=-\frac{\partial P}{\partial s}-\rho g\frac{dz}{ds}$$

In an analogous way it can be stated in n – direction

$$\rho\frac{U_s^2}{R}=-\frac{\partial P}{\partial n}-g\frac{dz}{dn}$$

where z in is to be chosen in the negative direction of the gravitational field. This equation expresses that for straight, non-curved stream filaments, i.e $R\to\infty$ the pressure variation vertical to the flow direction is given only by the gravitation. When the gravitational forces are negligible, the pressure over the cross-section of a non-curved stream filament is constant.

Stream Tube Theory

Starting from the general momentum equation in Euler's form, i.e. neglecting the molecular momentum-loss terms

$$\frac{\partial(\rho U_j)}{\partial t}+\frac{\partial(\rho U_i U_j)}{\partial x_i}=-\frac{\partial P}{\partial x_j}+\rho g_j$$

and integrating these over a control volume corresponding to the entire space of a stream filament, one obtains for stationary flow conditions:

$$\int_{F_c}\rho U_i U_j df = -\int_{F_c}\rho df_j + \int_{V_c}\rho g_j dV$$

For the special inflow and outflow conditions at the areas F_A and F_B of a stream filament it can thus be stated

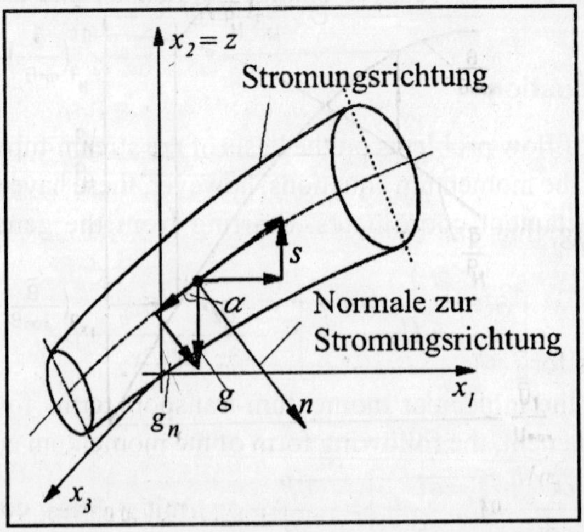

Fig. Considerations on the Momentum Equation for the Stream Filament

$$-\rho_A U_{s,A}^2 F_A + \rho U_{s,B}^2 F_B = +P_A F_A - P_B F_B - \int_V \rho g \frac{dz}{ds}(A\,ds)$$

or transcribed for $g = 0$, d.h. i.e. neglecting gravity:

$$\rho U_s^2 F_s = \text{const}$$

This form of momentum equation is employed in many problem solutions in flow mechanics.

Bernoulli Equation

When carrying out fluid-mechanical considerations, often the pressure and the velocity course/behaviour in flow direction are of interest. Such changes can, when resulting only from mechanical energy changes, be ascertained/determined from the "mechanical energy equation". This equation can be stated in general form as follows :

$$\rho\frac{D}{Dt}\left[\frac{1}{2}U_j^2+G\right]=-\frac{\partial(PU_j)}{\partial x_j}+P\frac{\partial U_j}{\partial x_j}-\frac{\partial(\tau_{ij}U_j)}{\partial x_i}+\tau_{ij}\frac{\partial U_j}{\partial x_i}$$

When carrying out considerations neglecting the molecular-dependent momentum -transport terms, i.e. for $\tau_{ij} = 0$, then one obtains for ρ = const and

thus $$\frac{\partial U_j}{\partial x_j} = 0$$

$$\rho \frac{D}{Dt}\left[\frac{1}{2}U_j^2 + G\right] = -\frac{\partial (PU_j)}{\partial x_j} = -U_j \frac{\partial P}{\partial x_j}$$

In the presence of stationary pressure fields it holds:

$$\frac{DP}{Dt} = \underbrace{\frac{\partial P}{\partial t}}_{=0} + U_i \frac{\partial P}{\partial x_i} = U_i \frac{\partial P}{\partial x_i}$$

so that under these conditions it holds:

$$\rho \frac{D}{Dt}\left[\frac{1}{2}U_j^2 + G\right] = -\frac{DP}{Dt}$$

For $G = -x_j g_j$ thus can be written:

$$\rho \frac{D}{Dt}\left[\frac{1}{2}U_j^2 + \frac{P}{\rho} - x_j g_j\right] = 0$$

which leads for $j = s$. to the statement of the Bernoulli equation:

$$\frac{1}{2}U_s^2 + \frac{P}{\rho} - g_s s = \text{const}$$

Considering that $-g_s s = gh$ one obtains the final form

$$\frac{1}{2}U_s^2 + \frac{P}{\rho} + gh = \text{const}$$

This equation can be interpreted physically such that the mass m, flowing into a stream filament per time unit, introduces the kinetic energy $\dot{m}\frac{1}{2}U_s^2$, the pressure energy $\dot{m}Pu = \dot{m}\frac{P}{\rho}$ and the potential energy $\Delta \dot{m} g h$ as total energy, whose sum along the stream filament cannot change, i.e. the total energy is constant along the stream filament. As at the same time m = const. holds, equation results from this.

The Total Energy Equation

The above considerations on the Bernoulli equation must experience expansion/extension when carrying out energy considerations for compressible media. The equation for the total energy has to be employed instead of the above-treated mechanical energy equation. According to equation it can be stated as follows:

$$\rho\frac{D}{Dt}\left[e+\frac{1}{2}U_j^2+G\right]=-\frac{\partial\dot{q}_j}{\partial x_i}-\frac{\partial(PU_j)}{\partial x_j}-\frac{(\tau_{ij}U_j)}{x_i}$$

Neglecting the molecular-dependent heat and momentum transport, i.e $\dot{q}_i = 0$ and $\tau_{ij} = 0$,it results for g ≈ 0:

$$\frac{\partial}{\partial t}\left[\rho e+\frac{1}{2}\rho U_j^2\right]+\frac{\partial}{\partial x_i}\left[U_i+\frac{1}{2}\rho U_j^2\right]=\frac{\partial}{\partial x_j}(PU_j)$$

For stationary flow processes then holds:

$$\frac{\partial}{\partial x_i}\left[\rho e+\frac{1}{2}\rho U_j^2\right]+\frac{\partial}{\partial x_i}\left[U_i\left(\rho e+\frac{1}{2}\rho U_j^2\right)\right]=\frac{\partial}{\partial x_j}(PU_j)$$

Introducing the enthalpy $h = e + P/\rho$ one obtains compressible media for the energy equation for stationary flows:

$$\left(h+\frac{1}{2}U_j^2\right)=const$$

When carrying out the above considerations for stream pipes, one obtains also the relation stated in equation, for area-averaged quantities though:

$$\tilde{h}+\frac{1}{2}\tilde{U}_j^2=const$$

The sum from the area-averaged enthalpy of a flowing fluid and the areaaveraged kinetic energy available per kg of the fluid is a constant for adiabatic flows free from viscosity in stream pipes, when gravity influences are negligible. With this the following equations for the computation of flows in stream pipes result.

Flows of incompressible fluids:

- Mass conservation: $\tilde{\rho}\tilde{F}_s\tilde{U}_s=const$
- Momentum conservation: $\tilde{\rho}\tilde{U}_s^2F_s+PF_s=const$
- Mechanical energy equation: $\left\{\frac{1}{2}\tilde{U}_s^2+\frac{\tilde{P}}{\tilde{\rho}}+g\tilde{h}\right\}=const$

Flows of compressible fluids:

- Mass conservation: $\tilde{\rho}\tilde{F}_s\tilde{U}_s=const$
- Momentum conservation: $\tilde{\rho}_s\tilde{U}_s^2\tilde{F}_s+PF_s=const$
- Energy equation: $\frac{1}{2}\tilde{U}_s^2+\tilde{h}=const$

INCOMPRESSIBLE FLOWS

Hydro-Mechanical Nozzle Flows

The flow problem diagrammed in figure can be solved, neglecting the friction forces, with the aid of the stream-tube theory, in order to obtain in this way a first overview of the flow processes taking place.

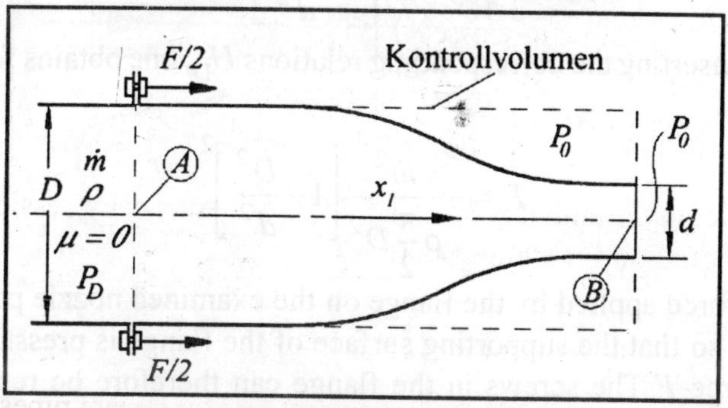

Fig. Nozzle Flow at the End of a Pipe

When applying one-dimensional flow computations, for the flow in cross-section A a velocity can be stated which is constant over the entire tube diameter.

$$U_D = \frac{\dot{m}}{\rho} \frac{1}{\frac{\pi}{4} D^2}$$

Because of the continuity equation holds:

$$\frac{\pi}{4} D^2 U_D = \frac{\pi}{4} d^2 U_d \rightsquigarrow U_d = \frac{D^2}{d^2} U_d$$

From the Bernoulli equation results:

$$\frac{P_D}{\rho} \frac{1}{2} U_D^2 = \frac{P_d}{\rho} + \frac{1}{2} U_2^2 = \frac{P_0}{\rho} + \frac{1}{2} U_2^2$$

From this P_D is computed as:

$$P_D = P_0 + \frac{\rho}{2}\left[U_d^2 - U_D^2\right] = P_0 + \frac{\rho}{2}\left[\left(\frac{D}{d}\right)^4 - 1\right] U_D^2$$

The flange force F on the control volume is computed from the integral momentum equation as follows:

$$-\rho U_D^2 \frac{\pi}{4} D^2 + \rho U_d^2 \frac{\pi}{4} d^2 - \underbrace{- P_D \frac{\pi}{4} D^2 + P_O \frac{\pi}{4} D^2}_{-\frac{\pi}{4} D^2 (P_D - P_O)} = F$$

or after rearrangement, in consideration of equiation

$$-\frac{\rho}{2}\frac{\pi}{4}D^2U_D^2\left[2-2\frac{d^2}{D^2}\frac{U_d^2}{U_D^2}\right]-\frac{\pi}{4}D^2\frac{\rho}{2}\left[\frac{D^2}{d^4}-1\right]U_D^2=F$$

and after further rearrangement

$$-\frac{\rho}{2}\frac{\pi}{4}D^2U_D^2\left[1-\frac{D^2}{d^2}\right]^2=F$$

When inserting the corresponding relations U_D, one obtains for the flange force F :

$$F=-\frac{\dot{m}^2}{\rho\frac{\pi}{2}D^2}\left[1-\frac{D^2}{d^2}\right]^2$$

DThe force applied by the flange on the examined nozzle part proves to be positive, so that the supporting surface of the flange is pressurized with a negative force F The screws in the flange can therefore be regarded to be force-free.

Sudden Cross-section Widening/Extension

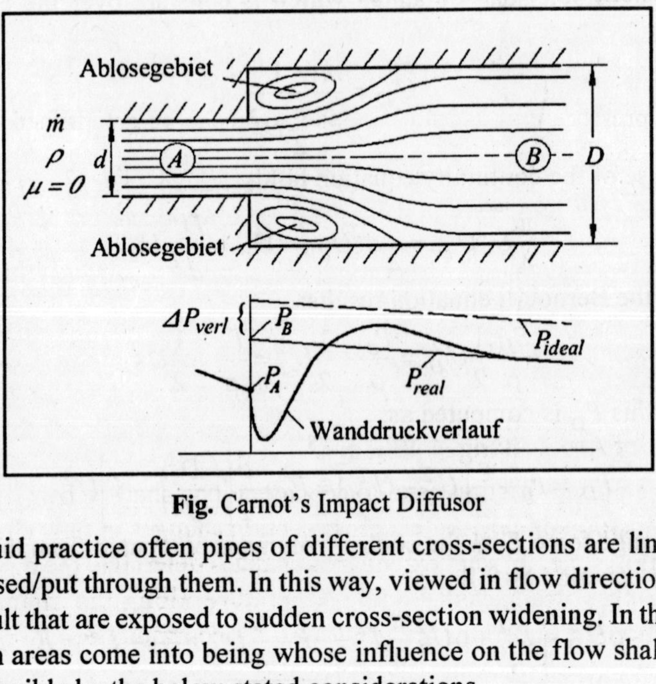

Fig. Carnot's Impact Diffusor

In fluid practice often pipes of different cross-sections are lined up and flows passed/put through them. In this way, viewed in flow direction, internal flows result that are exposed to sudden cross-section widening. In this manner separation areas come into being whose influence on the flow shall be made comprehensible by the below-stated considerations.

When carrying out considerations on flows between the planes A and B of the pipe flow, it yields from the continuity equation:

$$\frac{\dot{m}}{\rho} = U_d \frac{\pi}{4} d^2 = U_D \frac{\pi}{4} D^2$$

Thus the velocities U_d and U_D are known, they are determined by the given mass flow.

In the case of flows passing through the measuring length without losses, the following difference in pressure would result between A and B which can be computed from the Bernoulli equation:

$$\Delta P_{\text{ideal}} = (P_B - P_A)_{\text{ideal}} = \frac{\rho}{2} U_d^2 \left[1 - \frac{d^4}{D^4}\right]$$

Under real conditions, by the occurrence of the separation areas.

$$F = \rho U_d^2 \frac{\pi}{4} - \rho U_D^2 \frac{\pi}{4} D^2 + P_A \frac{\pi}{4} d^2 - P_B \frac{\pi}{4} D^2$$

When one neglects the contributions to the force F by momentum losses at the pipe walls, then the force F. is computed as the pressure force on the ring surface after the sudden widening, i.e. as

$$F = P_A \frac{\pi}{4} (D^2 = d^2)$$

Thus one obtains for the pressure difference

$$DP_{\text{real}} = (P_A - P_B)_{\text{real}} = \rho U_d^2 \frac{d^2}{D^2} \left(1 - \frac{d^2}{D^2}\right),$$

so that a pressure loss (Carnot's impact loss) can be determined as follows:

$$\Delta P_{\text{Verl.}} - \Delta P_{\text{ideal}} = \Delta P_{\text{real.}} = \frac{\rho}{2} U_d^2 \left(1 - \frac{d^4}{D^4}\right) = \frac{\rho}{2} \left(U_d^2 - U_D^2\right)$$

For D $\rightarrow \infty$ results as a maximum value for $\Delta P_{\text{Verl}} = \frac{\rho}{2} U_d^2$, the discharge pressure loss.

INCOMPRESSIBLE FLOWS

Influence of Area Changes on Flows

In the present section general considerations shall be carried out which give information on what effect cross-section changes in flow channels have on fluid flows, i.e. to what extent area changes determine the distribution of velocity, pressure, density and temperature along the channel. In the investigation the equations are employed, the continuity equation reads:

$$\tilde{\rho} \tilde{U}_1 F = const$$

Equation can be written in differential form:

$$\frac{dF}{F}+\frac{d\tilde{\rho}}{\tilde{\rho}}+\frac{d\tilde{U}_1}{\tilde{U}_1}=0$$

The variation of the velocity in flow direction can be described in a first approximation by Euler's equation, reduced for one-dimensional flows, i.e. it holds:

$$\tilde{\rho}\tilde{U}_1\frac{d\tilde{U}_1}{dx_1}=-\frac{d\tilde{P}}{dx_1}=-\frac{d\tilde{P}}{d\tilde{\rho}}\frac{d\tilde{\rho}}{dx_1}$$

On the basis of the energy equation for reversible adiabatic fluid flows, which for area-averaged quantities, by way of approximation, holds as follows:

$$\frac{\tilde{P}}{\tilde{\rho}^x}=const$$

the pressure diversion after the density is to be carried out under adiabatic conditions. As it holds, however:

$$\tilde{c}^2=\left(\frac{d\tilde{P}}{d\tilde{p}}\right)_{\text{ad}}$$

equation can also be written as follows:

$$\tilde{\rho}\tilde{U}_1\frac{\mathrm{d}\tilde{U}_1}{\mathrm{d}x_1}=-\tilde{c}^2\frac{d\tilde{\rho}}{\mathrm{d}x_1}$$

When one introduces the Mach number M*a* of the flow as:

$$\widetilde{M_\alpha}=\frac{\tilde{U}_1}{\tilde{c}}$$

equation can be written:

$$\frac{d\tilde{\rho}}{\tilde{\rho}}=-\widetilde{M_\alpha}^2\frac{d\tilde{U}_1}{\tilde{U}_1}$$

Inserting the result, we obtain:

$$\frac{dF}{F}-\widetilde{M_\alpha}^2\frac{d\tilde{U}_1}{\tilde{U}_1}+\frac{d\tilde{U}_1}{\tilde{U}_1}=0$$

or

$$\frac{d\tilde{U}_1}{\tilde{U}_1}=\frac{1-}{\left(1-\widetilde{M_\alpha}^2\right)}\frac{dF}{F}$$

When one takes into consideration that subsonic flows are given by $\widetilde{M_\alpha}<1$ and supersonic flows by $\widetilde{M_\alpha}>1$ the above relation expresses:

- In the presence of a subsonic flow ($\widetilde{M_\alpha}<1$). a decrease of the cross- sectional area of a flow channel in flow direction is linked to an increase of the flow velocity. An increase of the channel cross-sectional area in flow direction results in a decrease of the flow velocity.

- In the presence of a supersonic flow ($\widetilde{M}_\alpha > 1$) a decrease of the cross- sectional area of a flow channel in flow direction is liked to a decrease of the flow velocity. An increaseincrease of the flow cross-section in flow direction results in an increase of the flow velocity.

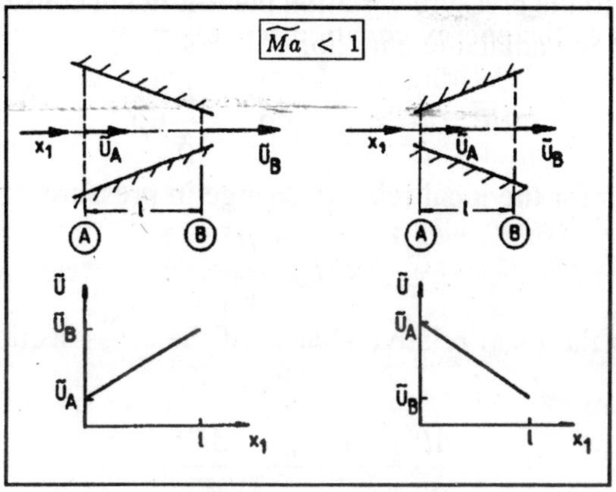

Fig. Influence of the Change of the Flow Cross-Section on a Subsonic Flow

Besides the changes of the flow velocity caused by changes of the cross-sectional areas, the changes in pressure, density and temperature of the flowing fluid are also of interest.

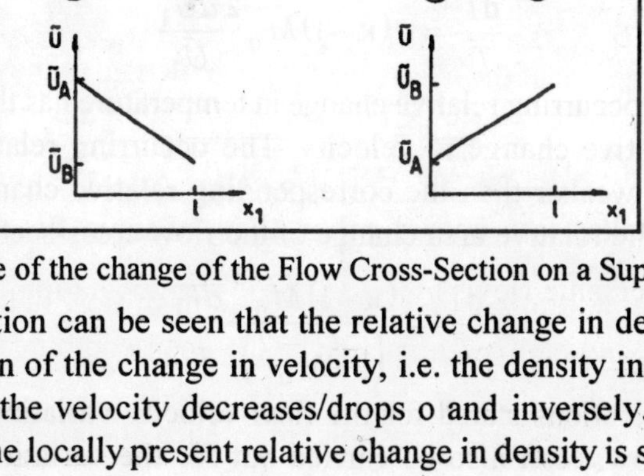

Fig: Influence of the change of the Flow Cross-Section on a Supersonic flow

From equation can be seen that the relative change in density has/owns the opposite sign of the change in velocity, i.e. the density increases in flow direction when the velocity decreases/drops o and inversely. In the area of subsonic flow the locally present relative change in density is smaller than the local relative change in velocity. In the area of supersonic flow the locally present relative change in density is larger than the relative change in velocity. As concerns the dependence from the cross- sectional area changes of the flow channel, it results for the change in density:

$$\frac{d\tilde{\rho}}{\tilde{\rho}} = \frac{\widetilde{M}_\alpha^{\,2}}{\left(1-\widetilde{M}_\alpha^{\,2}\right)} \frac{dF}{F}$$

With regard to the pressure variation the following considerations can be carried out. From the adiabatic equation follows:

$$d\tilde{P} = \frac{\tilde{P}}{\tilde{\rho}^{\kappa}} \kappa_{\tilde{\rho}(\kappa-1)} d\tilde{\rho} = \kappa \frac{\tilde{P}}{\tilde{\rho}} d\tilde{\rho}$$

Thus it holds for the local relative change in pressure

$$\frac{d\tilde{P}}{\tilde{P}} = \kappa \widetilde{M}_\alpha^{\,2} \frac{d\tilde{U}_1}{\tilde{U}_1}$$

or with regard to the local relative change of the cross-sectional area of the flow:

$$\frac{d\tilde{P}}{\tilde{P}} = \frac{\kappa \widetilde{M}_\alpha^{\,2}}{\left(1-\widetilde{M}_\alpha^{\,2}\right)} \frac{dF}{F}$$

Finally it is necessary to consider the variations in temperature. To this end the state equation for ideal gases is differentiated:

$$-\tilde{P}\frac{d\tilde{\rho}}{\tilde{\rho}^2} + \frac{d\tilde{P}}{\tilde{\rho}} = R d\tilde{T} \frac{dF}{F}$$

or transcribed:

$$-\frac{d\tilde{\rho}}{\tilde{\rho}} + \frac{d\tilde{P}}{\tilde{P}} = \frac{d\tilde{T}}{\tilde{T}}$$

Thus follows from the preceding relations

$$\frac{d\tilde{T}}{\tilde{T}} = -(\kappa-1)\widetilde{M}_\alpha^{\,2} \frac{d\tilde{U}_1}{\tilde{U}_1}$$

The locally occurring relative change in temperature has the opposite sign of the local relative change in velocity. The occurring relative changes in temperature are weaker than the corresponding relative changes in density. With regard to the relative area change of the flow cross-section it results:

$$\frac{d\tilde{T}}{\tilde{T}} = \frac{(\kappa-1)\widetilde{M}_\alpha^{\,2}}{\left(1-\widetilde{M}_\alpha^{\,2}\right)} \frac{dF_1}{F}$$

The considerations stated for the flow-velocity variation in supersonic and subsonic flows, can also be carried out for the variations in pressure, density and temperature with the aid of the above equations. Another important consideration can be stated through rearrangement of the above-derived relations such that it holds:

$$\frac{dF}{d\tilde{U}_1} = \frac{F}{\tilde{U}_1}\left(1 - \widetilde{M_\alpha}^2\right)$$

This relation expresses that the condition for achieving the sound velocity is given by $dF = 0$, i.e. $\widetilde{M_\alpha} = 1$ As for the second derivation holds:

$$\frac{d^2F}{d\tilde{U}_1^2} = \frac{F}{\tilde{U}_1^2}\widetilde{M_\alpha}^2\left(\widetilde{M_\alpha}^2 - 2\right)$$

for $\widetilde{M_\alpha} = 1$ holds a minimum of the flow cross-section.

Pressure-driven Compensating Flows through Converging Nozzles

In many technical plants flows of gases occur which are to be classified into the large group of compensating flows that can take place between reservoirs with differing pressure levels. Thus gases e.g. are often stored under high pressure in large storage reservoirs, in order to be led via correspondingly dimensioned/designed openings with connecting aggregates and discharge conduits to the intended purpose when need arises.

This discharge can idealized be understood as a compensating flow between two reservoirs or two chambers of which one represents the storage reservoir under pressure, while the environment represents the second reservoir. In the following considerations it is assumed that both reservoirs are very large so that constant reservoir conditions exist during the entire compensating flow under investigation. These are assumed to be known and are given by the pressure P_H, the temperature T_H etc. in the high-pressure reservoir, as well as through the pressure

P_N or. T_N for the low-pressure reservoir. The compensating flow shall take place via a continually converging nozzle, whose largest cross-section represents thus the discharge opening of the large reservoir, whereas the smallest nozzle cross-section represents the entrance/inlet opening into the low-pressure reservoir.

When one wants to investigate the fluid flows taking place in the above compensating flow more in detail, the final equations for flows through channels, pipes etc.

$$\tilde{\rho}\tilde{U}_1 F = const$$

$$\tilde{h} + \frac{1}{2}\tilde{U}_1^2 = \text{const}; \quad \frac{\tilde{P}}{\tilde{\rho}^\kappa} = \text{const}$$

$$\frac{\tilde{P}}{\tilde{\rho}} = R\tilde{T}$$

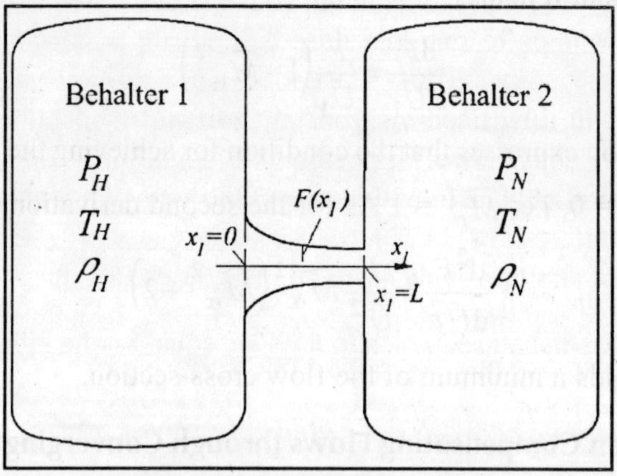

Fig. Compensating.Flow Between two Reservoirs through Converging Nozzle

With that a sufficient number of equations exists to determine the course of the area-averaged velocity and the area-averaged thermo-dynamical state quantities of the flowing gas during the process of the compensating flow, i.e. along the x_1 —axis.

When one considers that - based on the assumption of a large reservoir - in the interior of the high-pressure reservoir there is the constant pressure P_H and the velocity $(U_1)_H = 0$ then for the velocity U_1 at each point x_1 of the nozzle the following relation can be stated:

$$\tilde{h} + \frac{1}{2}\tilde{U}_1^2 = h_H$$

Taking into account that the enthalpy for an ideal gas can be stated as c_P T and that moreover the ideal gas equation holds, the above relation can be transcribed as follows:

$$c_P \frac{\tilde{P}}{R\tilde{\rho}} + \frac{1}{2}\tilde{U}_1^2 = \frac{\kappa}{\kappa-1}\frac{\tilde{P}}{\tilde{\rho}} + \frac{1}{2}\tilde{U}_1^2 = \frac{\kappa}{\kappa-1}\frac{P_H}{\rho_H}$$

The velocity U_1. is thus linked to the course of the pressure along the axis of the nozzle as follows:

$$\tilde{U}_1 = \sqrt{\frac{2\kappa}{\kappa-1}\left(\frac{P_H}{\rho_H} - \frac{\tilde{P}}{\tilde{\rho}}\right)}$$

The above equation indicates that for $\tilde{P} = 0$, i.e. for the outflow into a vacuum, a maximum flow velocity develops which is given by the state of the reservoir only:

$$U_{max} = \sqrt{\frac{2\kappa}{\kappa-1}\frac{P_H}{\rho_H}} = \sqrt{2c_P \cdot T_H}$$

Standardizing the flow velocity U_1, with U_{max}, existing at a Point x_1, one obtains:

$$\frac{\tilde{U}_1}{U_{max}} = \sqrt{1 - \frac{\tilde{P} \cdot \rho_H}{P_H \cdot \tilde{\rho}}}$$

or transcribed by means of the ideal gas equation:

$$\frac{\tilde{U}_1}{U_{max}} = \sqrt{1 - \frac{\tilde{T}}{T_H}}$$

Linking the adiabatic equation to the state equation the following relations:

$$\frac{\tilde{T}}{T_H} = \left(\frac{\tilde{\rho}}{\rho_H}\right)^{\kappa-1} \quad and \quad \frac{\tilde{T}}{T_H} = \left(\frac{\tilde{P}}{P_H}\right)^{\frac{\kappa-1}{\kappa}}$$

Thus the following equations hold:

$$\frac{\tilde{U}_1}{U_{max}} = \sqrt{\left[1 - \left(\frac{\tilde{\rho}}{\rho_H}\right)^{\kappa-1}\right]}$$

and

$$\frac{\tilde{U}_1}{U_{max}} = \sqrt{\left[1 - \left(\frac{\tilde{P}}{P_H}\right)^{\frac{\kappa-1}{\kappa}}\right]}$$

When choosing the standardized velocity $(\tilde{U}_1/U_{max})$ as a parameter for the representation of the flow in the nozzle, the course of pressure, density and temperature can be stated as follows:

$$\frac{\tilde{P}}{P_H} = \left[1 - \left(\frac{\tilde{U}_1}{U_{max}}\right)^2\right]^{\frac{\kappa}{\kappa-1}}$$

$$\frac{\tilde{\rho}}{P_H} = \left[1 - \left(\frac{\tilde{U}_1}{U_{max}}\right)^2\right]^{\frac{\kappa}{\kappa-1}}$$

$$\frac{\tilde{T}}{T_H} = \left[1 - \left(\frac{\tilde{U}_1}{U_{max}}\right)^2\right]$$

These relations are sated in Figure as functions of $(\tilde{U}_1/U_{max})$ Also stated is, along the $\left(\tilde{U}_1 / U_{max}\right)$–axis, the corresponding Mach number of the flow, which in consideration of the relation $c = \sqrt{(dP/d\rho)_{ad}} = \sqrt{\kappa R T}$ can be computed as follows:

$$\frac{\tilde{U}_1^2}{U_{max}^2} = \frac{\tilde{U}_1^2}{2c_P T_H} \frac{\kappa R \tilde{T}}{\kappa R \tilde{T}} = \widetilde{M}_{\alpha 1}^{-2} \frac{\kappa - 1}{2} \left(\frac{\tilde{T}}{T_H}\right)$$

When one considers the relation derived above for (T/T_H) equation, one obtains for the Mach number to be determined:.

$$\widetilde{M}_\alpha \sqrt{\frac{2}{\kappa - 1} \left[\frac{\left(\frac{\tilde{U}_1}{U_{max}}\right)^2}{\left[1 - \left(\frac{\tilde{U}_1}{U_{max}}\right)^2\right]}\right]}$$

Thus a Mach number of the flow is to be assigned to each statement of an area-averaged velocity standardized with the maximum velocity.

All quantities which are stated in the above equations can also be written as functions of the Mach number $\widetilde{M}_\alpha$ which in turn is to be considered as an area-averaged flow quantity describing the course of the flow along the x_1 – axis.

For the derivation of the dependency of the pressure, the density and the temperature from the Mach number of the flow, equation is written as follows:

$$c_P \tilde{T} + \frac{1}{2} \tilde{U}_1^2 = c_P T_H$$

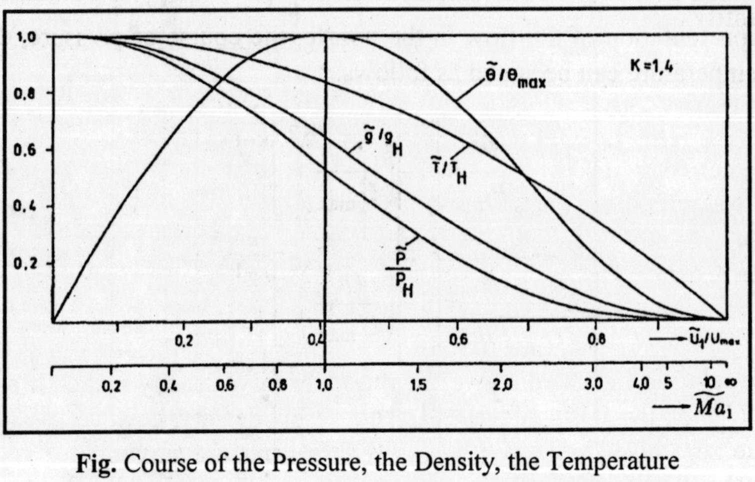

Fig. Course of the Pressure, the Density, the Temperature and the mass-flow Velocity in Pressure-Compensating Flows

By division with $c_p\tilde{T}$ one obtains:

$$\frac{T_H}{\tilde{H}} = 1 + \frac{\tilde{U}_1^2}{2c_p\tilde{T}}\frac{\kappa R}{\kappa R} = 1 + \frac{\kappa - 1}{2}\widetilde{M_{\alpha_1}}^2$$

or for the reciprocal:

$$\frac{\tilde{T}}{T_H} = \frac{2}{2 + (\kappa - 1)\widetilde{M\alpha_1^2}}$$

This equation makes it clear that a relation is given between the area-averaged temperature along the x_1 – axis and the Mach number existing at the same point of the flow.

With this for each x_1 – point the temperature can be computed, when the reservoir state is given and the Mach number of the flow known.

Taking into account the adiabatic equation, for the relation of pressure and reservoir pressure results:

$$\frac{\tilde{P}}{P_H} = \left(\frac{\tilde{T}}{T_H}\right)^{\frac{\kappa}{\kappa-1}} = \left[\frac{2}{2 + (\kappa - 1)\widetilde{M\alpha_1^2}}\right]^{\frac{\kappa}{\kappa-1}}$$

and

$$\frac{\tilde{\rho}}{P_H} = \left(\frac{\tilde{T}}{T_H}\right)^{\frac{\kappa}{\kappa-1}} = \left[\frac{2}{2 + (\kappa - 1)\widetilde{M\alpha_1^2}}\right]^{\frac{\kappa}{\kappa-1}}$$

The mass-flow density $\theta = \dot{m}/F = \tilde{\rho}\tilde{U}_1$, ..., i.e. the statement of the mass flowing per area and time unit through a flow cross-section. The course of this quantity can be written as follows, using the relations for $\tilde{U}_1$ and $\tilde{\rho}$:

$$\tilde{\rho}_1\tilde{U}_1 = \rho_H\left[1 - \left(\frac{\tilde{U}_1}{U_{\max}}\right)^2\right]^{\frac{\kappa}{\kappa-1}} \cdot \tilde{U}_1^2$$

or for the standardized mass-flow density:

$$\frac{\tilde{\rho}_1\tilde{U}_1}{\rho_H U_{\max}} = \frac{\tilde{U}_1}{U_{\max}}\left[1 - \left(\frac{\tilde{U}_1}{U_{\max}}\right)^2\right]^{\frac{\kappa}{\kappa-1}}$$

The relation indicated above for the mass-flow density makes it clear, that for $U_1 = 0$, $\theta = 0$ is achieved. The mass-flow density, however, assumes the value zero also for $U_1 = U_{\max}$ as with setting the maximally possible velocity the density of the fluid also contained in the mass-flow density has dropped to $\tilde{\rho} = 0$. Between these two minimal values the mass-flow density has to traverse a maximum which can be computed by differentiation of the above functions and by setting the derivation to zero. The value obtained by solving the resulting equation has to be inserted for ($\tilde{U}_1$ /U_{max}) in the above equation for the mass-flow density in order to achieve the maximal value. It is computed:

$$\theta_{max} = \rho_H \cdot U_{max} \cdot \sqrt{\frac{\kappa-1}{\kappa+1}} \left(\frac{2}{\kappa+1}\right)^{\frac{1}{\kappa-1}}$$

where for the velocity value it is obtained:

$$\frac{\tilde{U}_1}{U_{max}} = \sqrt{\frac{\kappa-1}{\kappa+1}} \quad \text{for } \theta = \theta_{max}$$

With this the mass-flow density standardized with the maximal value can be written as follows:

$$\frac{\theta}{\theta_{max}} = \sqrt{\frac{\kappa-1}{\kappa+1}} \cdot \frac{\tilde{U}_1}{U_{max}} \left[\frac{\kappa+1}{2}\left(1 - \frac{\tilde{U}_1^2}{U_{max}}\right)\right]^{\frac{1}{\kappa-1}}$$

The course of this quantity with $\tilde{U}_1 / U_{max}$ is also represented. The significance of the maximum of the mass-flow density for the course of compensating flows is dealt with more in detail further down. Its appearance prevents the steady increase of the mass flow with the increase of the pressure difference between pressure reservoirs when the compensating flow takes place via steadily converging nozzles.

A representation of the compensating flows through converging nozzles often regarded to be more simple is achieved by relating the quantities designating the flow to the corresponding quantities of the "critical state", which is designated by Mα = 1 To this state corresponds not only a certain Mach number, i.e. $\widetilde{M\alpha_1} = 1$, but also certain values of the thermo-dynamic state quantities: These can be determined from the equations by setting $\widetilde{M\alpha_1} = 1$ From this result the following values for thermo-dynamic state quantities of the fluid in critical state, i.e. for $\widetilde{M\alpha_1} = 1$:

$$\frac{\tilde{P}^*}{P_H} = \left(\frac{2}{\kappa+1}\right)^{\frac{\kappa}{\kappa-1}}$$

$$\frac{\tilde{P}^*}{\rho_H} = \left(\frac{2}{\kappa+1}\right)^{\frac{\kappa}{\kappa-1}}$$

$$\frac{\tilde{T}^*}{T_H} = \frac{2}{\kappa+1}$$

With these equations the pressure, density and temperature of a flowing medium can be determined in that cross-section of a converging nozzle in which the sound velocity occurs.

According to the considerations carried out a minimum of the cross-section has to exist at this point. As at this point the Mach number assumes the value $\widetilde{M\alpha_1} = 1$ the equation can be written as follows:

$$\frac{\tilde{U}_1^{*2}}{U^2_{\max}} = \frac{\kappa-1}{2}\left(\frac{\tilde{T}}{T_H}\right) = \frac{\kappa-1}{\kappa+1}$$

When comparing the values for $(\tilde{U}_1/U_{\max})$ of the relations, one finds that they are identical, i.e. the maximum mass-flow density can only occur in the narrowest cross-section of a nozzle, where the sound velocity then also takes place/sets in.

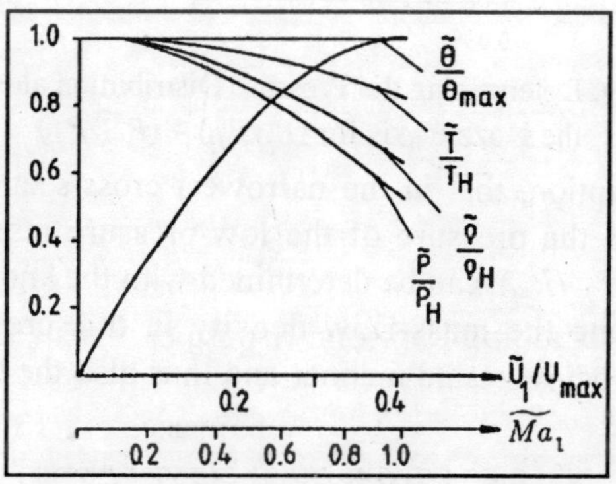

Fig. Course of the Pressure, the Density, the Temperature and the Massflow Density for Converging Nozzles

In accordance with the above-stated derivations of the basic equations for pressure-compensating flows between large reservoirs the flow shall be discussed which occurs in a steadily converging nozzle. The considerations shall be carried out in such a way that the mass flow is computed which results when a certain pressure relation (P_N/P_H) between the reservoirs comes about. Here two pressure ranges are of interest:

$\frac{P_N}{P_H} > \frac{P^*}{P_H}$ The relation of the reservoir pressures is larger than the critical pressure relation

$\frac{P_N}{P_H} < \frac{P^*}{P_H}$ The relation of the reservoir pressures is smaller than the critical pressure relation

When the pressure relation is larger than the critical value, a steady decrease of the relation of the reservoir pressures leads to a steady increase of the mass-flow density.

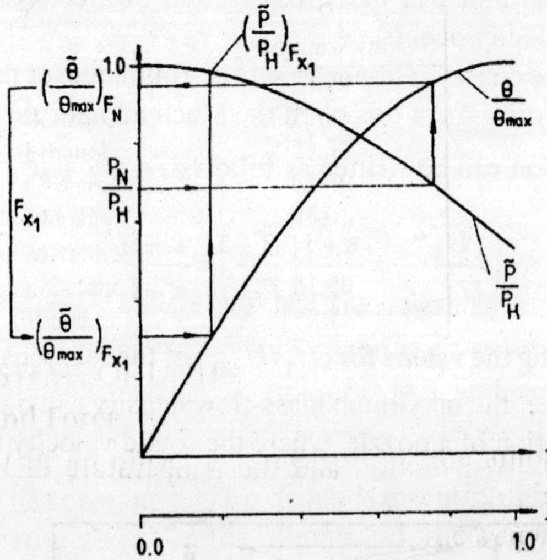

Fig. Determining the Pressure Distribution along the Nozzle axis for $(P_N/P_H) > (P^*_N/P_H)$

On the assumption, that in the narrowest cross-section of the steadily converging nozzle the pressure of the low-pressure reservoir sets in, the pressure relation (P_N/P_H). can be determined with the known values P_N. and P_H –. Via the same the mass-flow density in this cross-section can be determined in the below-stated manner and thus also the total mass flowing through the nozzle:

$$\dot{m}_H = F_H \tilde{\Theta}_H = F_H \left(\tilde{\rho}\tilde{U}_1\right)_H$$

For reasons of continuity this total mass flow is constant in all cross-section planes of the nozzle, so that it holds:

$$\dot{m}_H = \dot{m} \quad \text{d.h. } F_N \tilde{\Theta}_N = F_{x_1} \tilde{\Theta}_{x1}$$

Starting from the assumption that the specified/given distribution of the cross-section area of the nozzle along the x_1 – axis is known, then the massflow density distribution along the x_1 – axis can be determined. Via the same can then be computed, the pressure distribution along the nozzle, or the resulting distributions of the density and the temperature, but also of the Mach number and the flow velocity.

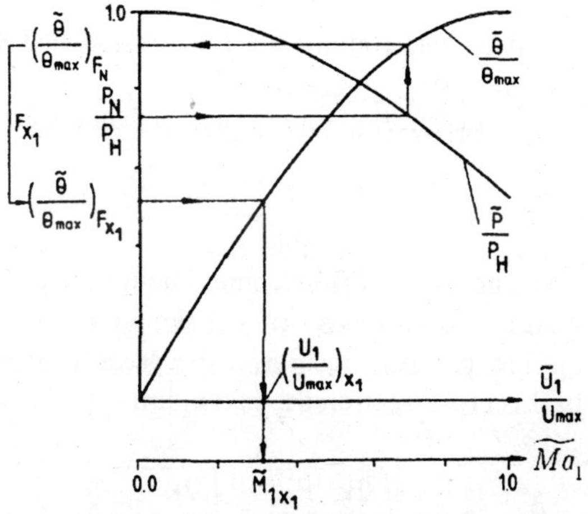

Fig. Determining the Mach number and the Velocity Distribution along a Converging nozzle for $(P_N/P_H) > (P^*_N/P_H)$

The way of proceeding in determining the pressure distribution along the nozzle, indicated in the above figure, can be transferred analogously also to defining the density distribution and the temperature distribution. For determining the distribution of the Mach number and the velocity, the way indicated in Figure holds.

From the above considerations follows that the velocity $(U_1)_N$ in the entrance cross-section of the nozzle is finite and that there the mass-flow density

$$\tilde{\theta}_H = \frac{F_N}{F_H}\left(\tilde{\rho}\tilde{U}_1\right)_N$$

is present. With this it is also said that in this cross-section a pressure, a density and a temperature are reigning which do not correspond to the values in the high-pressure reservoir. It is necessary to take this always into consideration when computing compensating flows through nozzles. The quantities designating the flows that exist at the nozzle entrance are to be determined via the above diagrams from the mass-flow density computed for the entrance cross-section.

When carrying out the above computations for determining the flow quantities and the thermo-dynamic quantities, it proves that with a decrease of the pressure relation (P_N/P_H) an increase of the mass-flow density in each cross-section of the nozzle is connected, as long as the pressure relation is larger than the critical value.

When the critical value itself is reached, i.e.:

$$\frac{P_N}{P_H}=\left(\frac{2}{\kappa+1}\right)^{\frac{\kappa}{\kappa-1}}=\frac{P^*}{P_H}$$

This value cannot be exceeded in the case of a further decrease of the pressure relation

(P_N/P_H) i.e. for all pressure relations smaller than the critical value:

$$\frac{P_N}{P_H}<\frac{P^*}{P_H}=\left(\frac{2}{\kappa+1}\right)^{\frac{\kappa}{\kappa-1}}$$

in the steadily converging nozzle a flow comes about which is identical for all pressure relations. At the exit cross-section of the nozzle, i.e. in the entrance cross-section to the low-pressure reservoir, the pressure P_N does not come about any more. In this cross-section the maximum mass-flow density rather is reached:

$$\tilde{\theta}_{\max}=\rho_H\sqrt{\frac{2\kappa}{\kappa-1}\frac{P_H}{\rho_H}}=\sqrt{\frac{2\kappa}{\kappa-1}P_H\rho_H}$$

or

$$\tilde{\theta}_{\max}=\rho_H\sqrt{2c_PT_H}$$

The total mass flow thus is computed:

$$\dot{m}=\dot{m}_{\max}=F_N\,\tilde{\theta}_{\max}$$

Starting again from the assumption that the nozzle form is known, then the mass-flow distribution existing along the axis can be computed via the continuity equation. When this distribution is known, the corresponding distributions of the pressure, the density, the temperature, the Mach number and the flow velocity can be determined. Of importance is that for all pressure relations (P_N/P_H), that are equal or smaller than the critical relation, one and the same flow comes about in the nozzle. In the exit cross-section of the nozzle for an area-averaged pressure exists which is larger than the pressure P_N, existing in the low-pressure reservoir.

$$\frac{P_N}{P_H}<\frac{\tilde{P}^*}{P_H}=\left(\frac{2}{\kappa+1}\right)^{\frac{\kappa}{\kappa-1}}$$

The pressure compensation takes place via fluid flows that form in the open-jet flow, stretching from the nozzle tip to the interior to the low-pressure reservoir.

Finally attention is drawn to important facts of the case which shall serve the comprehension of the occurring compensating flow. The above represen tations started from the state often existing in practice that compensating flows are controlled via pressure differences between reservoirs. This means that it was assumed that P_H, ρ_H or T_H are known and constant and that they have an influence on how the flow forms.

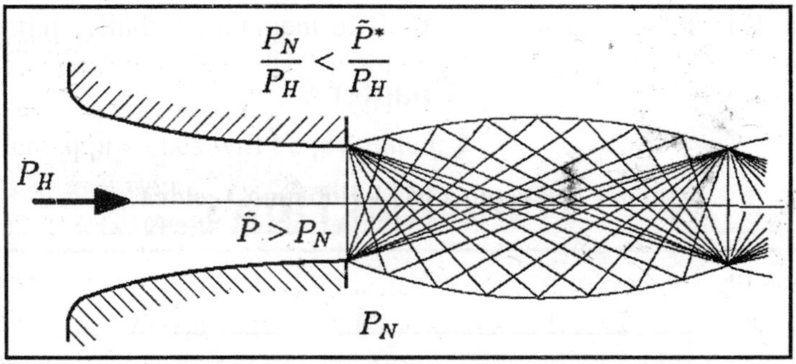

Fig. Pressure Compensation at the Nozzle exit Via Density Impacts

In the low-pressure reservoir it was only assumed that P_N is given and can be "forced upon" the flow in the narrowest cross-section of the nozzle (for (P_N/P_H) larger than the critical value (P^*/P_H)). The density of the flowing gas coming about for these conditions in the exit cross-section of the nozzle or the occurring temperature are not identical with the corresponding values of the fluid in the low-pressure reservoir.

A compensation of these values and the corresponding values of the lowpressure reservoir takes place in the open-jet flow following the nozzle flow. For pressure conditions:

$$\frac{P_N}{P_H} < \frac{P^*}{P_H} = \left(\frac{2}{\kappa+1}\right)^{\frac{\kappa}{\kappa-1}}$$

the compensation takes place between the pressure in the nozzle-exit crosssection and the pressure in the low-pressure reservoir, likewise in the open-jet flow following the nozzle flow.

Chapter 9

Potential Flows

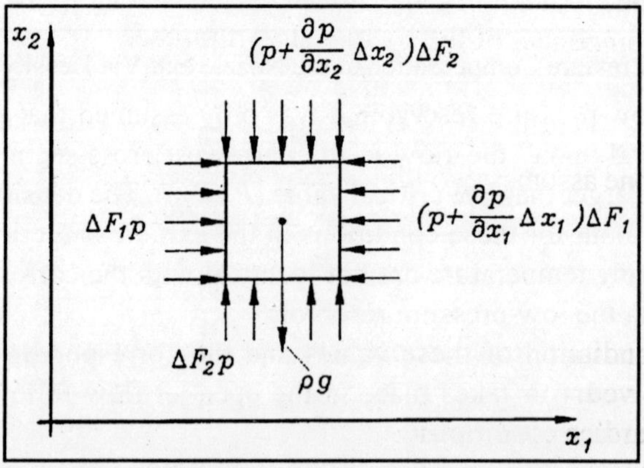

Fig. Graphic Representation of the Physical cause for Irrotationality of Ideal Flows (Kelvin's Theorem)

In order to make possible an integration of the partial differential-equation system of fluid mechanics with simple mathematical means, the introduction of the irrotationality of the flow field is necessary. The irrotationality is of importance for the computation of flow fields with simple means for the following reasons. A transport equation equivalent to the momentum equation had been derived for the rotational power which for viscosity-free flows is reduced to the simple form $D\omega/Dt = 0$.

From this equation two things follow. On the one hand it becomes evident that irrotational fluids fulfil automatically a simplified form of the momentum equation. On the other hand Kelvin's theorem results immediately, according to which all flows of viscosity-free fluids are irrotational, when at any point in time the irrationality of the flow field was detected. This can be understood graphically in that way that all surface forces acting on a liquid element attack normal to the surface and resultantly go through the centre of mass of the fluid element.

At the same time the inertia forces attack also in the centre of the mass, so that no resultant impulse comes about which can lead to a rotation. With this the conclusion is possible that rotating parts cannot receive an additional rotation by the pressure and inertia forces acting on ideal fluids. In addition to the above demand for irrotationality a further restriction shall now be made as regards the properties of the flows, namely the exclusive consideration of two-dimensional flows. This restriction is not a condition of the demand for irrotationality; one can on the contrary very well imagine three-dimensional flows of viscosity-free fluids that are irrotational. For twodimensional irrotational flows there exists, however, a very elegant solution method which is based on the employment of complex analytical functions and which is used exclusively in the following.

When occupying oneself with two – dimensional flow fields, the only remaining component of the rotational vector reads:

$$\omega_3 = \frac{1}{2}\left(\frac{\partial U_2}{\partial x_1} - \frac{\partial U_1}{\partial x_2}\right)$$

When one assumes two-dimensional flow fields to be irrotational, it holds $\omega = 0$ or:

$$\frac{\partial U_1}{\partial x_2} = \frac{\partial U_2}{\partial x_1}$$

This condition has to be fulfilled when two-dimensional flow problems are to be solved.

Disregarding singularities, for irrotational flow fields the above relation has to be fulfilled in all points of the flow field. This is tantamount to the statement that for two-dimensional irrotational flows a velocity potential driving the flow $\Phi(x_1, x_2)$ exists, to such an extent that the following relations hold:

$$U_1 = \frac{\partial \Phi}{\partial x_1} \text{ and } U_2 = \frac{\partial \Phi}{\partial x_2}.$$

Equation leads to the following relations:

$$\frac{\partial U_1}{\partial x_2} = \frac{\partial^2 \Phi}{\partial x_1 \partial x_2} = \frac{\partial U_2}{\partial x_1} = \frac{\partial^2 \Phi}{\partial x_1 \partial x_2},$$

which for irrotational flow fields, i.e. for $\omega_3 = 0$, confirm the reasonable introduction of a potential driving the velocity field. When one inserts the relations into the two-dimensional continuity equation for $\rho = \text{const}$ then one obtains the Laplace equation for the velocity potential:

$$\frac{\partial^2 \Phi}{\partial x_1^2} + \frac{\partial^2 \Phi}{\partial x_2^2} = 0.$$

For determining two-dimensional potential fields it is sufficient to solve equations, i.e. for determining the velocity field it is not necessary to solve the Navier-Stokes equation formulated in velocity terms. These equations have to be employed, however, for determining the pressure field.

The solution of the partial differential equations for the velocity potential requires at the boundary flow the boundary condition

$$\frac{\partial \Phi}{\partial n} = 0,$$

where n is the normal unit vector at each point of the boundary flow.

When the velocity potential or potential field Φ has been obtained as a solution of equation, the velocity components U_1 and U_2 can be determined for each point of the flow field by partial differentiations, according to the relations. After that determining the pressure via Euler's equations, i.e. via the momentum equations for viscosity-free fluids takes place. Determining the pressure can, however, also be done via the integrated form of Euler's equations, which leads to the "non-stationary Bernoulli equation".

The above representations make it clear that the introduction of the irrotationality of the flow field has led to considerable simplifications of the solution ansatz for the basic equations for flow problems. The equations that have to be solved for the flow field are linear and they can be solved decoupled from the pressure field.

The linearity of the equations to be solved is an essential property as it permits the superposition of individual solutions of the equations in order to obtain also solutions of complex flow fields. This solution principle will be used extensively in the following sections.

In the derivations of the above equations for two-dimensional potential flows the potential function was introduced in such a way that the irrotationality of the flow field was fulfilled identically. The introduction of the potential function Φi into the continuity equation then led to the two-dimensional Laplace equation; only such functions Φ, which fulfil this equation can be regarded as solutions of the basic equations of irrotational flows.

Via a procedure similar to the above procedure for the introduction of the potential function Φ "it is possible to introduce a second important function for two-dimsnional flows of incompressible fluids, the so-called stream function Ψ. The latter is defined in such a way that through the stream function the two-dimensional continuity equation is automatically fulfilled, i.e.:

$$U_1 = \frac{\partial \Psi}{\partial x_2} \text{ and } U_2 = -\frac{\partial \Psi}{\partial x_1}.$$

This relation inserted into the continuity equation shows directly that the stream function Ψ. introduced according to fulfils this equation; per definition this is the case for rotational and irrotational flow fields.

When one wants to define analytically or numerically the stream function of an irrotational flow, a potential flow, Ψ has to be a solution of the Laplace equation

$$\frac{\partial^2\Psi}{\partial x_1^2}+\frac{\partial^2\Psi}{\partial x_2^2}=0$$

as can be derived by inserting the equations into the condition for the irrotationality.

$$\frac{\partial U_2}{\partial x_1}-\frac{\partial U_1}{\partial x_2}=0$$

The stream function for two-dimensional potential flows fulfils the twodimensional Laplace equation, similar to the potential function Φ.

The stream function has a number of properties that prove useful for the treatment of two-dimensional flow problems. Lines of constant stream function values for example are path lines of the flow field when stationary conditions exist. This can be derived by stating the total differential of Ψ :

$$d\Psi=\frac{\partial\Psi}{\partial x_1}dx_1+\frac{\partial\Psi}{\partial x_2}dx_2.$$

For Ψ = const is $d\Psi = 0$ and thus results:

$$\left(\frac{dx_2}{dx_1}\right)_{\Psi=\text{const}}=-\frac{\dfrac{\partial\psi}{\partial x_1}}{\dfrac{\partial\Psi}{\partial x_2}}=\frac{U_2}{U_1}.$$

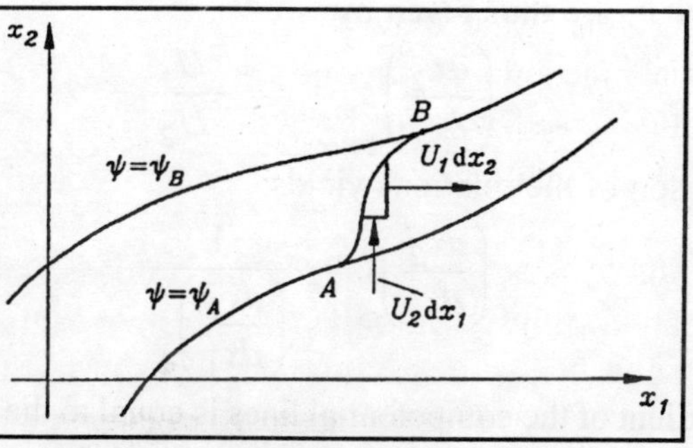

Fig. Schematic Representation of the Flow Between Flow Lines

This is the relation for the gradient of the tangent of the flow line, but also for the gradient of the path line of a fluid element.

Accordingly the total family of flow lines of a velocity field is described by all sorts of Ψ – values.

A further essential property of the stream function becomes clear by the fact that the difference of the stream-function values of two flow lines indicates the volume stream that flows between the flow lines. When computing the volume stream that passes the control area in flow direction for an area *AB* mit with the depth 1, one obtains:

$$\dot{Q} = \int_A^B U_1\, dx_2 - \int_A^B U_2\, dx_1 = \int_A^B (U_1\, dx_2 - U_2\, dx_1).$$

It holds, however: $d\Psi = U_1 dx_2 - U_2 dx_1$, so that can be written:

$$\dot{Q} = \int_A^B (U_1\, dx_2 - U_2\, dx_1) = \int_A^B d\Psi = \Psi_B - \Psi_A .$$

It is remarked that from the statement that Ψ = const are flow lines of the flow field follows of course immediately the condition that fixed walls have to run tangentially to lines Ψ = const From the orthogonality of equipotential and flow lines which is shown in the following section results at once that equipotential lines always have to stand vertically on fixed walls. When one considers the streamlines that can be stated for two-dimensional potential flows in connection with the potential lines of the same flow field, i.e. lines with Φ = const, one finds that these lie orthogonally to one another. This can be shown by stating the total differential $d\Phi$,

$$d\Phi = \frac{\partial \Phi}{\partial x_1} dx_1 + \frac{\partial \Phi}{\partial x_2} dx_2,$$

or writing same under consideration of equation as follows:

$$d\Phi = U_1 \cdot dx_1 + U_2 dx_2.$$

The lines Φ =. are thus given by

$$\left(\frac{\partial x_2}{\partial x_1}\right)_{\Phi=\text{const}} = -\frac{U_1}{U_2}.$$

A comparison of the relations yields:

$$\left(\frac{dx_2}{dx_1}\right)_{\Phi} = -\frac{1}{\left(\frac{dx_2}{dx_1}\right)_{\Psi}}$$

As the gradient of the equipotential lines is equal to the negative reciprocal of the gradient of the flow lines, these lines form an orthogonal net. The velocity along a stream can be computed as:

$$U_s = \left(\frac{\partial \Phi}{\partial s}\right)_{\Psi}$$

This relation is often used in investigations of flow fields for which values of flow lines and equipotential lines have been computed or obtained from measurements.

From the above derivations it is comprehensible that a stream function Ψ

can be computed when the potential function Φ is known and that also inversely the potential function Φ can be determined when the stream function Ψ is available. The procedure for determining one function from the other is to be regarded in accordance with the below-stated single steps for determining the stream function:

- The known potential function $\Phi(x, y)$ is examined with regard to whether it represents a solution of equation.
- By partial deriving to x_1 and x_2 the velocity components U_1 and U_2 are determined, in accordance with relations.
- From this can be determined the gradient of the equipotential lines,

$$\left(\frac{dx_2}{dx_1}\right)_\Phi = -\frac{U_1}{U_2}$$

- Equation follows for the gradient of the flow lines:

$$\left(\frac{dx_2}{dx_1}\right)_\Psi = \frac{U_2}{U_1}$$

- By integration of this relation the course of the flow line is conserved. These are lines of constant Ψ – values.

POTENTIAL AND COMPLEX FUNCTIONS

The representations have shown that the velocities U_1 and U_2 can be stated as partial derivations of the stream function and the potential function for irrotational two-dimensional flows of incompressible and viscosity-free fluids:

$$U_1 = \frac{\partial \Phi}{\partial x_1} = \frac{\partial \Psi}{\partial x_2}$$

and

$$U_2 = \frac{\partial \Phi}{\partial x_2} = -\frac{\partial \Psi}{\partial x_1}.$$

On the basis of their definition the stream and potential function satisfy the Cauchy-Rieman differential equations:

$$\frac{\partial \Phi}{\partial x_1} = \frac{\partial \Psi}{\partial x_2},$$

$$\frac{\partial \Phi}{\partial x_2} = -\frac{\partial \Psi}{\partial x_1}.$$

This gives expression to that a complex analytical function $F(z)$ can be stated in which $\Phi(x, y)$ represents the real part and $\Psi(x, y)$ the imaginary part of the function $F(z)$ is designated as the complex potential of the velocity field $F(z) = \Phi(x, y) + i\Psi(x, y)$,

where $x = x_1$ and $y = x_2$ can be seen and $z = x + iy$ indicates a point in the complex number plane. Inversely it can be said that for each analytical function holds that its real part represents automatically the potential of a flow field whose flow lines are described by the imaginary part of the complex function. As a consequence it results that each real part of an analytical function and also the imaginary part, each for itself, fulfil the two-dimensional Laplace equation. Analytical functions as they are dealt with in function theory can thus be employed for describing potential flows. When setting their real part $\Re\,(x, y)$ equal to the potential function $\Phi(x, y)$ and the imaginary part $\mathrm{Im}(x, y)$ equal to the stream function $\Psi\,(x, y)$, it is possible to state the equipotential and the flow lines. By proceeding in this way solutions to flow problems are obtained without partial differential equations having to be solved. The inverse way of proceeding that is thus sought here for the solution of flow problems, namely interpreting a known solution of the potential equation as a flow is regarded as acceptable because of the evident advantages of proceeding like this.

From a complex potential $F\,(z)$ a complex velocity can be derived by differentiation. As $F\,(z)$ represents an analytical function, thus is steady and steadily derivable, the derivation has to be independent of the direction in which it is determined, as is shown in the following. As because of the steadiness of $F\,(z)$ holds

$$\frac{dF}{dz} = \lim_{\Delta z \to 0} \frac{\Delta F}{\Delta z} = \lim_{\Delta z \to 0} \frac{\Delta F}{(z + \Delta z) - z}$$

$$= \lim_{\Delta z \to 0} \frac{\Delta F}{(x + \Delta x) + i\,(y + \Delta y) - (x + iy)}$$

and as one is free to choose the way on which Δz goes towards zero (the derivation has to be independent of the selected way), the following special ways can also be taken into consideration:

$$\Delta y = 0: \frac{dF}{dz} = \lim_{\Delta x \to 0} \frac{\Delta F}{(x + \Delta x) + iy - (x + iy)} = \lim_{\Delta x \to 0} \frac{\Delta F}{\Delta x} = \frac{\partial F}{\partial x}$$

$$\Delta x = 0: \frac{dF}{dz} = \lim_{\Delta y \to 0} \frac{\Delta F}{x + i\,(y + \Delta y) - (x + iy)}$$

$$= \lim_{\Delta y \to 0} \frac{\Delta F}{i\,\Delta y} = \frac{\partial F}{i\,\partial y} = -i\,\frac{\partial F}{\partial y}$$

The derivation of the complex potential $F(z)$ thus reads for $x = x_1$:

$$w\,(z) = \frac{dF\,(z)}{dz} = \frac{\partial \Phi}{\partial x_1} + i\,\frac{\partial \Psi}{\partial x_1}$$

or expressed in velocity components:

$$w\,(z) = U_1 - iU_2.$$

Based on the above considerations it also holds:

$$w(z)=\frac{dF(z)}{dz}=\frac{\partial\Phi}{i\,\partial x_2}+i\frac{\partial\Psi}{i\,\partial x_2}$$

or after transformation in consideration of $i^2=-1$

$$w(z)=\frac{\partial\Psi}{\partial x_2}-i\frac{\partial\Phi}{\partial x_2}=U_1-iU_2.$$

The above-stated relations are used in the following paragraphs to investigate diverse potential flows. Here occasionally use is made of that the complex number z can also be stated in cylinder coordinates (ρ,φ):

$$z=re^{(i\varphi)}=r\cos\varphi+ir\sin\varphi.$$

Between the velocity components in Cartesian coordinates and in cylinder coordinates the known relations:

$$U_1=U_r\cos\varphi-U_\varphi\sin\varphi$$
$$U_2=U_r\sin\varphi+U_\varphi\cos\varphi$$

hold. Thus for the complex velocity results:

$$w(z)=\frac{dF(z)}{dz}=U_1-iU_2=\left(U_r\cos\varphi-U_\varphi\sin\varphi\right)-i\left(U_r\sin\varphi+U_\varphi\cos\varphi\right)$$
$$=U_r(\cos\varphi-i\sin\varphi)-iU_\varphi(\cos\varphi-i\sin\varphi)$$
$$w(z)=\left(U_r-iU_\varphi\right)e^{(-i\varphi)}.$$

UNIFORM FLOW

Probably the simplest analytical function $F(z)$, disregarding a constant, is the function which is directly proportional to z proportional and whose proportional constant is a real number:

$$F(z)=U_0z=U_0(x+iy)$$

This analytical function describes a flow with the following potential and stream function:

$$\Phi(x,y)=U_0x \text{ and } \Psi(x,y)=U_0y.$$

Via the complex velocity one obtains:

$$w(z)=\frac{dF(z)}{dz}=U_0=U_1=iU_2$$

or $U_1=U_0$ and $U_2=0$ i.e. the complex potential $F(z)$ in equation describes a uniform flow parallel to the x_1 – axis or the x– axis.

This figure shows the flow line Ψ = const, where the arrows indicate the direction of the velocity. The potential lines Φ = const are not indicated in the figure. They represent the lines parallel to the x_2 – axis. When the proportionality constant is imaginary, i.e. it holds:

$$F(z) = iV_0 z = V_0(-y + ix)$$

then one obtains for the potential and stream functions:

$$\Phi(x, y) = -V_0 y \text{ and}$$
$$\Psi(x, y) = V_0 x$$

For the complex velocity it is computed:

$$w(z) = iV_0 = U_1 - iU_2$$

or $U_1 = 0$ and $U_2 = -V_0$., i.e. in this case the complex potential describes a flow parallel to the x_2 – axis or the y– axis which takes place in the direction of the negative axis.

When there is a flow in the direction indicated, the complex potential reads:

$$F(z) = (U_0 - iV_0)z$$
$$= (U_0 - iV_0)(x + iy).$$

From this result the following relations for $\Phi(x, y)$ and $\Psi(x, y)$:

$$\Phi(x, y) = U_0 x + V_0 y \text{ and}$$

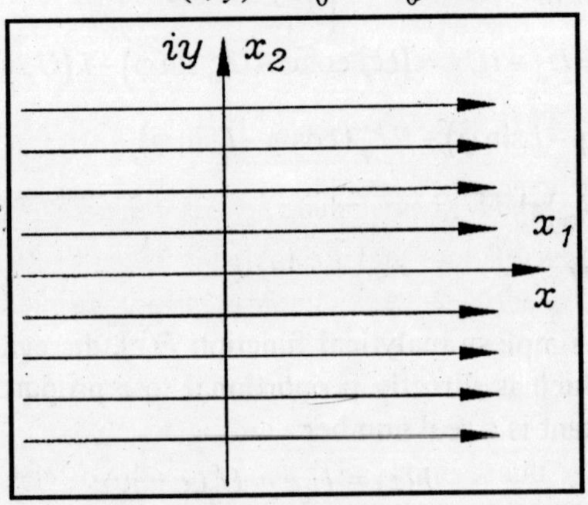

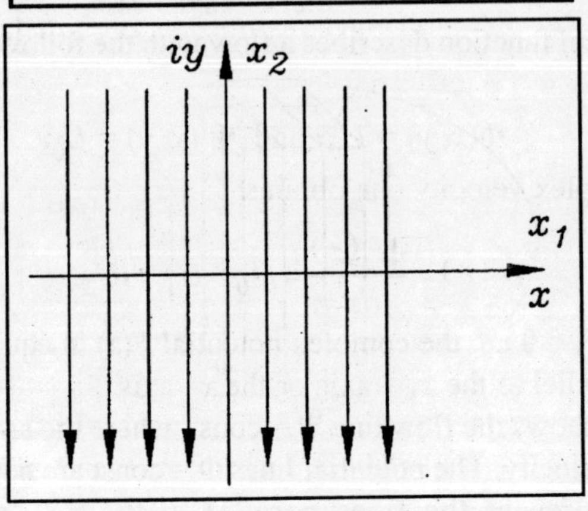

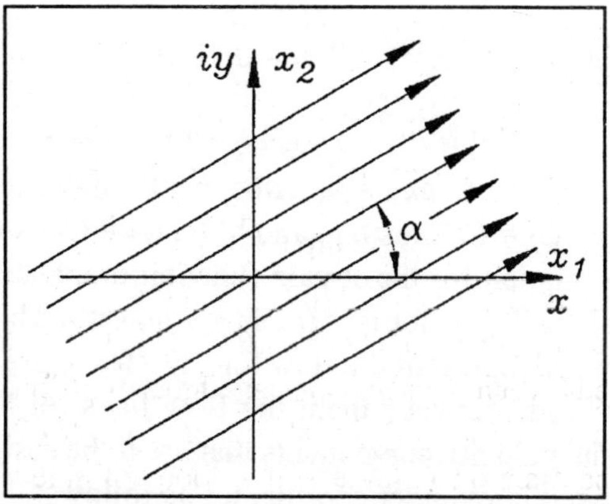

Fig. Uniform flow in (a) x_1–direction, (b) x_2-direction
(c) In direction of the angle a relative to the x_1-direction

$$\Psi(x, y) = U_0 y - V_0 x$$

Via the complex velocity it is computed:

$$w(z) = U_0 - iV_0 = U_1 - iU_2$$

$U_1 = U_0$ and $U_2 = V_0$, a velocity field.

CORNER AND SECTOR FLOW

Potential flows around corners and in sectors of angles are described by a complex potential which is proportional to z^n where for $n \leq 1$ flows around borders/corners are described, while for $n \geq 1$ flows in sectors of angles $\frac{\pi}{n}$ are obtained.

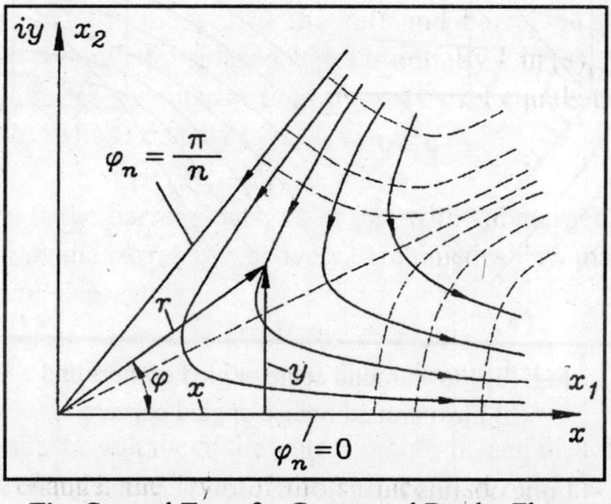

Fig. General Term for Corner and Border Flows

This shall be derived and explained in the following considerations. The derivations are based on the following complex potential:

$$F(z) = Cz^n.$$

When one replaces z by $z = re^{(\iota\varphi)}$ and divides the complex potential into real and imaginary parts, one obtains:

$$F(z) = C\,[r^n \cos(n\varphi) + ir^n \sin(n\varphi)].$$

After this potential and stream function can be stated as follows:

$$\Phi(r, \varphi) = Cr^n \cos(n\varphi) \text{ and } \Psi(r, \varphi) = Cr^n \sin(n\varphi).$$

The resulting relation for the stream function in makes clear, that $\Psi(r, \varphi)$ assumes the values $\Psi = 0$ for $\varphi = 0$ and for $\varphi = \pi/n$ This means that the lines $\varphi = 0$ and $\varphi = \pi/n$ represent the flow line $\Psi = 0$ and are regarded here as walls of the flow field. Between them the flow lines for $\Psi = r^n \sin(n\varphi) =$ const are stated. The velocity components that are to be assigned to this flow field can be expressed in cylinder coordinates as follows:

$$w(z) = \frac{dF(z)}{\mathrm{d}z} = nCz^{(n-1)} = nCr^{(n-1)}e^{\{i(n-1)\varphi\}}$$

or transcribed:

$$w(z) = \left[nCr^{(n-1)}\left(\cos(n\varphi) + i\sin(n\varphi)\right)\right]e^{(-i\varphi)},$$

so that can be stated :

$$U_r = nCr^{(n-1)}\cos(n\varphi) \text{ and } U_\varphi = -nCr^{(n-1)}\sin(n\varphi).$$

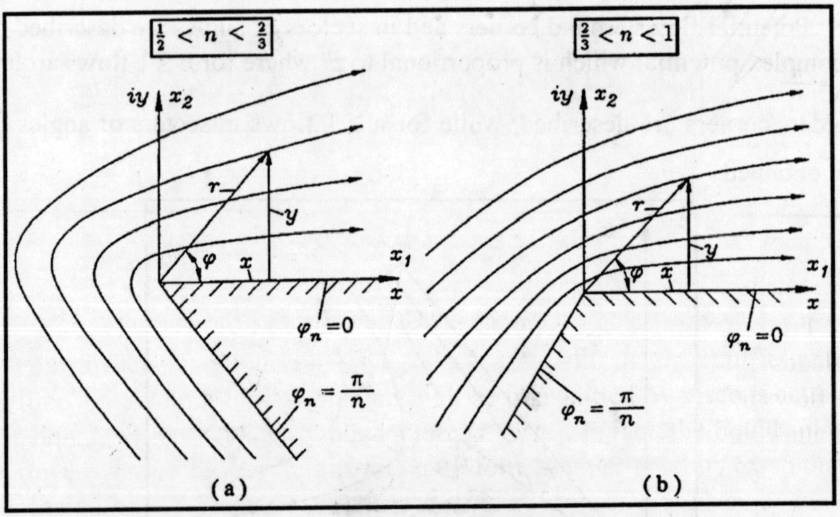

Fig. (a) Flow around acute-angled corners and
(b) Flow around obtuse-angled corners

For $\frac{1}{2} < n < 1$ one obtains fluid flows. Flows around borders are concerned here, which in general are designated as corner flows. For $\frac{2}{3} < n < 1$ flows around obtuse-angled corners are described and for $\frac{1}{2} < n \le \frac{2}{3}$. a representation is shown which comprises the flow around acute-angled corners.

For $1 < n < \infty$. result from the complex potential $F(z) = Cz^n$. flows in angle sectors, as they are sketched for obtuse-angled angle sectors $(1 < n < 2)$ and for acute-angled ones $(2 \le n \le \infty)$. As for $0 < \varphi < (\pi/2n)$ U_r. is always positive, while U_φ. assumes negative values in this domain, and as for $(\pi/2n) < \varphi < (\pi/n)$ U_r. becomes negative and U_φ. remains negative.

The planes $\varphi_n = 0$ and $\varphi_n = \pi/n$ represent a flow line. Along this flow line there are no velocity components in direction of the normal. The velocity changes along the boundary flow line. The flow in an angle sector with acute angle differs from the flow in an obtuse-angles flow domain only by the exponent n in the complex velocity potential.

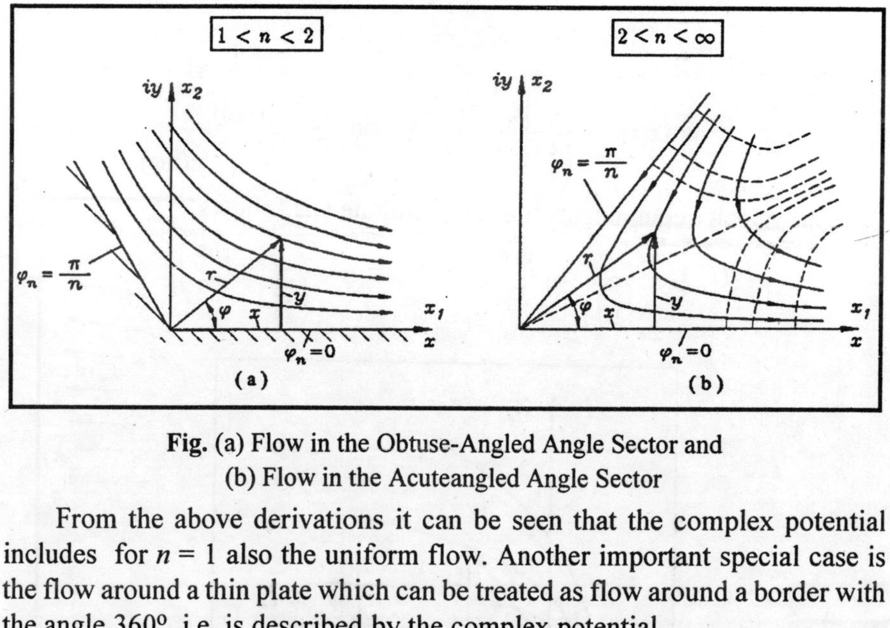

Fig. (a) Flow in the Obtuse-Angled Angle Sector and
(b) Flow in the Acuteangled Angle Sector

From the above derivations it can be seen that the complex potential includes for $n = 1$ also the uniform flow. Another important special case is the flow around a thin plate which can be treated as flow around a border with the angle 360°, i.e. is described by the complex potential

$$F(z) = Cz\left(\frac{1}{2}\right)$$

The proportionality constant is real and the angular area taken/occupied by the flow is

$$0 \le \varphi \le 2\pi.$$

In cylinder coordinates the complex potential can be written as follows:

$$F(z) = Cr^{\left(\frac{1}{2}\right)} e^{\left(i\frac{\varphi}{2}\right)}$$

The potential and stream function can be stated as follows:

$$\Phi(r,\varphi) = Cr^{\frac{1}{2}}\cos\left(\frac{\varphi}{2}\right) \text{ and } \Psi(r,\varphi) = Cr^{\frac{1}{2}}\sin\left(\frac{\varphi}{2}\right).$$

From the relation for the stream function can be derived that the lines φ = 0 and φ = 2π correspond to the flow line Ψ = 0 The flow lines for other φ–values are described by the stream function in equation. Also indicated are the equipotential lines which are also computable. The complex flow velocity is conserved by derivation of the complex potential:

$$w(z) = \frac{dF(z)}{dz} = \frac{C}{2z^{\left(\frac{1}{2}\right)}}$$

$$= \frac{C}{2r^{\left(\frac{1}{2}\right)}} e^{\left(-i\frac{\varphi}{2}\right)}$$

or transcribed:

$$w(z) = \frac{C}{2r^{\left(\frac{1}{2}\right)}}\left[\cos\left(\frac{\varphi}{2}\right) + i\sin\left(\frac{\varphi}{2}\right)\right]e^{(-i\varphi)}.$$

The velocity components thus are computed as follows:

$$U_r = \frac{C}{2r^{\left(\frac{1}{2}\right)}}\cos\left(\frac{\varphi}{2}\right) \text{ and } U_\varphi = -\frac{C}{2r^{\left(\frac{1}{2}\right)}}\sin\left(\frac{\varphi}{2}\right).$$

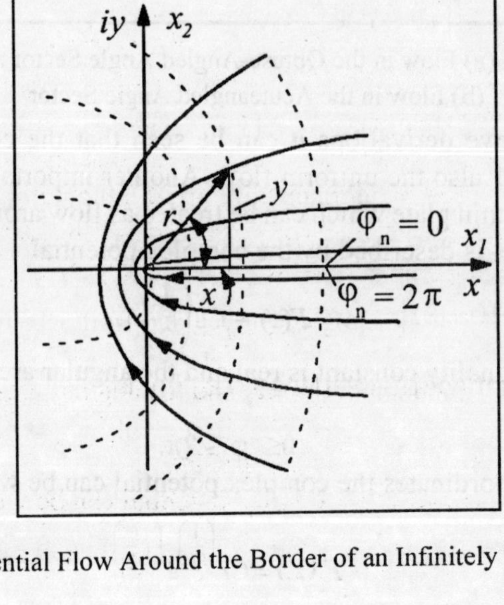

Fig. Potential Flow Around the Border of an Infinitely Thin Plate

These relations make it clear that the velocity component U_φ for $0 < \varphi < 2\pi$ is negative, while U_r for $0 < \varphi < \pi$ is positive and for $\pi < \varphi < 2\pi$ negative.

An important result of the above derivations it can be obtained that the velocity field possesses a singularity in its origin. It is caused by the flow around the plate border and characterized by extreme values of the velocity field. The values of both velocity components approach the value $r \to 0$ for ∞.

SOURCE OR SINK FLOWS AND POTENTIAL VORTEXES

When one chooses a complex potential $F(z)$, which is proportional to the natural logarithm of z., one obtains when selecting a real proportionality constant and depending on whether one chooses a positive or negative sign, the complex potential of a source or a sink flow:

$$F(z) = \pm C \ln z$$

or with $z = r \cdot e^{(i\varphi)}$

$$F(z) = \pm C\,[\ln r + i\varphi] = \Phi + i\Psi.$$

For the potential and flow function one obtains thus:

$$\Phi(r,\varphi) = \pm C\ In\ r \qquad \psi(r,y) = \pm C\varphi$$

$$\Phi(x,y) = \pm C\ In\sqrt{x^2+y^2} \quad \psi(x,y) = \pm C \arctan\frac{y}{x}$$

These equations show that the equipotential lines represent circles with r = const while the flow lines represent radial lines with φ = const. When computing the complex velocity:

$$w(z) = \frac{dF(z)}{dz} = \pm C\frac{1}{z} = \pm C\frac{(x-iy)}{x^2+y^2},$$

one obtains for the velocity components:

$$w(z) = \frac{\pm C}{x^2+y^2}(x-iy) = U_1 = -iU_2$$

$$\text{or } U_1 = \frac{\pm Cx}{x^2+y^2} \text{ and } U_2 = \frac{\pm Cy}{x^2+y^2}.$$

In equation can also be written:

$$w(z) = \frac{\pm C}{z} = \pm\frac{C}{r}e^{(-i\varphi)}$$

A comparison of equations shows that it holds:

$$U_r = \pm\frac{C}{r} \text{ and } U_\varphi = 0.$$

The velocity component U_r decreases with $1/r$, has, however a singularity in its origin $r = 0$.

A flow comes thus for the source flow and which is purely radial.

The volume flow released per time unit and unit depth of the source and which characterizes the strength of the source is given by:

$$\dot{Q} = \int_0^{2\pi} U_r r \, d\varphi = C\, 2\pi,$$

so that the complex potential for the source or the sink flow can be writtenas follows:

$$F(z) = \pm \frac{\dot{Q}}{2\pi} \ln z \qquad \begin{matrix} (+) = \text{source flow} \\ (-) = \text{sink flow} \end{matrix}$$

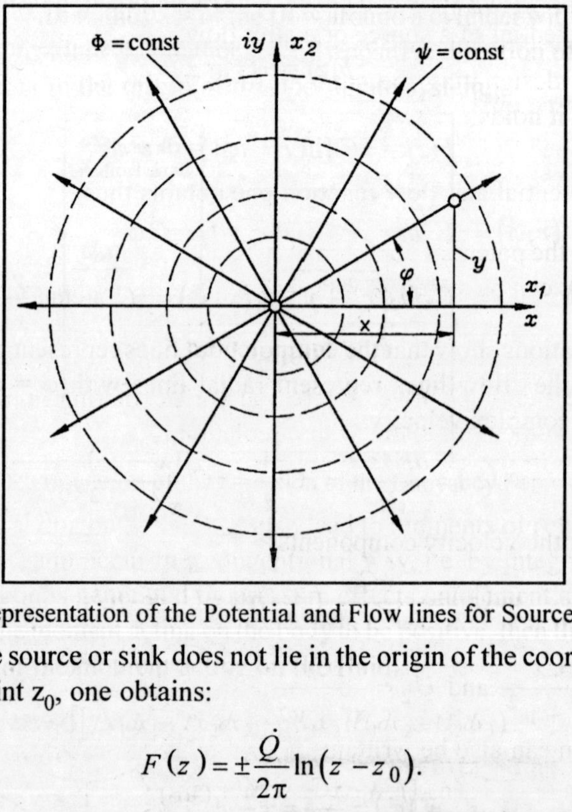

Fig. Representation of the Potential and Flow lines for Source Flows

When the source or sink does not lie in the origin of the coordinate system but in the point z_0, one obtains:

$$F(z) = \pm \frac{\dot{Q}}{2\pi} \ln(z - z_0).$$

When considering a potential z proportional to the natural logarithm, in which the proportionality constant is imaginary, one obtains *F*(z) of a potential vortex:

$$F(z) = iC \ln z = C(-\varphi + i \ln r).$$

For the potential and stream function can be stated from this:

$$\Phi(r, \varphi) = -C\varphi \text{ and } \Psi(r, \varphi) = C \ln r$$

or.

$$\Phi(z,y) = -C \arctan\frac{y}{x} \text{ and } \Psi(x,y) = C \ln\sqrt{x^2+y^2}$$

These relations show that the equipotential radial lines represent $\varphi =$ const while the flow lines are circles with $r =$ const For the complex velocity can be stated:

$$w(z) = \frac{dF(z)}{dz} = iC\frac{1}{z} = i\frac{C}{r}e^{(-i\varphi)}.$$

By comparing the equations, it results:

$$U_r = 0 \text{ and } U_\varphi = -\frac{C}{r}$$

The resulting flow field is that of a potential vortex with the characteristic decrease of the circumferential velocity with the distance from the vortex centre. When designating the strength/force of the potential vortex by the circulation Γ, it holds:

$$\Gamma = \oint U_s d = \int_0^{2\pi} U_\varphi r\, d\varphi = -2\pi C$$

With this the potential vortex rotating in mathematically positive direction (Γ is positive), can be stated as follows owing to its complex potential:

$$F(z) = -\frac{\Gamma}{2\pi} i \ln z$$

When the sign is positive, a potential vortex rotating in mathematically negative direction comes about with circulation.

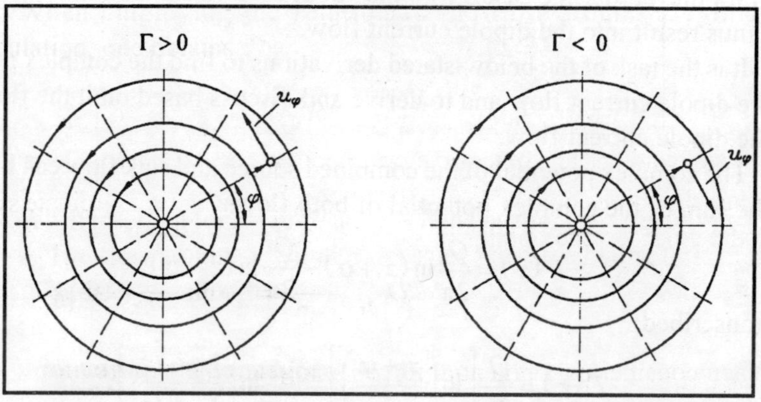

Fig. Flow Lines and Equipotential lines of the Potential Vertex

A strict distinction has to be made between the potential vortex and vortex motions whose flow fields are linked to rotations, like e.g. vortexes where the entire flow field forms analogously to the rotation of a solid/body. The flow field of the potential vortex is irrotational. The entire circulation is limited to the vortex-centre line where the total rotation is located.

DIPOLE CURRENT FLOW

In this section a potential flow shall be discussed which is defined as dipole current flow and results as a borderline case of the superposition of a source flow with a sink flow. Considered is a source with the strength $\dot{Q}$ which is located on the x–-axis in the distance (–*a*). from the origin of a coordinate system and a sink of the same strength which has been arranged on the *x*-axis in the distance (+*a*).

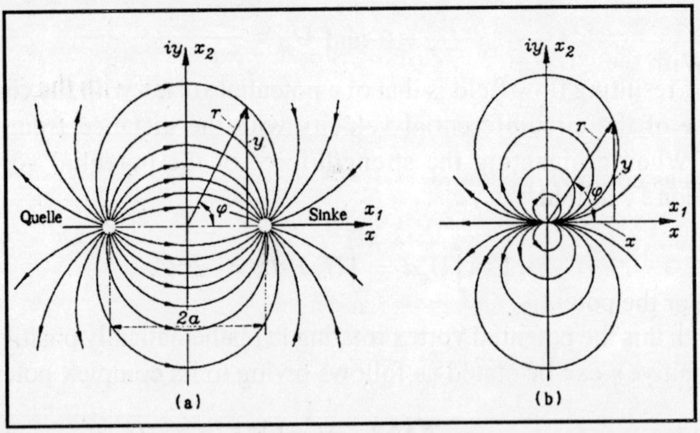

Fig. Flow Lines of (*a*) Source and Sink Flows as well as (b) A Dipole Current Flow

When the distance a is reduced, source and sink move closer together until for the borderline case $a \to 0$ they both coincide in the coordinate origin and thus result into the dipole current flow.

It is the task of the below-stated derivations to find the complex potential of the dipole current flow and to derive and discuss based on it the flow field of the dipole current flow.

The complex potential of the combined source and sink flow can be stated as the sum of the complex potential of both flows:

$$F(z) = \frac{\dot{Q}}{2\pi} \ln(z + \alpha) - \frac{\dot{Q}}{2\pi} \ln(z - \alpha)$$

or transcribed:

$$F(z) = \frac{\dot{Q}}{2\pi}\left[\ln\left(\frac{z+\alpha}{z-\alpha}\right)\right] = \frac{\dot{Q}}{2\pi} \ln\left[\frac{1+\alpha/z}{1-\alpha/z}\right].$$

When carrying out a series expansion for the term $\left(\frac{1}{1-\alpha/z}\right)$ one obtains:

$$F(z) = \frac{Q}{2\pi} \ln\left[\left(1+\frac{\alpha}{z}\right)\left(1+\frac{\alpha}{z}+\frac{\alpha^2}{z^2}+\frac{\alpha^3}{z^3}+\ldots\right)\right]$$

or after performed multiplication:

$$F(z)=\frac{Q}{2\pi}\ln\left(1+2\frac{\alpha}{z}\right)$$

When one carries out another series expansion:

$$\ln\left(1+2\frac{\alpha}{z}\right)=2\frac{\alpha}{z}-2\frac{\alpha^2}{z^2}+\frac{8\alpha^3}{3z^3}\mp\cdots$$

one obtains for small values (a/z)

$$F(z)=\frac{\dot{Q}}{2\pi}2\frac{\alpha}{z}.$$

With the strength/force of the dipole current flow:

$$D=\frac{\dot{Q}\alpha}{\pi}$$

results as complex potential:

$$F(z)=\frac{D}{z}=\frac{D}{(x+iy)}.$$

For the potential and stream function the following terms can be derived:

$$\Phi(r,\varphi)=\frac{Dx}{x^2+y^2}\text{ and }\Psi(r,\varphi)=\frac{-Dy}{x^2+y^2}$$

$$\Phi(r,\varphi)=\frac{D}{r}\cos\varphi\text{ and }\Psi(r,\varphi)=\frac{-D}{r}\sin\varphi$$

The flow lines and equipotential lines are indicated. For the complex velocity can be derived:

$$w(z)=\frac{\mathrm{d}F(x)}{\mathrm{d}z}=-\frac{D}{z^2}=-\frac{D}{r^2}e^{(-i2\varphi)}$$

or transcribed:

$$w(z)=-\frac{D}{r^2}(\cos\varphi-i\sin\varphi)e^{(-i2\varphi)}$$

From this result the velocity components:

$$U_r=-\frac{D}{r^2}\cos\varphi\text{ and }U_\varphi=-\frac{D}{r^2}\sin\varphi.$$

The signs of these velocity components confirm the direction of the flow indicated.

POTENTIAL FLOWS AROUND A CYLINDER

The significance of the dipole current flow lies in the fact that its complex potential can be superimposed with the complex potential of the uniform flow parallel to the x-axis and that thus a complex potential arises which describes the flow around a cylinder. The simple superposition of the $F(z)$-functions of flows is admissible as the partial differential equations derived from the basic equations of flow mechanics are linear for the potential and stream function. By addition of the complex potentials for the even flow parallel to the x-axis and for the dipole current flow one obtains the following relation:

$$F(z) = U_0 z + \frac{D}{z} = U_0 r e^{(i\varphi)} + \frac{D}{r} e^{(-i\varphi)},$$

which is equivalent to

$$F(z) = U_0 r(\cos\varphi + i\sin\varphi) + \frac{D}{r}(\cos\varphi - i\sin\varphi).$$

For the potential and stream function thus the following terms can be found:

$$\Phi(r.\varphi) = \left(U_0 r + \frac{D}{r}\right)\cos\varphi \text{ and } \Psi(r,\varphi) = \left(U_0 r - \frac{D}{r}\right)\sin\varphi.$$

When one inserts now the radius $r = R$ of a cylinder, the stream function along a cylinder wall results as:

$$\Psi(r,\varphi) = \left(U_0 R - \frac{D}{R}\right)\sin\varphi.$$

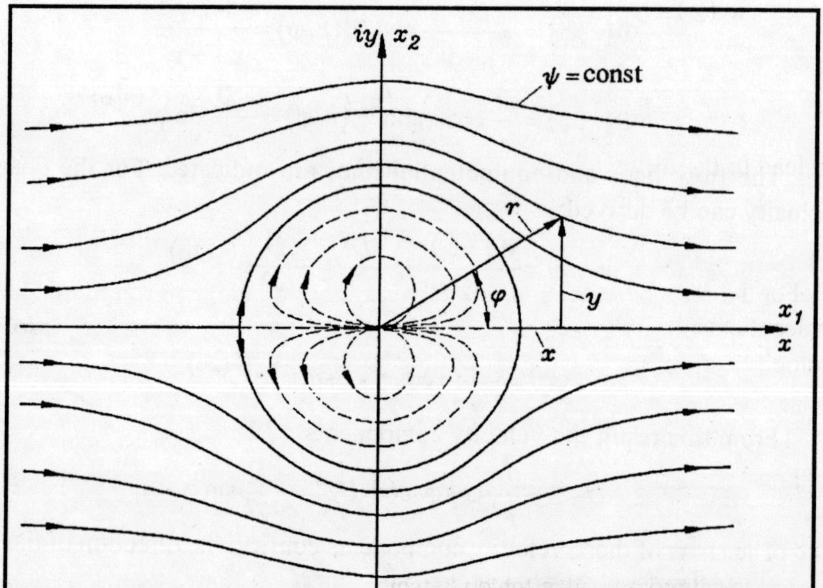

Fig. Flow Lines of the Flow Around a Cylinder

When choosing the strength of the dipole current flow $D = (U_0 R^2)$, one obtains for the stream function $\Psi = 0.$ along the cylinder wall ($r = R$ for all φ). From this representation can be learnt that the outer cylinder line is a dividing line between an internal flow caused by the dipole current flow and an external flow coming from the flow parallel to the x–axis. We thus have a flow whose external flow can be interpreted as the flow resulting from a two-dimensional flow of an incompressible viscose-free fluid around a cylinder.

When one takes into consideration the relation $D = U_0R^2$, derived for the strength of the dipole current flow, for the complex potential of the flow around a cylinder $r \geq R$ can be stated:

$$F(z) = U_0\left(z + \frac{R^2}{z}\right).$$

In addition, for the potential and stream function holds:

$$\Phi(r,\varphi) = U_0\left(r + \frac{R^2}{r}\right)\cos\varphi \text{ and } \Psi(r,\varphi) = U_0\left(r - \frac{R^2}{r}\right)\sin\varphi.$$

For the complex velocity can be derived:

$$w(z) = \frac{dF(z)}{dz} = U_0\left(1 - \frac{R^2}{z^2}\right) = U_0\left[1 - \frac{R^2}{r^2}e^{(-i2\varphi)}\right].$$

Further conversions yield:

$$w(z) = U_0\left[e^{(i\varphi)} - \frac{R^2}{r^2}e^{(-i\varphi)}\right]e^{(-i\varphi)},$$

$$= U_0\left[(\cos\varphi + i\sin\varphi) - \frac{R^2}{r^2}(\cos\varphi - i\sin\varphi)\right]e^{(-i\varphi)},$$

and lead to the following velocity components:

$$U_r = U_0\left(1 - \frac{R^2}{r^2}\right)\cos\varphi \text{ and } U_\varphi = -U_0\left(1 + \frac{R^2}{r^2}\right)\sin\varphi.$$

For the outer cylinder area ($r = R$) results: $U_r = 0$; along the circumferential surface there is only a flow in circumferential direction. For the latter results a velocity component:

$$U_\varphi = -2U_0\sin\varphi \text{ for } r = R$$

For $\varphi = \dfrac{\pi}{2}$ thus a velocity component exists which is equal to twice the value of the inflow velocity.

The indicated potential flow around a cylinder results in a solution having flow-o conditions which are equal to the inflow conditions, so that no force resulting from the flow acts on the cylinder. This can also be derived from the solution for the velocity field itself. As concerns the quantity of the U_j–component, there exists a symmetry to the x-axis, so that the pressure distribution is also symmetrical and therefore no resulting buoyancy force comes about. Because of a likewise existing symmetry of the pressure distribution is also symmetrical and therefore no resulting buoyancy force comes about. Because of a likewise existing symmetry of the pressure distribution to the y–axis no resulting resisting force is produced either. As this result is contradictory to our experience (d'Alambert's paradox), this investigation shows clearly the significance of the viscosity terms in the basic equations of fluid mechanics. When these terms are not considered/included in fluidtechnical considerations for obtaining relevant information with regard to fluid physics, fluid forces on bodies can only be dealt with to some extent.

FLOW AROUND A CYLINDER WITH CIRCULATION

When repeating the procedure where the complex potential of two potential flows were added, in order to obtain after determining a free parameter, the strength D of the dipole current flow, the potential flow around a cylinder, one can e.g. state the following complex potential:

$$F(z)=U_0\left(z+\frac{R^2}{z}\right)+\frac{i\Gamma}{2\pi}\ln z+Ci.$$

This complex potential results from the summation of the potential of the flow around cylinder and the potential vortex, where the centres of both flows lie in the origin of the coordinates. The constant C was included in the above equation to be able to choose the quantity of the stream function again in a way that $\Psi = 0$ when $r = R$ i.e. the outer cylinder is to represent the flow line $\Psi = 0$ in the final relation. For determining now the constant C we insert in the above equation $z = re^{(i\varphi)}$

$$F(z)=U_0\left(re^{(i\varphi)}+\frac{R^2}{r}e^{(-i\varphi)}+\frac{i\Gamma}{2\pi}\ln\left(re^{(i\varphi)}\right)\right)+Ci,$$

Making use of the relation $e^{i\varphi} = \cos\varphi + i\sin\varphi$ it results:

$$F(z)=U_0\left[\left(r+\frac{R^2}{r}\right)\cos\varphi+i\left(r-\frac{R^2}{r}\right)\sin\varphi\right]-\frac{\Gamma}{2\pi}\varphi+i\frac{\Gamma}{2\pi}\ln r+C_i,$$

from which follow the relations for the potential and stream function:

$$\Phi(r,\varphi)=U_0\left(r+\frac{R^2}{r}\right)\cos\varphi-\frac{\Gamma}{2\pi}\varphi$$

and

$$\Psi(r,\varphi)=U_0\left(r-\frac{R^2}{r}\right)\sin\varphi+\frac{\Gamma}{2\pi}\ln r+C$$

In order to obtain $\Psi = 0$ for $r = R$ and all φ one has to choose the constant

$C=-\left(\frac{\Gamma}{2\pi}\right)\ln R$ Thus results for the complex potential of the flow around a cylinder with circulation:

$$F(z)=U_0\left(z+\frac{R^2}{z}\right)+i\frac{\Gamma}{2\pi}\ln\frac{z}{R}.$$

This potential describes the even flow parallel to the x–axis, in connection with a dipole current flow and a potential vortex located in the origin of the coordinate system. For this flow the potential and stream function can be stated as follows:

$$\Phi(r,\varphi)=U_0\left(r+\frac{R^2}{r}\right)\cos\varphi-\frac{\Gamma}{2\pi}\varphi.$$

$$\Psi(r,\varphi)=U_0\left(r-\frac{R^2}{r}\right)\sin\varphi+\frac{\Gamma}{2\pi}\ln\frac{r}{R}.$$

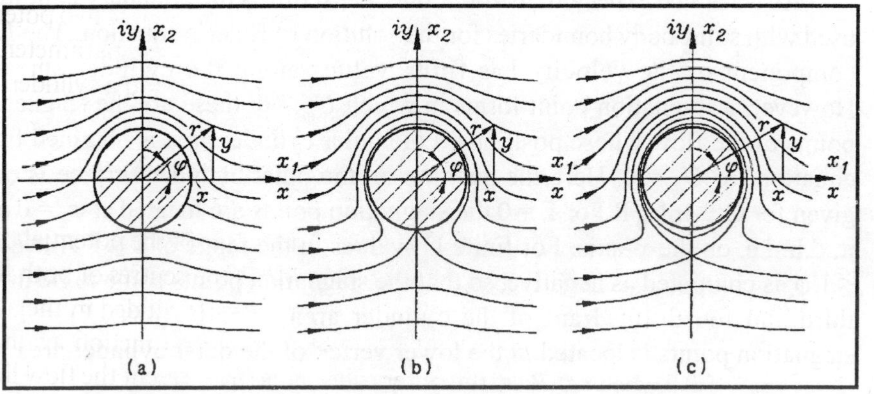

Fig. Flow Lines for the Flow Around a Cylinder with Rotation

(α) Normal Circulation $0\le\frac{\Gamma}{4\pi U_0 R}\le 1;$

(b) Normal Circulation $\frac{\Gamma}{4\pi U_0 R}=1;$

(c) Normal Circulation $\frac{\Gamma}{4\pi U_0 R}>1.$

The corresponding flow and equipotential lines for three typical domains of circulation. The velocity components of the flow field can be computed via the complex velocity:

$$w(z)=U_0\left[1-\frac{R^2}{r^2}e^{(-i2\varphi)}\right]+\frac{i\Gamma}{2\pi r}e^{(-i\varphi)}$$

$$=\left[U_0\left(e^{(i\varphi)}-\frac{R^2}{r^2}e^{(-i\varphi)}\right)+i\frac{\Gamma}{2\pi r}\right]e^{(-i\varphi)}.$$

By comparing this relation with equation the following velocity components result:

$$U_r = U_0\left(1+\frac{R^2}{r^2}\right)\cos\varphi \text{ and } U_\varphi = -U_0\left(1+\frac{R^2}{r^2}\right)\sin\varphi - \frac{\Gamma}{2\pi r}.$$

For $\Gamma = 0$ the equations stated result for the potential flow around a cylinder without circulation.

When setting in the $r = R$, in the above relation, one obtains the velocity components U_r and U_φ along the circumferential area of the cylinder:

$$U_r = 0 \text{ and } U_\varphi = -2U_0\sin\varphi - \frac{\Gamma}{2\pi R}.$$

As was to be expected, the flow line $\Psi = 0$ fulfils the boundary condition used with solid/body boundaries for the solution of Euler's equation. The U_φ–component of the velocity has finite values along the cylinder surface. However, a stagnation point forms in which $U_\varphi = 0$; these are the stagnation points of the flow whose position on the outer cylinder area is obtained from equation for $U_\varphi = 0$. Here the position on the outer cylinder surface is only given for G £ 4pU$_0$ R For $\Gamma = 0$ the stagnation points are located at $\varphi_s = 0$ and π, d.h. i.e. on the x–axis. For finite Γ–values in the range of $0 < \Gamma/(4\pi U_0 R) < 1$ φ_s is computed as negative, so that the stagnation points come to lie in the third and fourth quadrant of the cylinder area. For $\Gamma/(4\pi U_0 R) = 1$ the stagnation points is located in the lower vertex of the outer cylinder area: for this value $\varphi_s = -\frac{\pi}{2}$ is computed and $\frac{3}{2\pi}$.

When the circulation of the flow is increased further, so that $\Gamma > (4\pi U_0 R)$ holds, the stagnation point of the flow cannot form any more along the outer cylinder area; the formation of a "free stagnation point" in the flow field comes about. The position of this point for $U_r = 0$ and $U_\varphi = 0$ can be computed from the above equations for the velocity components, i.e. from:

$$U_0\left(1-\frac{R^2}{r_s^2}\right)\cos\varphi_s = 0$$

and

$$U_0\left(1+\frac{R^2}{r_s^2}\right)\sin\varphi_s = -\frac{\Gamma}{2\pi r_s}.$$

As $r_s \neq R$, i.e. the formation of the free stagnation point on the circumferential area is excluded, the first of the above two equations can only be fulfilled for $\varphi_s = \frac{\pi}{2}$ or $\frac{3}{2\pi}$ Thus the second conditional equation for the position coordinate of the "free stagnation point" reads:

$$U_0\left(1+\frac{R^2}{r_s^2}\right) = \mp\frac{\Gamma}{2\pi r_s}$$

As $\Gamma > 0$ can be assumed in the above equation, and as the left side of the equation can only adopt positive values, only the positive sign of the above equation with the requirements concerning the flow yields consistent values, i.e. the conditional equation for r_s reads:

$$U_0\left(1+\frac{R^2}{r_s^2}\right)=\frac{\Gamma}{2\pi r_s}$$

or transcribed

$$r_s^2-\frac{\Gamma}{2\pi U_0}r_s+R^2=0.$$

As a solution of this equation one obtains:

$$r_s=\frac{\Gamma}{4\pi U_0}\pm\sqrt{\left(\frac{\Gamma}{4\pi U_0}\right)^2-R^2}.$$

With this the position coordinates of the free stagnation point result as:

$$\varphi_s=\frac{3\pi}{2} \text{ and } \frac{r_s}{R}=\frac{\Gamma}{4\pi U_0 R}\left[1+\sqrt{1-\left(\frac{4\pi U_0 R}{\Gamma}\right)^2}\right].$$

The negative sign of the root in the solution for r_s was omitted in the statement of the position coordinates for the free stagnation point, as this would lead to a radius which is located within the outer cylinder area. As only the flow around the cylinder is of concern, this second solution of the square equation for r_s holds no interest.

Moreover, it was also excluded from the solution for the position coordinates of the free stagnation point that the angle φ_s has also a solution for $\frac{\pi}{2}$ The reason for this lies in the fact that for $\frac{\Gamma}{4\pi U_0 R}=1$ the stagnation point appears as a solution only in the lower vertex of the outer cylinder area. An inclusion of the solution for $\varphi_s=\frac{\pi}{2}$ would mean that a small increase of the circulation, to an extent that the standardized circulation is given a value larger than 1, would lead to a jump of the stagnation point from the lower to the upper vertex. Considerations on the stability of the position of the stagnation points show, however, that only the lower stagnation point, i.e $\varphi_s=\frac{3\pi}{2}$ can exist as a stable solution. Because of the superposition of the flow around a cylinder with a potential vortex a flow field has come about, which again is symmetrical concerning the *y*–axis. With this it is in turn determined that owing to the flow the outer cylinder area obtains no resulting force acting in flow direction, i.e. no resisting force occurs because of the flow. Owing to the circulation an asymmetrical flow in relation to the x-axis has come about,however, and this leads to a buoyancy, i.e. to a resulting force on the cylinder, directed upwards. As the velocity component on the upper side of the cylinder is larger than on the underside, because of the Bernoulli equation an excess pressure results prevailing on the underside, which causes a flow force directed upwards. The quantitative determination of this force requires integral relations.

SUMMARY OF IMPORTANT POTENTIAL FLOWS

In the preceding representations a number of potential flows was discussed which are known as basic flows and whose treatment gives an insight into the occurring flow processes. In the following table further analytical functions are stated, in addition to the already extensively discussed examples, which can be used for the derivation of potential and stream functions and the corresponding velocity fields of potential flows. By equating the indicated potential or stream-function values to a constant, the equipotential or flow lines of the potential flow can be stated. The procedure concerning the derivations of fluid-mechanically interesting quantities shall be represented her once gain briefly with the aid of the sourcesink flow taken from the table.

Example: $F(z) = \frac{\dot{Q}}{2\pi}.\ln z = \frac{\dot{Q}}{2\pi}(\ln r + i\varphi); \quad z = x + iy = re^{i\varphi}$

Potential: $\Phi = \frac{\dot{Q}}{2\pi}\ln r = \frac{\dot{Q}}{2\pi}\ln\sqrt{x^2 + y^2}$

Stream function: $\Psi = \frac{\dot{Q}}{2\pi}\varphi = \frac{\dot{Q}}{2\pi}\arctan\frac{y}{x}$

Velocity: $u = \frac{\partial\Phi}{\partial x} = \frac{\dot{Q}}{2\pi}\frac{x}{x^2 + y^2} = \frac{\partial\Psi}{\partial y}$

$$u = \frac{\partial\Phi}{\partial y} = \frac{\dot{Q}}{2\pi}\frac{y}{x^2 + y^2} = -\frac{\partial\Psi}{\partial x}$$

$$c = \sqrt{u^2 + v^2} = \frac{\dot{Q}}{2\pi}\sqrt{\frac{x^2 + y^2}{\left(x^2 + y^2\right)^2}} = \frac{\dot{Q}}{2\pi r}$$

Equipotential lines: $y = \sqrt{e^{\frac{2\pi}{c}\cdot K\Phi} - x^2}$

$$\Phi = K_\Phi$$

Stream Lines: $y = x\tan\left(\frac{2\pi}{\dot{Q}}\right)K_\Psi$

$$\Psi = K_\Psi$$

Complex potential $F(z)$	Potential $\Phi(x, y)$	Stream function $\Psi(x, y)$	Velocity u	v	c	Flow lines $\Psi = \text{const}$
$(u_\infty - iv_\infty)z$ parallel flow	$u_\infty x + v_\infty y$	$u_\infty y - v_\infty x$	u_∞	v_∞	$c_\infty = \sqrt{u_\infty^2 + v_\infty^2}$	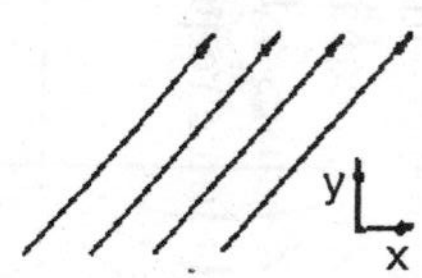
Translation flow $u_\infty z$ in x-direction	$u_\infty x$	$u_\infty y$	u_∞	0	u_∞	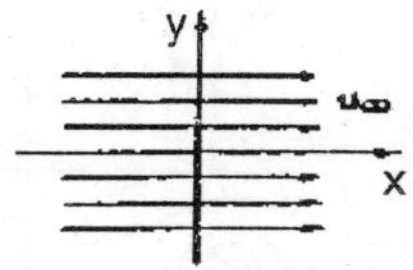
$\frac{a}{2}z^2$ Stagnation-point flow Corner flow a real > 0	$\frac{a}{2}(x^2 - y^2)$	axy	ax	$-ay$	ar	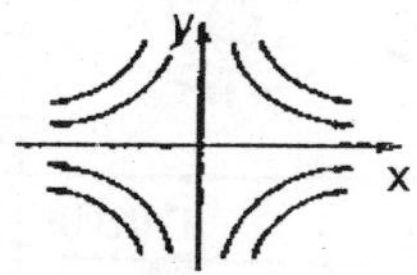
$\frac{\dot{Q}}{2\pi} \ln z$ Source $\dot{Q} > 0$, Sink $\dot{Q} < 0$	$\frac{\dot{Q}}{2\pi} \ln r = \frac{\dot{Q}}{2\pi} \ln \sqrt{x^2 + y^2}$	$\frac{\dot{Q}}{2\pi}\varphi = \frac{\dot{Q}}{2\pi} \arctan \frac{y}{x}$	$\frac{\dot{Q}}{2\pi} \frac{x}{x^2+y^2}$	$\frac{\dot{Q}}{2\pi} \frac{y}{x^2+y^2}$	$\frac{\dot{Q}}{2\pi r}$	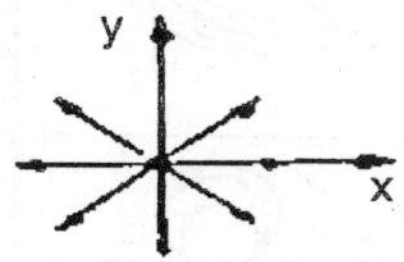
Vortex $\frac{\Gamma}{2\pi} i \ln z$ $\Gamma > 0$ Right-handed $\Gamma < 0$ Left handed	$-\frac{\Gamma}{2\pi} \arctan \frac{y}{x}$	$\frac{\Gamma}{2\pi} \ln \sqrt{x^2 + y^2}$	$\frac{\Gamma}{2\pi} \frac{y}{x^2+y^2}$	$-\frac{\Gamma}{2\pi} \frac{x}{x^2+y^2}$	$\frac{\Gamma}{2\pi r}$	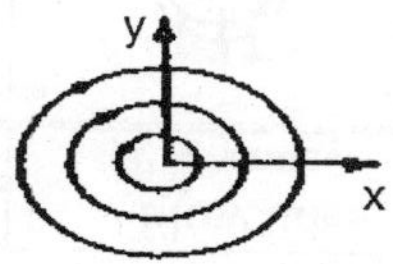

Complex potential $F(z)$	Potential $\Phi(x,y)$	Stream function $\Psi(x,y)$	Velocity u	Velocity v	Velocity c	Flow lines $\Psi = \text{const}$
$\frac{m}{z}$ Dipol	$\frac{mx}{x^2+y^2}$	$-\frac{my}{x^2+y^2}$	$m\frac{y^2-x^2}{(x^2+y^2)^2}$	$-m\frac{2xy}{(x^2+y^2)^2}$	$-\frac{m}{r^2}$	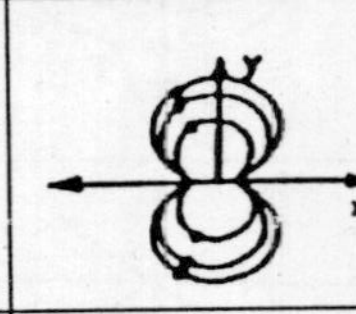
$u_\infty z + \frac{\dot{Q}}{2\pi}\ln z$ Parallel flow + source/sink	$u_\infty x + \frac{\dot{Q}}{2\pi}\ln r$	$u_\infty y + \frac{\dot{Q}}{2\pi}\varphi$	$u_\infty + \frac{\dot{Q}}{2\pi}\frac{x}{x^2+y^2}$	$\frac{\dot{Q}}{2\pi}\frac{y}{x^2+y^2}$		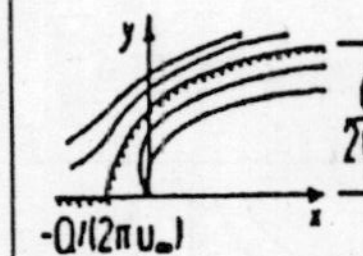
Parallel flow + dipole $u_\infty\left(z + \frac{R^2}{z}\right)$ Flow around a cylinder	$u_\infty x\left(1 + \frac{R^2}{x^2+y^2}\right)$	$u_\infty y\left(1 - \frac{R^2}{x^2+y^2}\right)$	on the cylinder: $2u_\infty \sin^2\varphi$	$-2u_\infty \sin\varphi\cos\varphi$	$2u_\infty \sin\varphi$	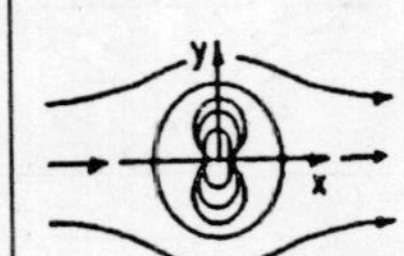
$u_\infty\left(z + \frac{R^2}{z}\right) + \frac{\Gamma}{2\pi}i\ln z$ Flow around a cylinder + vortex	$u_\infty x\left(1 + \frac{R^2}{x^2+y^2}\right) - \frac{\Gamma}{2\pi}\varphi$	$u_\infty y\left(1 - \frac{R^2}{x^2+y^2}\right) + \frac{\Gamma}{2\pi}\ln r$	on the cylinder: $2u_\infty \sin^2\varphi + \frac{\Gamma}{2\pi R}\sin\varphi$	$-2u_\infty \sin\varphi\cos\varphi - \frac{\Gamma}{2\pi R}\cos\varphi$	$2u_\infty \sin\varphi + \frac{\Gamma}{2\pi R}$	
Parallel flow + vortex	$u_\infty x + \frac{\Gamma}{2\pi}\varphi$	$u_\infty y + \frac{\Gamma}{2\pi}\ln r$	$u_\infty + \frac{\Gamma}{2\pi}\frac{y}{x^2+y^2}$	$-\frac{\Gamma}{2\pi}\frac{x}{x^2+y^2}$		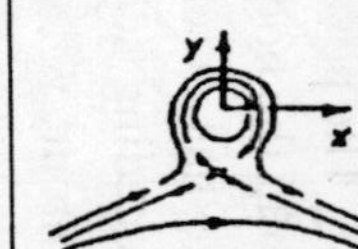

FLOW FORCES ON BODIES

The possibility was mentioned to compute from the pressure distribution along the body contour the forces acting on bodies via potential flows. When one has determined the velocity field of a potential flow according to the preceding sections, the velocity distribution is also present along the body contour, which represents a flow line of the flow field. In each point of the flow holds the Bernoulli equation in the following form:

$$P+\frac{\rho}{2}\left(U_s^2+U_n^2\right)=\text{const}$$

For the flow line and thus the body contour holds $U_n = 0$, i.e.:

$$P+\frac{\rho}{2}U_s^2=\text{const}$$

The quantity U_s^2 can be computed from U_1 and U_2 or. U_r and U_φ as follows:

$$U_s^2=U_1^2+U_2^2=U_r^2+U_\varphi^2$$

Along the contour of a flow body the following integrations can be carried out

$$K_1=-\oint P\cos\varphi ds=-\oint P dx_2 \text{ and } K_2=-\oint P\sin\varphi ds=-\oint P dx_1,$$

in order to conserve the flow forces in the x_1-direction or the x_2-direction of a Cartesian coordinate system (here φ. is the angle between body contour and x_2-axis). When referring the forces to the inflow direction and choosing the latter such that it is identical with the x_1-direction K_1. results in the resisting force on the body, while K_2 yields the buoyancy.

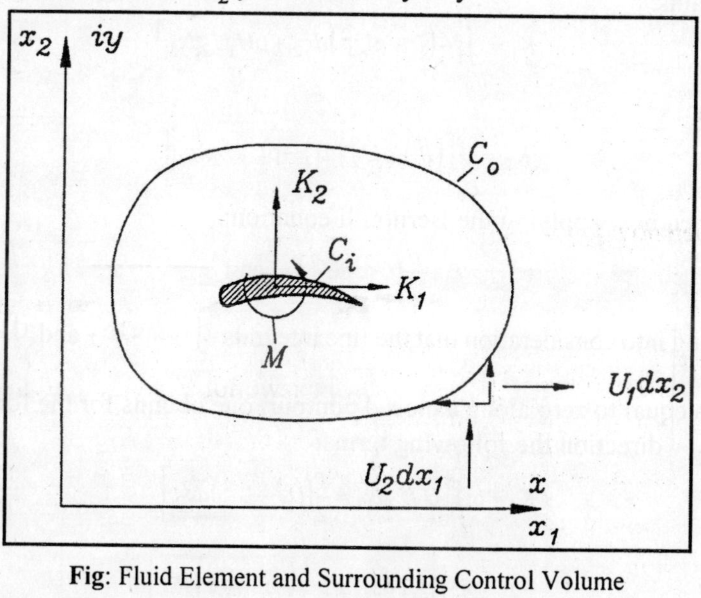

Fig: Fluid Element and Surrounding Control Volume

The attempt shall be made to conserve the forces directly via the computation bases which use the complex velocity. To this effect a control volume around the flow body with the height 1 vertical to the flow is assumed.

In this way a control room comes about which is determined by an internal and an external contour. The fluid forces attacking in the centre of gravity of the flow body and in direction of the x_1-axis and x_2-axis are indicated likewise. Also stated is the moment which a body can get by occurring flow forces.

When now applying to the control room the momentum equations in integral form, it can be expressed in words that the increase of x_1 – or. x_2 – momentum of the flow can only be caused by the flow forces attacking the body in x_1 – or x_2 – direction. In x_1 – direction results:

$$-K_1 - \oint_{C_0} P dx_2 = \oint_{C_0} \rho U_1 (U_1 dx_2 - U_2 dx_1).$$

This relation considers that the internal contour of the control room represents the surface of a flow body, so that the fluid does not flow through it. The pressure forces acting on the internal contour C_i in x_1 -direction were combined into the resulting force K_1 The force acts in positive direction on the body and thus in negative direction on the fluid; this explains the negative sign in front of K_1.

A similar relation can be written for the x_2 – direction:

$$-K_2 + \oint_{C_0} P dx_1 = \oint_{C_0} \rho U_2 (U_1 dx_2 - U_2 dx_1).$$

When solving the two equations in terms of the forces and transforms, one obtains:

$$K_1 = \oint_{C_0} \left[-\left(P + \rho U_1^2\right) dx_2 + \rho U_1 U_2 dx_1 \right]$$

and

$$K_2 = \oint_{C_0} \left[\left(P + \rho U_2^2\right) dx_1 - \rho U_1 U_2 dx_2 \right].$$

When now applying the Bernoulli equation:

$$P + \frac{\rho}{2}\left(U_1^2 + U_2^2\right) = \text{const}$$

and taking into consideration that the line integrals $\oint_{C_0} (\text{const}) dx_1$ and $\oint_{C_0} (\text{const}) dx_2$ are both equal to zero along a closed contour, one obtains for the forces in x_1 – and x_2 – direction the following terms:

$$K_1 = \rho \oint_{C_0} \left[U_1 U_2 dx_1 = \frac{1}{2}\left(U_1^2 - U_2^2\right) dx_2 \right]$$

$$K_2 = -\rho \oint_{C_0} \left[U_1 U_2 dx_2 + \frac{1}{2}\left(U_1^2 - U_2^2\right) dx_1 \right].$$

When now considering the quantity:

$$i\frac{\rho}{2}\oint_{C_0} w^2(z)\,\mathrm{d}z = i\frac{\rho}{2}\oint_{C_0}(U_1 - iU_2)^2(dx + idy)$$

one obtains:

$$i\frac{\rho}{2}\oint_{C_0} w^2(z)\,\mathrm{d}z = \rho\oint_{C_0}\left[\left(U_1U_2dx_1 - \frac{1}{2}\left(U_1^2 - U_2^2\right)dx_2\right)\right.$$

$$\left. + i\left(U_1U_2dx_2 + \frac{1}{2}\left(U_1^2 - U_2^2\right)dx_1\right)\right].$$

This equation shows that the flow forces in x_1 – and x_2 – direction that act on a body can be computed as follows:

$$i\frac{\rho}{2}\oint_{C_0} w^2(z)\,\mathrm{d}z = K_1 - iK_2.$$

Via this relation, the Blasius integral for flow forces, the flow forces on bodies lying in potential flows can be computed easily.

When employing the above relation to compute the resulting force components on the cylinder with circulation, one obtains, beginning with the complex potential:

$$F(z) = U_0\left(z + \frac{R^2}{z}\right) + i\frac{\Gamma}{2\pi}\ln\frac{z}{R}$$

for the complex velocity:

$$w(z) = \frac{\mathrm{d}F(z)}{dz} = U_0\left(1 - \frac{R^2}{z^2}\right) + \frac{i\Gamma}{2\pi z}.$$

For $w^2(z)$ is computed:

$$w^2(z) = U_0^2 - \frac{2U_0^2R^2}{z^2} + \frac{U_0^2R^4}{z^4} + \frac{iU_0\Gamma}{\pi z} - \frac{iU_0\Gamma R^2}{\pi z^3} - \frac{\Gamma^2}{4\pi^2 z^2}$$

or transcribed:

$$w^2(z) = U_0^2 + \frac{U_0^2R^4}{z^4} - \frac{1}{z^2}\left(2U_0^2R^2 + \frac{\Gamma^2}{4\pi^2}\right) - i\left[\frac{U_0\Gamma R^2}{\pi z^3} - \frac{U_0\Gamma}{\pi z}\right].$$

Inserted into the relation for the components K_1 and K_2 of the flow force, one obtains for the integration along the outer cylinder area:

$$K_1 = iK_2 = i\frac{\rho}{2}\oint w^2(z)dz = i\frac{\rho}{2}\oint\left[U_0^2 + \frac{U_0^2R^4}{z^4} - \frac{1}{z^2}\left(2U_0^2R^2 + \frac{\Gamma}{4\pi^2}\right)\right.$$

$$\left. - i\left(\frac{U_0\Gamma R^2}{\pi z^3} - \frac{U_0\Gamma}{\pi z}\right)\right]dz.$$

When introducing into this integral $z = re^{(i\varphi)}$ and considering that for the outer cylinder area $r = R$. holds, then the integration can be carried out and leads to the result:

$$K_1 - iK_2 = -i\rho U_0\Gamma$$

or $K_1 = 0$ and $K_2 = \rho U_0 \Gamma$. This is the Kutta-Joukowski law. This law says that the flow force occurring with a potential flow around a cylinder is equal to zero, when there is no circulation.

When there is circulation, no resisting force occurs but a buoyancy which per linear measure is proportional to the fluid density of the inflow velocity and the circulation:

$$K_2 = \rho U_0 \Gamma.$$

As the sign of this force is positive, there is buoyancy. The inflow direction, the direction of rotation of the vortex and the direction of the resulting buoyancy represent the directions of the axes of a rectangular coordinate system oriented to the right.

The positive force in the case of the flow around a cylinder with circulation comes about as a result of the mathematically positive direction of rotation of the potential vortex in the origin of the coordinate system.

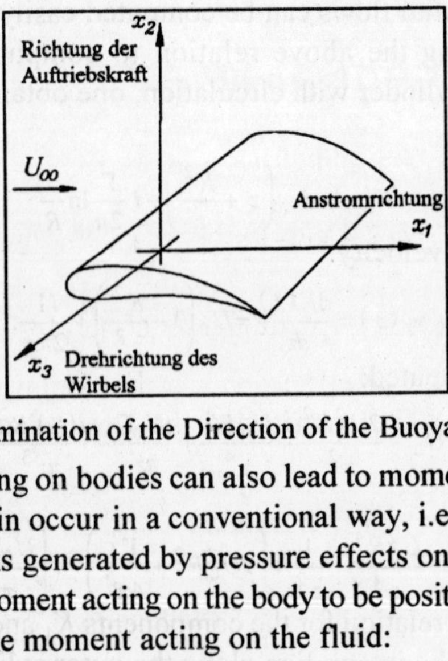

Fig. Determination of the Direction of the Buoyancy Forces

Flow forces acting on bodies can also lead to moments of rotation whose computation can again occur in a conventional way, i.e. by integration of the moment contributions generated by pressure effects on element areas. When again assuming the moment acting on the body to be positive, then the following equation holds for the moment acting on the fluid:

$$M = \oint_{C_0} \left[Px_1 dx_1 + Px_2 dx_2 + \rho U_1 x_2 \left(U_1 dx_2 - U_2 dx_1 \right) \right.$$
$$\left. - \rho U_2 x_1 \left(U_1 dx_2 - U_2 dx_1 \right) \right] = 0.$$

When solving in terms of M one obtains:

$$M = -\oint_{C_0} \left[Px_1 dx_1 + Px_2 dx_2 + \rho \left(U_1^2 x_2 dx_2 + U_2^2 x_1 dx_1 \right. \right.$$
$$\left. \left. - U_1 U_2 x_2 dx_1 - U_1 U_2 x_1 dx_2 \right) \right].$$

When eliminating the pressure via the Bernoulli equation

$$P+\frac{\rho}{2}\left(U_1^2+U_2^2\right)=\text{const}$$

and considering that the integrals are

$$\oint_{C_0}(\text{const})\,x_1dx_1=\oint_{C_0}(\text{const})\,x_2dx_2=0$$

one obtains:

$$M=\frac{\rho}{2}\oint_{C_0}+\left[\left(U_1^2-U_2^2\right)\left(x_1dx_1-x_2dx_2\right)+2U_1U_2\cdot\left(x_1dx_2+x_2dx_1\right)\right]$$

and it can be shown that the following holds:

$$M=\frac{\rho}{2}\text{Re}\left(\oint_{C_0}zw^2(z)\,\text{d}z\right).$$

An evaluation of the integral yields:

$$M=\text{Re}\left[\frac{\rho}{2}\oint_{C_0}(z+iy)\left(U_1-iU_2\right)^2(\text{d}x+i\text{d}y)\right]$$

considering that $x_1=x$ and $x_2=y$ one obtains:

$$M=R\text{e}\left\{\frac{\rho}{2}\oint\left[\left(U_1^2-U_2^2\right)\left(x_1\text{d}x_1-x_2\text{d}x_2\right)+2U_1U_2\cdot\left(x_1\text{d}x_2+x_2\text{d}x_1\right)\right]\right.$$
$$\left.+i\left[\left(U_1^2-U_2^2\right)\left(x_1\text{d}x_2+x_2\text{d}x_1\right)-2U_1U_2\left(x_1\text{d}x_1-x_2\text{d}x_2\right)\right]\right\}$$

The real part of equation corresponds to the term, which was to be proved. When employing the relation to the flow around a cylinder with circulation, one obtains :

$$M=R\text{e}\left[\frac{\rho}{2}\oint_{C_0}\left(U_0^2z-\frac{2U_0^2R^2}{z}+\frac{U_0^2R^4}{z^3}+\frac{iU_0\Gamma}{\pi}-\frac{iU_0\Gamma R^2}{\pi z^2}-\frac{\Gamma^2}{4\pi^2 z}\right)\text{d}z\right]$$

When inserting for $z=re^{(i\varphi)}$ and $r=R$ in equation, one obtains as a solution $M=0$. The flow around a cylinder does not furnish a hydrostatic moment on the cylinder, even when the flow has circulation.

Chapter 10

Wave Motions in Fluids Free from Viscosity

GENERAL CONSIDERATIONS

Fluid flows were considered whose analytical treatment was possible by employing simplified forms of the generally valid basic equations of fluid mechanics. The solution methods required for this are known, i.e. they are at disposal and it is known that they can be employed. The application of methods was shown which permit the solutions of the basic equations of fluid mechanics in order to obtain one-dimensional and two-dimensional flows.

In particular, potential flows were dealt with whose given properties were chosen such that methods of the theory of functions can be employed to treat analytically two-dimensional and irrotational flow problems.

The special properties of potential flows made it possible to take over a fully developed domain of mathematics into flow mechanics and to employ it for computing potential and streamline fields. From these fields velocity fields of the treated potential flows could be derived and the employment of the mechanical energy equation and its integral form, respectively finally lead to pressure distributions in the considered flow fields.

The latter again lead to the computation of forces and moments for control volume which are of particular interest for the solution of engineering problems. Simplifications of the flow properties by introducing two-dimensionality and irrotationality thus have permitted a closed treatment of flow problems with known mathematical methods.

The treatment of wave motions of fluids is planned. As with all mechanical wave motions, motions are concerned that occur around a medium rest position, i.e. the fluid particles involved in the wave motion experience no change of position in the time average.

Thus in the case of wave motions in fluids only the vibration state characterizing a wave continues and not the fluid itself. This holds independently from whether with wave motions in fluids longitudinal or transversal waves form shows the oscillation motion of fluid particles for both wave modes.

One can infer that the considered wave motions are periodical, with regard to space as well as time. Oscillations on the other hand are periodical either concerning time or space.

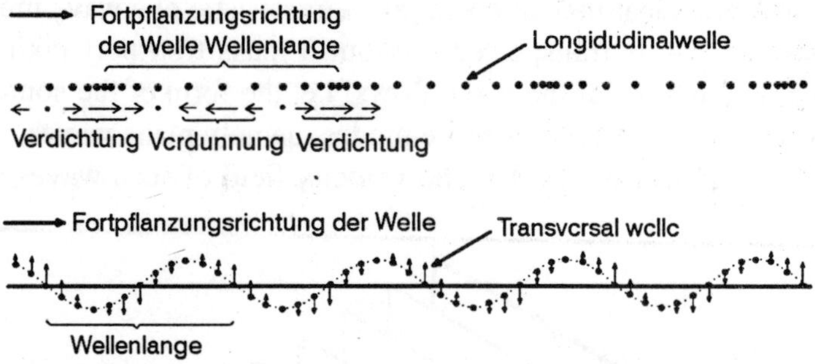

Fig. Instantaneous Image of a Progressing Longitudinal and Transversal Wave

Mechanical longitudinal waves, which are characterized by condensations and dilutions, i.e. by specific changes in volume or density, can exist in all matters having "volume elasticity", i.e. react with elastic counter forces in case of occurring volume changes. Such counter forces form in gases, as volume changes are coupled to pressure changes, so that for an ideal gas at T = const holds

$$P \cdot dv \quad = -v \cdot dP,$$

and thus longitudinal waves can occur in isothermal gases, which are not possible in thermodynamically ideal fluids because of $\rho = 1/v$ = const. makes also clear that the formation of transversal waves is dependent on the presence of "shear forces", i.e. lateral forces, in order to permit the wave motion of "particles" diagonally to the direction of propagation. With this mechanical transversal waves only occur in solid matters which can build up elastic transversal forces in the case of corresponding stress. In purely viscose fluids no transversal waves are possible. At first sight his statement seems to be a contradiction to observations of water waves whose development and propagation can be observed easily when one throws an object in a water container.

A transversal wave develops which, however, proves to be a wave motion restricting itself to a small area on the water surface. In the interior of the fluid the wave motion cannot be observed. Moreover can be seen that the observed wave on the surface does not form owing to "shear forces" but that the presence of gravity or the occurrence of surface tensions are responsible for the wave motion.

In fluids the most different wave motions are possible, whose initiation and conservation are connected to an energy input into the fluid. For the generation of a wave and its conservation a certain work e ort is necessary which then propagates in the space as energy of the wave. With this it is essential that two different energy modes are possible, and also occur, between which an exchange of energy can take place in periodical sequence.

This makes it clear that an essential characteristic of a wave motion in a fluid is that energy is transported without a mass transport taking place. Depending on the form of the wave flows, i.e. the form of the source of the wave motion, one distinguishes wave modes, namely plane waves, spherical waves and cylindrical waves. For the velocity field of such waves it can be stated:

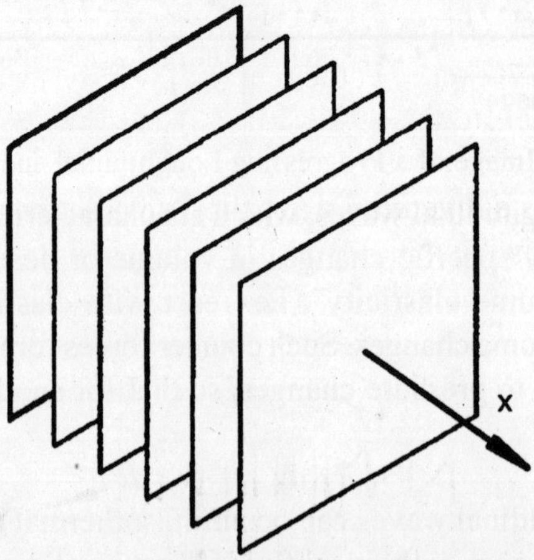

Fig. Diagram of a Two-dimensional Wave and Statements Concerning the Nomenclature

Plane waves:

$$u'(x, t) = u_A \cdot \sin\left[2\pi\left(\frac{t}{T} \mp \frac{x}{\lambda}\right)\right]$$

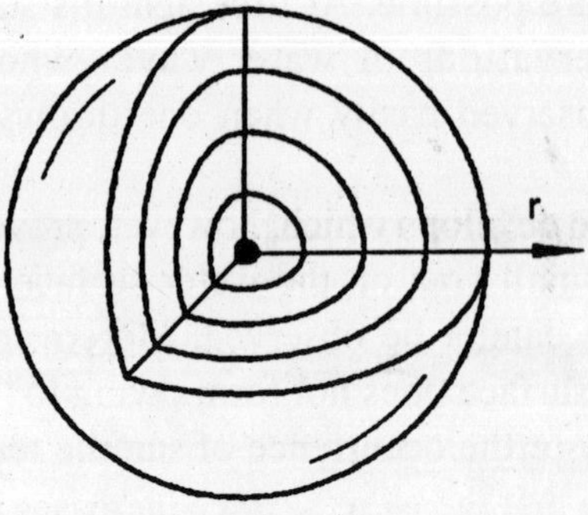

Fig. Diagram of a Spherical wave and Statements Concerning the Nomenclature

In the case of a plane wave the mean energy density is constant, as a considered surface of a wave does not change along the propagation direction *x*. In the above relation *T* is the oscillation period of the wave motion and l is the wave length. The periodicity of the plane wave in the propagation direction *x* and the time *t* can be seen from the sinusoidal term.

Spherical wave:

$$u'(x, t) = \frac{u_A}{r} \cdot \sin\left[2\pi\left(\frac{t}{T} \mp \frac{r}{\lambda}\right)\right]$$

As concerns spherical waves in fluids, the energy density decreases with the square of the distance of point $r = 0$ ab, as the surface increases with the square of the distance.

In point $r = 0$ the exciter of the spherical wave is located; the entire energy of the wave is concentrated at this place. Here the above-stated equation of the wave only holds for $r \neq 0$.

Term indicates a diverging wave that runs/moves away from the exciter centre and the positive sign indicates a diverging wave running/moving towards the exciter centre.

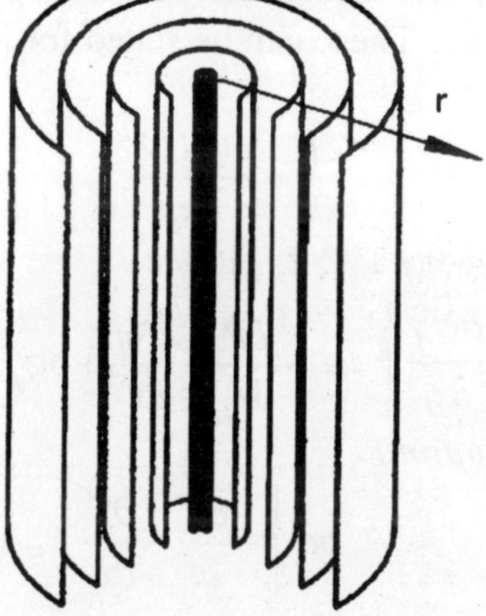

Fig. Diagram of a Cylindrical Wave and Statements Concerning the Nomenclature

Cylindrical wave:

$$u'(x, t) = \frac{u_A}{\sqrt{r}} \cdot \sin\left[2\pi\left(\frac{t}{T} \mp \frac{r}{\lambda}\right)\right]$$

Cylindrical waves propagate radially from the exciter line located in the centre , i.e. $r = 0$, such that the wave surface increases in a linear way with the distance *r* Thus the energy density decreases in a linear way with the distance *r* Therefore then the amplitude of the wave is in inverse ratio proportional to the root of the distance *r* from the exciter line. Again the negative sign in front of the r/λ-Term indicates a wave moving from the exciter line in positive r-direction, whereas the positive sign describes a wave moving towards the exciter line.

Many general properties of wave motions known from physics can be transferred to wave propagations in fluids. Nevertheless, in a book meant as an introduction into fluid mechanics special considerations are required, in particular is it necessary to arouse the deeper comprehension of the causes of the considered wave motions and to show how to deal with them on the basis of the Navier-Stokes equations.

The derivations of properties of theses wave motions shall show in which way to proceed in fluid mechanics to derive the properties from the basic equations of fluid mechanics. The aim of the descriptions/explications thus is not a broad consideration of different wave motions in fluids, but an introduction into the mathematical treatment of longitudinal and transverse waves in fluids.

LONGITUDINAL WAVES: SOUND WAVES IN GASES

In order to be able to deal theoretically with the properties of longitudinal sound waves in ideal gases. These can be stated for ideal gases as follows:

Continuity equation:

$$\frac{\partial \rho}{\partial t}+\frac{\partial(\rho U_i)}{\partial x_i}=0$$

Momentum equations: (j = 1, 2, 3)

$$\rho\left[\frac{\partial U_j}{\partial t}+U_i\frac{\partial U_j}{\partial x_i}\right]=-\frac{\partial P}{\partial x_j}-\frac{\partial \tau_{ij}}{\partial x_i}+\rho g_j$$

Thermal energy equation:

$$\rho c_v\left[\frac{\partial T}{\partial t}+\frac{\partial T}{\partial x_i}\right]=-\frac{\partial q_i}{\partial x_i}-P\frac{\partial U_i}{\partial x_i}-\tau_{ij}\frac{\partial U_i}{\partial x_i}$$

State equation:

$$\frac{P}{\rho}=RT \text{ and } e=c_v\cdot T$$

The above system of partial differential equations and thermo-dynamic state equations comprises 7 unknowns, namely U_1, U_2, U_3, P, ρ, e and T, for the determination of which 7 equations are at disposal. Thus we have a closed equation system which, with sufficient starting and boundary conditions, can be solved in principle, therefore allows one to treat fluid motions.

The above-indicated equation system is considerably simplified when one neglects the molecule-dependent heat and momentum transport, so that all terms of the momentum and energy equation provided with q_i and t_{ij}. can be dropped and mass forces are neglected. When maintaining the Tensor way of writing, the equations can, after introduction of this neglect, be written as follows:

Continuity equation:

$$\frac{\partial \rho}{\partial t} + \frac{\partial(\rho U_i)}{\partial x_i} = 0$$

Momentum equations: (j = 1, 2, 3)

$$\rho \frac{DU_j}{Dt} = \rho \left[\frac{\partial U_j}{\partial t} + U_i \frac{\partial U_j}{\partial x_i} \right] = \frac{\partial P}{\partial x_i}$$

Energy equation:

$$\rho \frac{De}{Dt} = \rho C_v \frac{DT}{Dt} = \rho C_v \left[\frac{\partial T}{\partial t} + U_i \frac{\partial T}{\partial x_i} \right] = -P \frac{\partial U_i}{\partial x_i}$$

State equation:

$$\frac{P}{\rho} = RT \text{ and } e = c^v \cdot T$$

Taking into consideration that the continuity equation can be written as follows:

$$\frac{\partial \rho}{\partial t} + U_i \frac{\partial \rho}{\partial x_i} + \rho \frac{\partial U_i}{\partial x_i} = \frac{D\rho}{Dt} + \rho \frac{\partial U_i}{\partial x_i} = 0$$

the following relation holds:

$$\frac{\partial U_i}{\partial x_i} = -\frac{1}{\rho} \frac{D\rho}{Dt}$$

Inserting into the energy equation and considering the state equation, the energy equation can be written in the following form:

$$\rho c_v \left[\frac{D\rho}{Dt} \left(\frac{P}{\rho R} \right) \right] = P \left[\frac{1}{\rho} \frac{D\rho}{Dt} \right]$$

or:

$$\frac{1}{P} \frac{DP}{Dt} = \left(\frac{R + c_v}{c_v} \right) \frac{1}{\rho} \frac{D\rho}{Dt}$$

or considering $R = (c_P - c_v)$ and $k = (c_P/c_v)$ transcribed:

$$\frac{1}{P} \frac{DP}{Dt} = k \frac{1}{\rho} \frac{D\rho}{Dt}$$

Equation allows the following generally holding solution:

$$\frac{D}{Dt}(\ln P) = \frac{D}{Dt}(\ln \rho^k)$$

or,

$$\frac{D}{Dt}\left[\ln\left(\frac{P}{\rho^k}\right)\right] = 0 \rightsquigarrow \frac{P}{\rho^k} = \text{const.}$$

The above relation gives expression to that the energy equation can be reduced to the adiabatic equation of thermodynamics, assuming that no molecule-dependent heat and momentum transport takes place and taking into account the continuity equation and the state equation for ideal gases. Along a streamline of a flow thus the following relation holds for the indicated conditions:

$$\frac{P}{\rho^k} = \text{const.}$$

There exist now a number of fluid-mechanic processes in compressible media that can be dealt with by means of equations, which result by further simplifications from the above equations.

Assuming that there are flow processes that preferably take place such that the velocity field only depends on a position coordinate, then it holds $U_1 = U_1\,(x_1)$, $U_2 = U_2\,(x_1)$ as well as$U_3 = U_3 = (x_1)$ Moreover it shall hold that $\frac{\partial U_2}{\partial x_1} \ll \frac{\partial U_1}{\partial x_1}$ and $\frac{\partial U_3}{\partial x_1} \ll \frac{\partial U_1}{\partial x_1}$ so that the following equations can be stated:

Continuity equation:

$$\frac{\partial \rho}{\partial t} + \frac{\partial}{\partial x_1}(\rho U_1) = 0$$

Momentum equation:

$$\rho\left[\frac{\partial U_1}{\partial t} + U_1\frac{\partial U_1}{\partial x_1}\right] = -\frac{\partial P}{\partial x_1}$$

Energy equation:

$$\frac{P}{\rho^k} = \text{const.}$$

By sound waves one understands the propagation of small disturbances in gases.

The sound velocity thus is the velocity with which small disturbances propagate in a fluid at rest. While for an incompressible fluid an infinitely large propagation velocity results for small fluids, for compressible fluids results a finite propagation velocity defined by the kind of gas and its temperature.

The quantitative determination of the propagation velocity of small disturbances in fluids at rest can be derived as stated below via the equations which hold for non-stationary one-dimensional flows of viscosity-free compressible fluids:

When computing the pressure variation $(\partial P/\partial x_1)$. in consideration of equation, one obtains:

$$\frac{\partial P}{\partial x_1} = \left(\frac{dP}{d\rho}\right)_{ad} \frac{\partial \rho}{\partial x_1},$$

as the pressure with the allowance of adiabatic conditions is only a function of a thermodynamical quantity, like the density. The derivation $(dP/d\rho)$ has to be formed for the adiabatic system condition required by equation. The continuity equation and the momentum equation can thus be written for one-dimensional fluid flows:

$$\frac{\partial \rho}{\partial t} + U_1 \frac{\partial \rho}{\partial x_1} + \rho \frac{\partial U_1}{\partial x_1} = 0$$

$$\frac{\partial U_1}{\partial t} + U_1 \frac{\partial U_1}{\partial x_1} + \frac{1}{\rho}\left(\frac{dP}{d\rho}\right)_{ad} \frac{\partial \rho}{\partial x_1} = 0$$

With this two conditional equations are available for the two unknowns U_1 and ρ whose analytical solution is sought on the hypothesis that small pressure and density fluctuations exist, i.e. that it holds:

$$U_1 = 0 + u'(x_1, t) \quad P = P_0 + p'(x_1, t)$$

$$\rho = \rho_0 + \rho'(x_1, t)$$

This relation inserted in the above conditional equations leads to:

$$\frac{\partial}{\partial t}(\rho_0 + \rho') + u' \frac{\partial}{\partial x_1}(\rho_0 + \rho') + (\rho_0 + \rho')\frac{\partial u'}{\partial x_1} = 0$$

$$\frac{\partial u'}{\partial t} + u' \frac{\partial u'}{\partial x_1} + \left(\frac{dP}{d\rho}\right)_{ad} \frac{1}{(\rho_0 + \rho')} \frac{\partial}{\partial x_1}(\rho_0 + \rho') = 0$$

When considering that the disturbance variables ρ' and u' depend on place and time, while the quantities P_0 and r_0 depend neither on place nor time, the following partial differential equation system results, on the hypothesis that products of fluctuation quantities are negligible with reference to linear terms:

$$\frac{\partial \rho'}{\partial t} + \rho_0 \frac{\partial u'}{\partial x_1} = 0 \text{ and } \frac{\partial u'}{\partial t} + \left(\frac{dP}{d\rho}\right)_{ad} \frac{1}{\rho_0} \frac{\partial \rho'}{\partial x_1} = 0$$

These now are two conditional equations for ρ' and u' that can be brought to a solution. In order to obtain this solution for the propagation of sound waves, the first of the above two equations is derived with respect to t:

$$\frac{\partial^2 \rho'}{\partial t^2} + \rho_0 \frac{\partial^2 u'}{\partial x_1 \partial t} = 0$$

The second equation multiplied by ρ_0 and differentiated to x_1 yields:

$$\rho_0 \frac{\partial^2 u'}{\partial x_1 \partial t} + \left(\frac{dP}{d\rho}\right)_{\text{ad}} \frac{\partial^2 \rho'}{\partial x_1^2} = 0$$

The subtraction of equations from equation results in the conditional equation for ρ':

$$\frac{\partial^2 \rho'}{\partial t^2} - \left(\frac{dP}{d\rho}\right)_{\text{ad}} \left(\frac{\partial^2 \rho'}{\partial x_1^2}\right) = 0$$

Furthermore, the derivation of equation to x_1 yields

$$\frac{\partial^2 \rho'}{\partial t \partial x_1} + \rho_0 \frac{\partial^2 u'}{\partial x_1^2} = 0$$

and multiplication of equation by $\left(\rho \Big/ \left(\frac{dP}{d\rho}\right)_{\text{ad}}\right)$ and differentiation to t leads to:

$$\frac{\partial^2 \rho'}{\partial t \partial x_1} + \frac{\rho_0}{\left(\frac{dP}{d\rho}\right)_{\text{ad}}} \frac{\partial^2 u'}{\partial t^2} = 0$$

The subtraction of equation from and multiplication by $\frac{1}{\rho}\left(\frac{dP}{d\rho}\right)_{\text{ad}}$ results in

$$\frac{\partial^2 u'}{\partial t^2} - \left(\frac{dP}{d\rho}\right)_{\text{ad}} \frac{\partial^2 u'}{\partial x_1^2} = 0,$$

of the conditional equation for the velocity fluctuation u'.

When comparing the conditional equations for ρ' and u', one sees that both are described by one and the same form of conditional equation, i.e. ρ' and u will show the same dependency on place and time. The solution is stated by the one-dimensional wave equation, i.e. there exists a wave propagation velocity with a propagation velocity which generally reads:

$$\sqrt{\left(\frac{dP}{d\rho}\right)_{\text{ad}}} = \sqrt{k\rho^{k-1}\frac{P}{\rho^k}} = \sqrt{k\frac{P}{\rho}} = \sqrt{kRT}$$

The general solutions of the differential equations for $\rho'(x_1, t)$ and $u(x_1, t)$ can be stated as follows:

$$\rho' = f_\rho(x_1 - ct) + g_\rho(x_1 + ct) \quad \text{and} \quad u' = f_u(x_1 - ct) + g_u(x_1 + ct)$$

$f_{\rho', u}(x_1 - c_t)$ represents the respective wave which propagates in x_1-direction and $g_{\rho, u}(x_1 + ct)$ the wave moving in the negative x_1-direction.

Further considerations on the propagation of disturbances in compressible media at rest can now be made on the basis of the above results. To this effect one computes from the general solution for u (wave in positive x_1-direction):

$$\frac{\partial u'}{\partial t} = -c\left(\frac{\partial f}{\partial \eta}\right) \text{ mit } \eta = x_1 - ct$$

and

$$\frac{\partial u'}{\partial t} = -c\frac{\partial u'}{\partial x_1}$$

From the momentum equation follows:

$$\frac{\partial u'}{\partial t} = -\frac{c^2}{\rho_0}\frac{\partial \rho'}{\partial x_1} = -c\frac{\partial \rho'}{\partial x_1}$$

or transcribed

$$\frac{\partial u'}{\partial x_1} = \frac{c}{\rho_0}\frac{\partial \rho'}{\partial x_1} \quad \Rightarrow \quad \frac{u'}{c} = \frac{\rho'}{\rho_0}$$

When we have a disturbance in the form of a compression wave, i.e. $\rho' > 0$, then also $u' > 0$,and this means that the fluid particles move in the direction of the disturbance when a compression disturbance occurs. When on the other hand an expansion disturbance occurs, i.e. $\rho' < 0$, then also $u' < 0$, and in this case the fluid particles move opposite to the direction of the propagation of the disturbance.

The most important result of the above derivations was that small disturbances in non-viscose and compressible fluids at rest propagate with sound velocity that can be computed as follows:

$$c = \sqrt{\left(\frac{dP}{d\rho}\right)_{ad}} = \sqrt{kRT}$$

TRANSVERSAL WAVES: SURFACE WAVES

General Solution Set-up

On the free surfaces of fluids wave appearances can occur, i.e. propagation of transversal waves owing to introduced disturbances. These can be two- or three-dimensional, however, the analytical treatment of surface waves presented here concentrates on two-dimensional surfaces. By linearization of the basic equations written in potential form one obtains the partial differential equations solved normally for surface waves. These indicate that the field of propagation of surface waves belongs to the potential theory.

Their treatment takes place separately nevertheless, as a special problem is concerned, i.e. a special class of flow appearances whose treatment correspondingly requires a special methodology. The latter is shown below in an introducing way. The relations stated in the following can again be derived from the basic equations, which can be stated as follows for a fluid-mechanically ideal fluid, i.e. a fluid free from viscosity:

$$\frac{\partial U}{\partial t} + U_i \cdot \frac{\partial U_j}{\partial x_i} = -\frac{1}{\rho}\frac{\partial P}{\partial x_j} + g_j$$

When integrating this equation over a period of time τ, one obtains

$$\bar{U}_j + \int_0^\tau U_i \frac{\partial U_j}{\partial x_i} dt = -\frac{1}{\rho}\frac{\partial}{\partial x_j}\int_0^\tau P\, dt + \int_0^\tau g_j\, dt$$

This equation can now be interpreted with $\pi = \int_0^\tau P\, dt$ as the pressure impulse during the time interval τ, for small time intervals τ as follows for ρ = const:

$$\bar{U}_j = \frac{\partial}{\partial x_j}\frac{P}{\rho} \text{ mit } \int_0^\tau U_i \frac{\partial U_j}{\partial x_i} dt \approx 0 \quad \text{and} \quad \int_0^\tau g_j \cdot dt \approx 0$$

Thus the fluid motion generated as a result of pressure impulses on free surfaces is described by a velocity potential, by $\bar{U}_j = U_j$:

$$\bar{U}_j = -\frac{\partial \phi}{\partial x_j} \quad \text{mit} \quad \phi = \frac{P}{\rho}$$

The motion thus is irrotational. Strictly speaking all this holds only at the free surface and the determination of ϕ in the entire flow area requires further considerations still. The continuity equation can be written as follows for ϕ.

$$\frac{\partial^2 \phi}{\partial x_i \cdot \partial x_i} = 0 = \frac{\partial^2 \phi}{\partial x_1^2} + \frac{\partial^2 \phi}{\partial x_2^2} + \frac{\partial^2 \phi}{\partial x_3^2}$$

The momentum equation can be written as stated below:

$$\frac{DU_j}{Dt} = -\frac{1}{\rho} \cdot \frac{\partial P}{\partial x_j} + g_j / \cdot U_j$$

or can be transcribed after multiplication by U_j as follows:

$$\frac{D}{Dt}\left(\frac{1}{2}U_j^2\right) = -\frac{1}{\rho}\frac{DP}{Dt} - \frac{1}{\rho}\frac{\partial P}{\partial t} - \frac{DG}{Dt}$$

With $g_j = -\rho \frac{DG}{Dt}$ for $\frac{\partial g}{\partial t} = 0$,

$$\frac{D}{Dt}\left(\frac{1}{2}U_j^2\right) = -\frac{1}{\rho}\frac{DP}{Dt} - \frac{1}{\rho}\frac{\partial P}{\partial t} - \frac{DG}{Dt}$$

or transcribed:

$$\frac{\partial \phi}{\partial t} + \frac{P}{\rho} + \frac{1}{2}U_j^2 + G = F(t)$$

The function *F (t)* introduced by the integration can be included into the potential ϕ, so that it holds:

$$\frac{\partial \phi}{\partial t} + \frac{P}{\rho} + \frac{1}{2}U_j^2 + G = 0$$

Represents a two-dimensional surface wave whose deflection, measured from the position of rest $x_2 = 0$ can be stated a follows:

$$x_2 = y = h(x_1, t) = h(x, t)$$

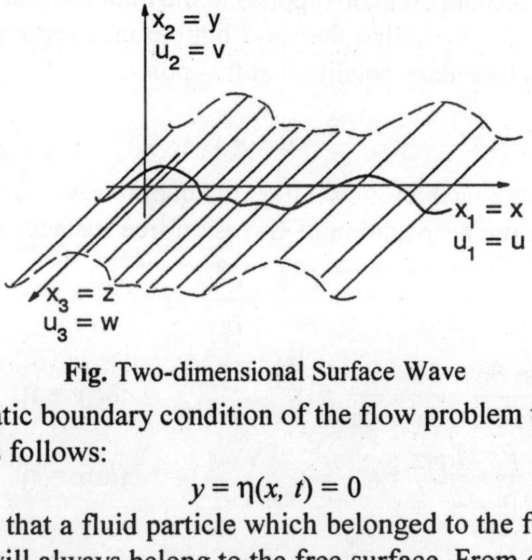

Fig. Two-dimensional Surface Wave

The kinematic boundary condition of the flow problem to be solved can thus be stated as follows:

$$y = \eta(x, t) = 0$$

This means that a fluid particle which belonged to the fluid surface at a point in time *t* will always belong to the free surface. From equation results with u_i as fluid velocity of the considered wave motion

$$\frac{D}{Dt}(y-\eta) = 0 = \frac{\partial}{\partial t}(y-\eta) + u_i\frac{\partial}{\partial x_i}(y-\eta) = 0$$

or the deflections carried out:

$$-\frac{\partial \eta}{\partial t} - u_1\frac{\partial \eta}{\partial x_1} + u_2 - u_3\frac{\partial \eta}{\partial x_3} = 0$$

When introducing now the potential function ϕ with

$$u_1 = \frac{\partial \phi}{\partial x_1},\quad u_2 = \frac{\partial \phi}{\partial x_2} \quad \text{and} \quad u_3\frac{\partial \phi}{\partial x_3}$$

the following relation results for the free surface with $x_1 = x$, $x_2 = y$ and $x_3 = z$:

$$\frac{\partial \phi}{\partial y} = \frac{\partial \eta}{\partial x} \cdot \frac{\partial \eta}{\partial x} + \frac{\partial \phi}{\partial z} \cdot \frac{\partial \eta}{\partial z}$$

In the entire area of the flow the potential function fulfills the continuity equation which thus can be stated in two-dimensional form as follows:

$$\frac{\partial^2 \phi}{\partial x^2} + \frac{\partial^2 \phi}{\partial y^2} = 0$$

On the prerequisite of absence of viscosity the Bernoulli equation can be employed in the form indicated by equation.

$$\frac{\partial \phi}{\partial t} + \frac{P}{\rho} + \frac{1}{2} U_j^2 + G = 0$$

This is equivalent to the assumption that typically the pressure along the free surface is constant and corresponds to the atmospheric pressure over the surface. When now including the solid bottom in a certain position $y = -h$, one obtains as a boundary condition at this point:

$$\frac{\partial \phi}{\partial y} = 0 \text{ for } y = -h$$

Thus one obtains the following set of equations, which are to be fulfilled in order to treat the propagation of waves on free surfaces analytically.

$$\frac{\partial^2 \phi}{\partial x^2} + \frac{\partial^2 \phi}{\partial y^2} = 0$$

$$\frac{\partial \eta}{\partial t} + \frac{\partial \phi}{\partial x}\frac{\partial \eta}{\partial x} + \frac{\partial \phi}{\partial z} + \frac{\partial \eta}{\partial z} = \frac{\partial \phi}{\partial y} \qquad \text{for } y = \eta$$

$$\frac{\partial \phi}{\partial t} + \frac{P}{\rho} + \frac{1}{2} U_j^2 + g\eta = 0 \qquad \text{for } y = \eta$$

$$\frac{\partial \phi}{\partial y} = 0 \qquad \text{for } y = -h$$

Here the last equations are to be understood as boundary conditions. Thus it becomes clear that the problem when solving wave problems for fluids with free surfaces is characterized by the imposed kinematic and dynamical boundary conditions.

It proves to be a peculiarity here that the main problem when solving problems concerning the wave motion in fluids with free surfaces is the introduction of the boundary conditions and not the solution of the differential equations describing the fluid motion.

Considerable simplifications of the equation system result further for the assumption of surface waves of small amplitudes. Assuming that the amplitude of the wave is smaller than all other linear dimensions of the problem, i.e.

smaller than the depth of the water h, and the wavelength λ, it results that η is small and $\frac{\partial \eta}{\partial x}$ also can be assumed to be small. The latter is the gradient of the course of the water surface. Moreover it hold that can be assumed to be small, as surface waves have no high frequencies and the assumption of small amplitudes is also valid here. Thus it holds for two-dimensional waves:

$$\frac{\partial \eta}{\partial t} = \frac{\partial \phi}{\partial \eta} \quad \text{for } y = \eta$$

This equation still contains the problem that the boundary condition under lying it for surface waves has to be imposed at the point $y = \eta$ However, when one expands $\frac{\partial \phi}{\partial \eta}$ in a Taylor series

$$\frac{\partial \phi}{\partial y}(x,\eta,t) = \frac{\partial \phi}{\partial y}(x,0,t) + \eta \frac{\partial^2 \phi}{\partial \eta^2}(x,0,t) + \cdots$$

it can be seen that the second term on the right side can be neglected because of the assusmed small η-values. In an analogous way it holds

$$\frac{\partial \phi}{\partial y}(x,\eta,t) + \frac{P(x,t)}{\rho} + g \cdot \eta(x,t) = F\,(t)$$

and for small velocities the following relation:

$$\frac{\partial \phi}{\partial t}(x,0,t) + \frac{P(x,t)}{\rho} + g \cdot \eta(x,t) = 0$$

where the function $F\,(t)$ was included into the potential $\phi(x,\,y,\,z)$. The differentiation of 11.64 to t yields:

$$\frac{\partial^2 \phi}{\partial t^2} + \frac{1}{\rho}\frac{\partial P}{\partial t} + g\frac{\partial \eta}{\partial t} = \frac{\partial^2 \phi}{\partial t^2} + (x,0,z) + \frac{1}{\rho}\frac{\partial P(x,t)}{\partial t} + g \cdot \frac{\partial \phi}{\partial y}(x,0,z) = 0$$

so that one obtains the following simplified set of equations for the treatment of surfaces of small amplitudes:

$$\frac{\partial^2 \phi}{\partial x^2} + \frac{\partial^2 \phi}{\partial y^2} = 0$$

$$\frac{\partial \phi}{\partial y}(x,0,t) = \frac{\partial \eta}{\partial t}(x,t) \qquad (\text{for } y = \eta)$$

$$\frac{\partial^2 \phi}{\partial t^2}(x,0,t) + \frac{1}{\rho}\frac{\partial P(x,t)}{\partial t} + g\frac{\partial \phi}{\partial y}(x,0,t) = 0 \qquad (\text{for } y = \eta)$$

$$\frac{\partial \phi}{\partial y}(x,-h,t) = 0 \qquad (\text{for } y = -h)$$

With the above listed equations gravitational waves and capillary waves can be treated, which usually represent small amplitudes.

PLANE STANDING WAVES

When considering wave motions, where the fluid particles move only in parallel to the $x_1 - x_2$–plane, i.e. where the pressure P and the velocity U_j are independent of x_3, so that the fluid motions in all areas parallel to the $x_1 - x_2$–plane take place in the same way, a plane wave motion with the following potential results:

$$\phi(x, y, t) = \phi(x, y)\cos(\varphi t + \epsilon)$$

For the case of a standing wave to be dealt with in this section it stated:

$$\phi(x, y) = \frac{P(y)}{\rho}\sin[k(x-\xi)]$$

The potential ϕ fulfills the Laplace equation.

$$\frac{\partial^2\phi}{\partial x^2} + \frac{\partial^2\phi}{\partial y^2} = 0$$

$$\text{Mit } \rho\frac{\partial^2\phi}{\partial x^2} = -P(y)k^2\sin[k(x-\xi)] \text{ and } \rho\frac{\partial^2\phi}{\partial y^2} = P\frac{d^2P}{dy^2}\sin[k(x-\xi)]$$

one obtains the following differential equation:

$$\frac{d^2P}{dy^2} - k^2P = 0$$

the solution of which is as follows:

$$P(y) = C_1\exp(ky) + C_2\exp(-ky)$$

$$\text{fur } -\infty \le y \le 0 \quad -\infty \le +x \le +\infty$$

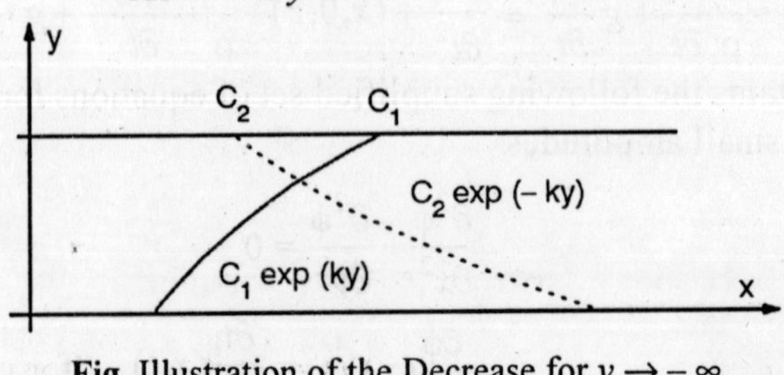

Fig. Illustration of the Decrease for $y \to -\infty$

The integration constant C_2, when carrying out more precise considerations, results as $C_2 = 0$, as otherwise for large depths $y \Longrightarrow -\infty$ the P (y)-term would become very big, so that the solution:

$$\phi(x, y) = \frac{C_1}{\rho}\exp(ky)\sin[k(x-\xi)]$$

can be obtained, or

$$\phi(x, y, t) = \frac{C_1}{\rho}\exp(ky)\sin[k(x-\xi)]\cos(\varphi t+\epsilon)$$

When starting from the assumption that the occurring fluid motion is slow, the equation

$$\frac{\partial\phi}{\partial t}+\frac{P}{\rho}+\frac{1}{2}U_j^2+g\eta = 0 \qquad \text{for } y=\eta$$

can be written as follows:

$$\frac{\partial\phi}{\partial t}+\frac{P}{\rho}+g\eta = 0 \text{ for } y=\eta$$

A derivation to the time yields the relation indicated below, as the pressure along the free surface does not change:

$$\frac{\partial^2\phi}{\partial t^2}+g\frac{\partial\eta}{\partial t} = \frac{\partial^2\phi}{\partial t^2}+gu_2$$

With $u2 = \frac{\partial\phi}{\partial y}$ it results:

When employing to one obtains:

$$\frac{\partial^2\phi}{\partial t^2} = -\frac{C_1}{\rho}\exp(ky)\sin[k(x-\xi)]\varphi^2\cos(\varphi+\epsilon)$$

and

$$\frac{\partial\phi}{\partial y} = +\frac{C_1}{\rho}\exp(ky)\sin[k(x-\xi)]\cos(\varphi+\epsilon)$$

$$\varphi^2 = kg$$

Thus for the remaining considerations the following fluid motion has to be examined which for the sake of simplicity is considered for $\xi = 0$ and $\epsilon = 0$.

$$\phi(x, y, t) = \frac{C_1}{\rho}\exp(ky)\sin(kx)\cos(\varepsilon t)$$

For the free surface it is computed from equation:

$$\eta = -\frac{1}{g}\frac{\partial\phi}{\partial t} = -\frac{1}{g}\frac{\partial}{\partial t}\phi(x,0,t)$$

or

$$\eta = \frac{C_1\varphi}{g}\sin(kx)\sin(\varphi t)$$

With $A = \frac{C_1\varphi}{\rho g}\sin(\varphi t)$ it holds:

$$\eta = A \sin(kx)$$

For $x = \frac{m \cdot \pi}{k}$ result for m = 0, ±1, ±2, ± · · · nodal points of a standing wave. In the middle between these nodes are the "antinodal points" of the wave motion. The wavelength of the sinusoidal fluid motion can be computed as:

$$\lambda = \frac{2\pi}{k}$$

The amplitude of the wave motion is $\frac{C_1 \varphi}{g} \cdot \sin(\varphi t) = A$ quency of the wave motion it holds:

$$f = \frac{\varphi}{2\pi} = \frac{1}{T}$$

Taking into consideration and one obtains:

$$T = \frac{1}{f} = \sqrt{\frac{2\pi k}{g}} \quad \text{bzw.} \quad \lambda = \frac{\gamma \tau^2}{2\pi}$$

or

$$\lambda = \frac{g}{2\pi f^2}$$

The above relations show that the wavelength of standing fluid waves decreases with increasing frequency of the motion.

PLANE PROGRESSING WAVES

For the considerations stated below it is assumed that the considered fluid takes up the space as follows:

$$-\infty \le y \le 0 \quad \text{and} -\infty \le x \le +\infty$$

and at the point $y = 0$ occupies a free surface which for the considerations carried out possesses a finite surface. The equations required for the treatment of progressing waves can be stated as follows:

$$\frac{\partial^2 g}{\partial x^2} + \frac{\partial^2 \phi}{\partial y^2} = 0$$

With $y = \eta(x_1, t)$ for the free surface it holds:

$$\frac{D}{Dt}(y - \eta) = 0 \rightsquigarrow u_2 = \left(\frac{\partial}{\partial t} + u_1 \frac{\partial}{\partial x_1} \right) \eta$$

Neglecting the term of second order it results:

$$\frac{\partial \phi}{\partial y} = \frac{\partial \eta}{\partial t}$$

For the pressure at the free surface it can be stated:

$$P = -\sigma\left[\frac{1}{R_1}+\frac{1}{R^2}\right]$$

where R_1 and R_2 represent the main radiuses of curvature of the disturbed surface and σ indicates the surface tension. Linearized it can be written:

$$P = -\sigma\frac{\partial^2\eta}{\partial x^2}$$

where the pressure above the free surface is formulated with $P = 0$ Otherwise P is to be replaced by $(P = P_0)$.

For plane progressing waves the following potential can be stated:

$$\phi(x, y, t) = C\exp(ky)\cos[k(x - ct)]$$

with $\phi = 0$ for $y = -\infty$ The formulation for. fulfills the continuity equation in the form.

The Bernouilli equation can be stated as follows:

$$\frac{P}{\rho} = -\frac{\partial\phi}{\partial t} - gy.$$

Or transcribed:

$$\frac{\partial\phi}{\partial t} = -\frac{P}{\rho} - g\eta = \frac{\sigma}{\rho}\frac{\partial^2\eta}{\partial x^2} - g\eta.$$

From this follows:

$$\frac{\partial^2\phi}{\partial t^2} = \frac{\sigma}{\rho}\frac{\partial^2}{\partial x^2}\left(\frac{\partial\eta}{\partial t}\right) - g\frac{\partial\eta}{\partial t}.$$

and in consideration of

$$\frac{\partial^2\phi}{\partial t^2} = \left[\frac{\sigma}{\rho}\frac{\partial^2}{\partial x^2} - g\right]\frac{\partial\phi}{\partial g}$$

For the left side of it can be written with:

$$\frac{\partial^2\phi}{\partial t^2} = -Ck^2c^2\exp(ky)\cos[k(x - ct)] = -k^2c^2\phi$$

so that it holds:

$$k^2c^2\phi = \left[g - \frac{\sigma}{\rho}\frac{\partial^2}{\partial x^2}\right]\frac{\partial\phi}{\partial y}$$

With $\frac{\partial\phi}{\partial y} = k\phi$ and $\frac{\partial^2\phi}{\partial x^2} = -k^2\phi$ results from for the velocity of the progressing wave:

$$c^2 = \frac{g}{k} + \frac{k\sigma}{\rho}$$

With $k = \frac{2\pi}{\lambda}$ it can be seen that for long waves the influence of the gravity dominates:

$$c = \sqrt{\frac{g}{k}} \quad \text{shear waves}$$

For waves with small wavelengths the capillary e ects dominate:

$$c = \sqrt{\frac{k\sigma}{\rho}} \quad \text{capillary waves}$$

Concerning wavelengths, often the path curves of the fluid particles are also of interest, which occur at the water surface or in certain depths below the water surface. In this respect the following considerations can be carried out, where x_0 and y_0 are introduced as the coordinates which

$$u_x = \frac{\partial\phi}{\partial x} = \frac{dx}{dt} = Ck\exp(ky)\sin[k(x-ct)]$$

$$u_y = \frac{\partial\phi}{\partial y} = \frac{dy}{dt} = Ck\exp(ky)\cos[k(x-ct)]$$

From this result:

$$x = x_0 + C \cdot k\exp(ky)\cos[k(x-ct)]\left(\frac{+1}{Ck}\right)$$

$$y = y_0 + ck\exp(ky)\sin[k(x-ct)]\left(\frac{-1}{ck}\right)$$

Or transcribed:

$$(x-x_0)^2 + (y-y_0)^2 = \left(\frac{C}{c}\right)^2 \exp(2ky_0)$$

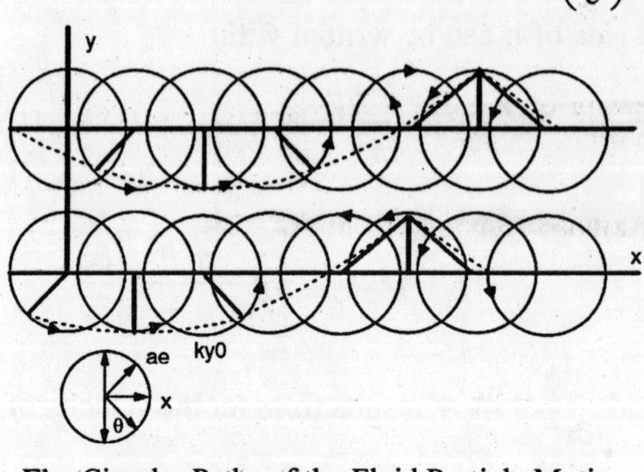

Fig. Circular Paths of the Fluid Particle Motion

The path curves of the fluid particles are computed as circles whose radius gets smaller with increasing water depth. For the water surface the radius of the circular path is equal to the amplitude of the surface wave, while in a depth of a wavelength it has already decreased to a 1/535th of the wave amplitude at the water surface. This makes it clear that the considered wave motion remains limited to an area in immediate proximity of the water surface.

Shows the circular paths described by fluid particles. These will run in an anticlockwise direction, so that it holds:

$$(x - x_0) = a \exp(ky_0) \sin \Theta$$
$$(y - y_0) = -a \exp(ky_0) \cos \Theta$$

so that for Θ it holds:

$$\Theta \quad = kx_0 + kct \cos (yt + \in)$$

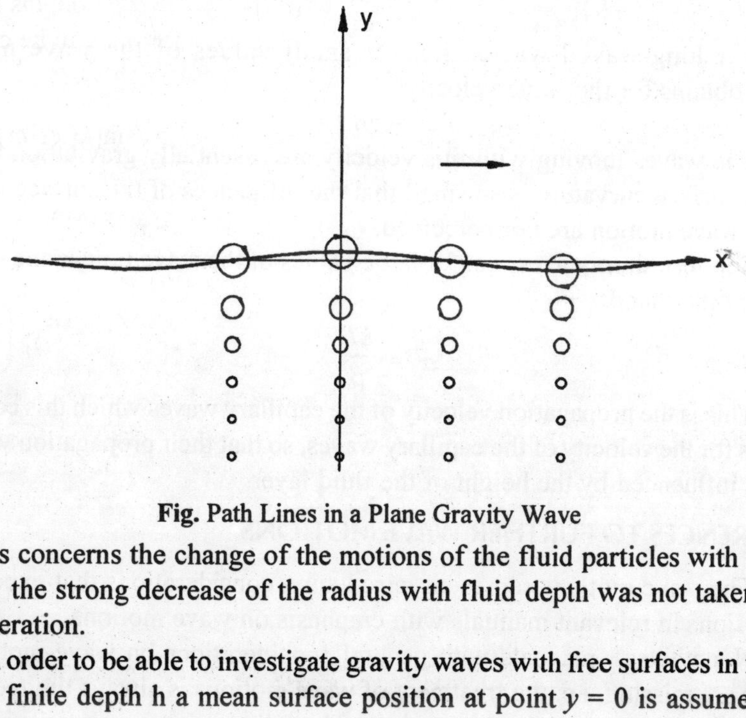

Fig. Path Lines in a Plane Gravity Wave

As concerns the change of the motions of the fluid particles with water depth, the strong decrease of the radius with fluid depth was not taken into consideration.

In order to be able to investigate gravity waves with free surfaces in fluids with a finite depth h a mean surface position at point $y = 0$ is assumed. At position $y = -h$ a wall is considered as being given, so that a mean fluid film thickness with the height h occurs. To fulfill now the continuity equation:

$$\frac{\partial^2 \phi}{\partial x^2} + \frac{\partial^2 \phi}{\partial y^2} = 0$$

by a wave with the wave number k., the following potential formulation is carried out:

$$\phi = C \cosh k(y + h) \cos k(x - ct)$$

which does not only fulfill the continuity equation, but permits also to fulfill the boundary condition at the bottom of the fluid layer:

$$\frac{\partial \phi}{\partial y} = 0 \text{ for } y = -h$$

It then results a condition for the free surface that can be stated as below indicated:

$$kc^2 \cosh kh = \left(g + \frac{k^2 T}{\rho} \right) \sinh kh,$$

Or resolved to the wave velocity one obtains:

$$c^2 = \left(g + \frac{k^2 T}{\rho} \right) \frac{\tanh kh}{k}$$

For long-waved waves, i.e. for small values of the wave number k one obtains for the wave velocity:

$$c^2 = gh.$$

The waves moving with this velocity are essentially gravitation waves, as the surface curvature is so small that the influences of the surface tension at the wave motion are not perceived.

For very short waves, i.e. for large values of the wave time k one obtains on the other hand:

$$c^2 = \frac{kT}{\rho}$$

This is the propagation velocity of the capillary waves which this equation shows for the velocity of the capillary waves, so that their propagation velocity is not influenced by the height of the fluid layer.

REFERENCES TO FURTHER WAVE MOTIONS

The wave motions represent introducing considerations that experience extensions in relevant manuals with emphasis on wave motions.

Nonetheless, manuals with general considerations on wave motions in fluids are missing, i.e. the treatment of wave motions is always limited to the treatment of very special wave motions. The following wave motions in fluids are dealt with:

- Gerstner waves
- Solitar waves
- Rossby-waves
- Stokes-waves
- Cnoidal-waves
- Axisymmetric waves

So if one wants to find the introducing literature on the wave motion observed in nature, it is necessary to perceive the physical causes o the considered wave motion.

Thus one observes for example that a long body which is moved transversely to its linear expansion near the free surface forms waves mainly in its wake. In front of the body one observes viewed from the amplitude smaller surface waves, when the dimensions of the bodies in flow direction are smaller as compared to$(\sigma/\rho g)^{1/2}$ Otherwise the gravity waves occurring behind the body dominate and the capillary waves that can be observed in front of the body are negligible. So when one has recognized the nature of the observed wave motions.

Index